The Open University

Science: Level 2

S278
EARTH'S PHYSICAL RESOURCES
ORIGIN, USE AND
ENVIRONMENTAL IMPACT

E

FOSSIL F
AND

ST

The S278 Course Team

Chair

Peter Webb

Course managers

Jessica Bartlett
Annemarie Hedges

Authors

Tom Argles
Steve Drury
Peter Sheldon
Sandy Smith
Peter Webb

Course assessor

Professor David Manning
(*University of Newcastle*)

Production Team

Jessica Bartlett (*Indexer*)
Gerry Bearman (*Editor*)
Steve Best (*Graphic artist*)
Kate Bradshaw (*Software designer*)
Roger Courthold (*Graphic artist*)
Becky Graham (*Editor*)
Sarah Hack (*Graphic artist*)
Liz Lomas (*Course team assistant*)
Judith Pickering (*Project manager*)
Jane Sheppard (*Graphic designer*)
Andy Sutton (*Software designer*)
Pam Wardell (*Editor*)
Damion Young (*Software designer*)

Acknowledgements

The S278 Course Team gratefully acknowledges the contributions of members of the S268 *Physical Resources and Environment* Course Team (1995) and of its predecessor, S238 *The Earth's Physical Resources* (1984).

This publication forms part of an Open University course S278 *Earth's Physical Resources: Origin, Use and Environmental Impact*. The complete list of texts which make up this course can be found on the back cover. Details of this and other Open University courses can be obtained from the Student Registration and Enquiry Service, The Open University, PO Box 197, Milton Keynes, MK7 6BJ, United Kingdom: tel. +44 (0)870 333 4340, email general-enquiries@open.ac.uk

Alternatively, you may visit the Open University website at http://www.open.ac.uk where you can learn more about the wide range of courses and packs offered at all levels by The Open University.

To purchase a selection of Open University course materials visit http://www.ouw.co.uk, or contact Open University Worldwide, Michael Young Building, Walton Hall, Milton Keynes MK7 6AA, United Kingdom for a brochure. tel. +44 (0)1908 858785; fax +44 (0)1908 858787; email ouwenq@open.ac.uk

The Open University
Walton Hall, Milton Keynes
MK7 6AA

First published 2006.

Edited, designed and typeset by The Open University.

Printed and bound in the United Kingdom at the University Press, Cambridge.

ISBN 0 7492 6997 9

1.1

CONTENTS

AN INTRODUCTION TO ENERGY

Understanding energy resources involves considering all types of energy source from various scientific and technological standpoints, with a focus on the uses, limitations and consequences of using energy that is available to humanity. The central theme of this book is how geoscientific information helps us to understand the formation and distribution of useful repositories of energy. This chapter sets the scene by considering how much energy human society uses and the basic concepts of energy, work, power and efficiency, then briefly investigates the different types of energy available, their sources and renewability.

1.1   Energy use

Until about 8000 years ago humans relied on hunting and gathering for food, and burning wood to keep warm. Their exact energy demands can at best only be estimated but to survive they probably needed about as much energy as it takes to run a couple of ordinary domestic light bulbs continuously. Later, agriculture developed, and although wood was still the chief fuel, animal power, animal dung and charcoal were also used. Even today, such energy sources based on natural biomass dominate the lives of human populations in the so-called 'Third World' or 'developing countries'. The 19th century heralded a large increase in energy use in what were to become industrialized countries (Figure 1.1), particularly the use of coal. Homes and other buildings were heated; factories and railways were powered by steam engines (Figure 1.2); mining and chemical industries developed and agriculture became more mechanized. The emergence of technological societies in the 20th century resulted in an even larger increase in energy use for manufacturing, agriculture, transport and a host of other

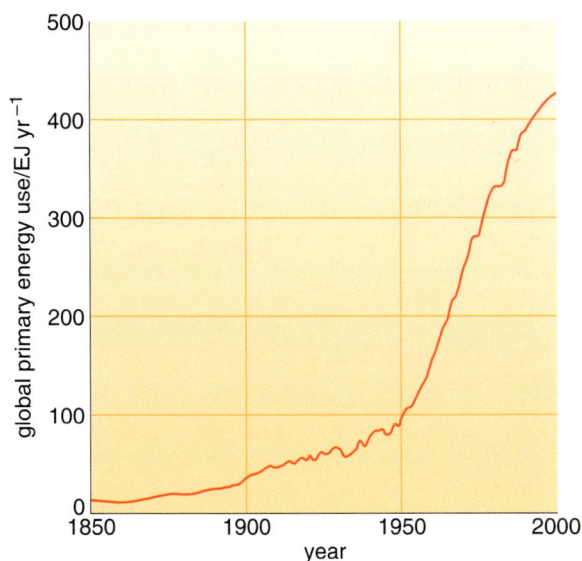

Figure 1.1 Growth in global primary energy use 1850–2000. The EJ unit of energy will be explained in Box 1.1.

Figure 1.2 A model of the *Rocket* steam locomotive that was designed and built by George Stephenson in 1829.

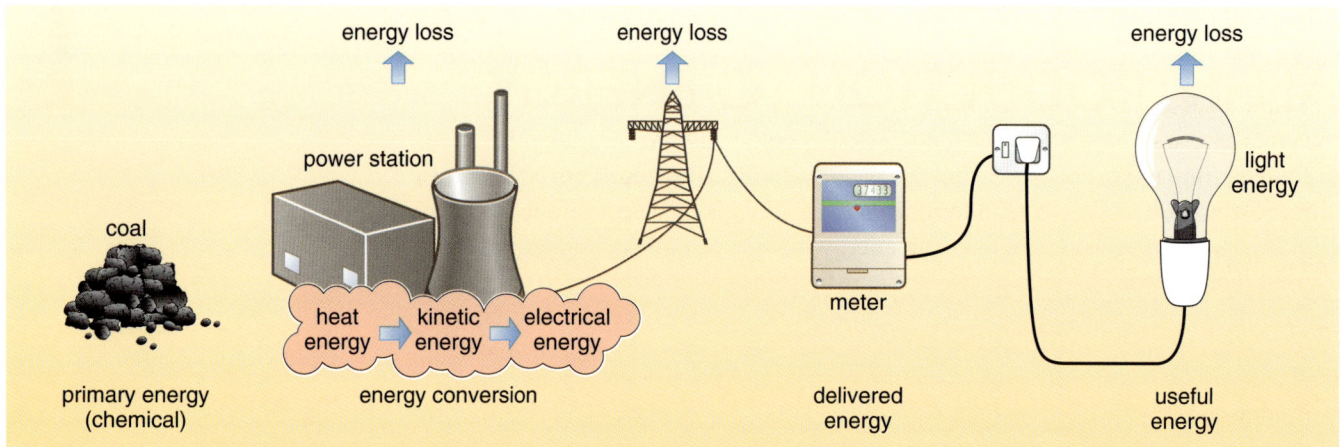

applications. In technologically advanced countries the largest increases have been in using gas for heating, oil products for transport, and electricity as a convenient means of transferring energy generated by a variety of sources (Figure 1.3).

It is important to remember that the **primary energy** released by all forms of energy resources is not the amount that performs useful tasks: it is *the total amount of energy released by human activity*. Energy use and conversion, as you will see, can never be fully efficient. For that reason, the energy *consumed* usefully by society is considerably less than primary energy released: you will find that we refer to both primary energy and energy consumption (sometimes demand), depending on the context.

World population rose from some 5 million 10 000 years ago, through 1 billion in the 19th century to 6.5 billion in 2005. This rapid increase in population, together with a sharp increase in the demand that each person in the developed world has for energy, led to the dramatic rise in global energy consumption (Figure 1.1).

All the Earth's physical resources, for example metals in ores, water supplies and building stone, depend on using energy to extract, process and transport them. In effect, the ability to extract and use the Earth's physical resources depends on whether there is a ready supply of energy at the right price. If there were a limitless supply of cheap energy we could turn the entire stock of all physical resources into reserves. One aim of this book, and indeed the whole course, is to examine the limits that exist in reality: some are governed by physical laws, others depend on economics and there are also limits posed by sustaining the Earth's environmental conditions on which life depends.

Figure 1.3 Energy conversion in an electrical supply system whose primary energy source is coal. Useful electrical energy delivered to the home is much less than the primary energy used in generating it. This is due to losses through inefficiency at the power station, in transmission and in running household appliances.

1.2 Energy, work, power and efficiency

In everyday speech we often refer colloquially to the *powerful* politician, the *energetic* child, the *working* mother and the *efficient* administrator. We use these terms imprecisely, and often wrongly, compared with their scientific definitions.

1.2.1 Some basic concepts

Energy is defined as *the capacity to do work*, and **work** is even more precisely defined as *a force acting on an object that causes its displacement*, and is calculated from *force × distance*. Work is therefore the foundation for scientific study of motion and change.

The unit of work is the same as that for energy: the joule (J). Yet a joule, unlike the units of mass, length or time, is not a fundamental unit. Working out the joule in fundamental units takes us to the root of the physics involved. The definition of work (force × distance) shows that one joule is actually one newton metre (N m).

Force is *mass × acceleration*, so one **newton** *is the force that gives a mass of one kilogram (kg) an acceleration of one metre per second per second (m s⁻²), and is therefore equivalent to* 1 kg m s⁻².

However, work is not just mechanical movement, such as moving sacks of flour around or turning a wheel. It is also involved in heating a substance (vibrating its molecules), changing its state (melting and boiling) and compressing it, along with many other phenomena.

As an example of the connection between work and energy consider the energy bound up with a photon of light that strikes the photoelectric cell in a solar-powered pocket calculator. Some of that photon's energy becomes an electrical current that contributes to the calculator's microchip executing a calculation. However, most of the original light energy goes to heating up the cell and the liquid crystal display that shows the answer, and there might even be a beep of sound when the calculation is complete. During this process a tiny amount of work is done, but the *form* of the energy has changed, from the otherwise limitless potential of light to travel an infinite distance through a vacuum, never losing its 'capacity to do work', to the motion of particles of matter. The last change is eventually expressed by a minuscule rise in temperature in and around the calculator, which ultimately dissipates to help heat up the rest of the Universe! Energy can neither be created or destroyed (the Law of Conservation of Energy) so the energy of the photon doesn't disappear but is spread far and wide. Although eternal, in practice it is beyond recovery: *useful* energy in a high *grade* form is *degraded* through the material work that it has done.

Energy takes many natural forms: light, heat, sound, mechanical movement (**kinetic energy**), that gained by position in a gravitational field (**potential energy**), the movement of electrons (electricity), chemical energy, that released through Einstein's famous matter–energy conversion $E = mc^2$ (nuclear energy) and a great many more. In this book we deal mainly with the forms of energy that are available at sufficiently high grade to be able to do useful work. An implicit theme is that all energy sources are themselves products of work done naturally:

- Chemical energy — 'fossil fuels', such as coal and petroleum, and 'biofuels', such as wood, are the products of conversion of solar energy into the energy of chemical combination through photosynthesis by plants (Chapters 2, 3 and part of Chapter 5);

- Nuclear energy — nuclear fuels (Chapter 4) are heavy radioactive isotopes produced by thermonuclear fusion in long-dead stars that became supernovae;

- Geothermal energy (part of Chapter 5) comes from heat produced by the natural decay of radioactive isotopes distributed at very low concentrations in the Earth;

- Solar energy (part of Chapter 5) is emitted by thermonuclear fusion within the Sun;

- Tidal energy (part of Chapter 5) is essentially the work done by the gravitational fields of the Moon and Sun on the oceans;

- Wind and wave energy (part of Chapter 5) are solar energy converted into work done by the Earth's atmosphere and oceans;

- Hydro energy (part of Chapter 5) relies on work done by solar energy in evaporating water that falls as rain or snow at high topographic elevations, which in turn has gravitational potential energy.

Every application that 'uses' energy is in fact converting one form of energy into other forms. Some of this energy conversion is useful, some is not. A kettle boils water, but it may also 'sing' (sound energy) and heat up itself. A hydroelectric power station turns the kinetic energy of moving water into electricity through generators, but at the same time friction between the moving parts of the generators produces heat and sound. The heat and sound are degraded and less useful forms of energy.

● Which energy changes take place when an electric kettle boils water?

○ Electrical energy is first converted into heat energy as it increases the vibrational movement of atoms in the metal heating element. Heat is conducted through the element into the water, making the water molecules vibrate more, thereby raising its temperature. Finally, heat energy converts water into steam.

The point to be grasped here is that *changes* of energy from one form to another are commonplace in everyday life.

Power is the *rate at which energy is delivered or work is done*. It is worth noting the difference between power and energy: energy is *an amount* of work done over an indefinite time interval and power is the *rate* at which energy is converted, i.e. the amount used or made available per second. The units for measuring energy and power are summarized in Box 1.1.

Every three months or so most people in the UK receive an electricity bill, which states the number of 'Units' of electricity they have used, as measured by their electricity meter. The 'Units' quoted are kilowatt hours (kW h). But what, in purely physical terms, does 1 kW h represent? One kilowatt hour is the energy delivered at a power of 1 kW for one hour. This is equivalent to 1 kJ per second for an hour, i.e. 3600 kJ.

In practical terms, you might think that one kW h was a measure of all the work done around your home by various appliances in an hour. You would be very disappointed, because not one of them is perfectly efficient. In fact over 70% of each 'Unit' would ultimately have been lost to heating the rest of the Universe without doing any useful work. As you will see below and in later chapters, generating and then transmitting the electrical energy to your home will have been inefficient too, so a high proportion of your bill had been spent on work that was of no practical use to you.

Box 1.1   Units of energy and power

All forms of energy are measured in the same unit, the **joule** (J). A joule is the same amount of energy irrespective of which form (light, sound, electrical, and so on) that energy takes.

Recalling the earlier definitions in this section, when 1 kg falls 1 m at the acceleration due to gravity at the Earth's surface (9.81 m s^{-2}), the work done (force × distance) is mass in kg × acceleration due to gravity in m s^{-2} × height in m, i.e.:

$$\text{work done} = 1\,\text{kg} \times 9.81\,\text{m s}^{-2} \times 1\,\text{m} = 9.81\,\text{N m}$$
$$= 9.81\,\text{J}$$

So a joule is equivalent to the work done when a mass of (1/9.81) kg or 0.102 kg falls through a metre: a mass about the size of the apple reputed to have fallen on Isaac Newton's head to inspire his theory of gravity! A joule is a small amount of energy so when considering national or international energy use, large multiples of the joule are needed, and these are named in the metric (SI) system using standard prefixes (Table 1.1).

There is another energy unit that will be used occasionally in this book, the **tonne oil equivalent (toe)**. This is often used by energy statisticians as a convenient means of comparing amounts of energy available from fuels other than oil. One toe is the *chemical energy contained in one tonne of oil*, and equates approximately to 4.19×10^{10} J (i.e. 41.9 GJ).

Being a rate, power is measured in joules converted per second (J s^{-1}), called **watts** (W). One watt is *equivalent to one joule per second*. Note that if you are hit on the head by a falling apple, even though its energy is only about 1 J, the impact is almost instantaneous, so its power is high. Just as a joule is a small unit of energy, so a watt is a small unit of power. Larger quantities of power are quoted using the same prefixes as energy, as given in Table 1.1.

An electric supply of 1 kW will run a microwave oven, or around ten light bulbs. Working really hard, the human body can deliver about 1 kW for a very short period of time.

Table 1.1   Units of energy and power. In Section 1.3.3 and Chapter 5 you will also find the milliwatt (mW or 10^{-3} watts) used.

Energy		Number in powers of ten	Power	
Unit	Abbreviation		Unit	Abbreviation
kilojoule	kJ	10^3	kilowatt	kW
megajoule	MJ	10^6	megawatt	MW
gigajoule	GJ	10^9	gigawatt	GW
terajoule	TJ	10^{12}	terawatt	TW
exajoule	EJ	10^{18}	exawatt	EW

When we convert energy from one form to another, the output that is useful to us is never as much as the energy input. The ratio of the useful output to the input (i.e. the *ratio of useful work done to the energy supplied*) is called the **energy efficiency** of the process and is usually expressed as a percentage. Every natural process involves a change in grade of energy through the work that it does, but one joule always ends up as one joule, because of the Law of Conservation of Energy. Ignoring this transformation in grade, energy conversion could be said to be 100% efficient. But that misses the point. A perfect energy-changing machine would get out as much useful energy as was put in, so it would be 100% efficient, but such a machine does not and probably cannot exist.

Efficiency can be as high as 90% in a water turbine, but only around 35–40% in a coal-fired power station. The energy delivered to do useful work is considerably less than the primary energy consumed in electricity generation.

1.2.2   Present-day energy use

Global annual consumption of all forms of primary energy increased more than tenfold during the 20th century (Figure 1.1), and by the year 2002 reached an estimated 451 EJ. About three-quarters of this energy came from coal, oil and gas (Figure 1.4).

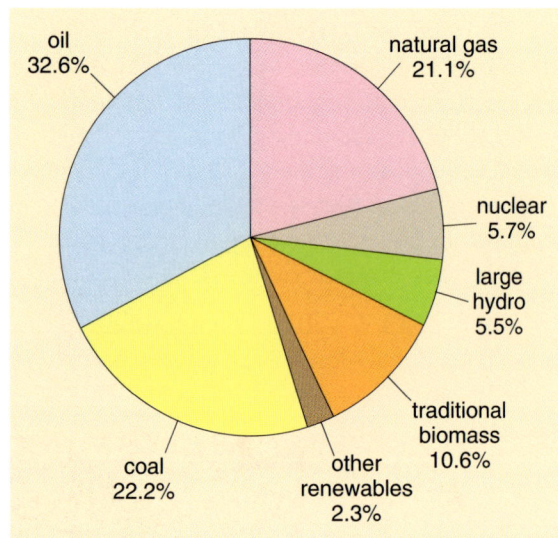

Figure 1.4 Contributions of various energy sources to global primary energy consumption in 2002. 'Renewables' and 'biomass' will be defined in Sections 1.6.2 and 1.7.

If the global annual energy consumption is 451 EJ, what would be the average rate of consumption each second of every day, i.e. the global power demand?

$$\frac{451 \times 10^{18}\ \text{J}}{(60 \times 60 \times 24 \times 365)\ \text{s}} = 1.43 \times 10^{13}\ \text{J s}^{-1} = 14.3\ \text{TW}$$

Make a note of these equivalent amounts, as you will be comparing them with the energy and power available from various sources later.

With a global population of 6.5 billion, each person's 'drain' on primary energy is, on average, around 73 GJ per year. But globally, there are major regional differences in energy consumption (Figure 1.5a). Developed countries, with industrial as well as domestic demands, use energy in vast quantities and at alarming rates. In North America it is around 350 GJ per person per year, nearly five times the global average, and totalling around 28% of global energy use by about 4.5% of world population. People in Europe and the former Soviet Union use about double the global average. Figure 1.5b, which shows the amount of lighting seen from space at night, gives a graphic picture of the inequalities of energy use.

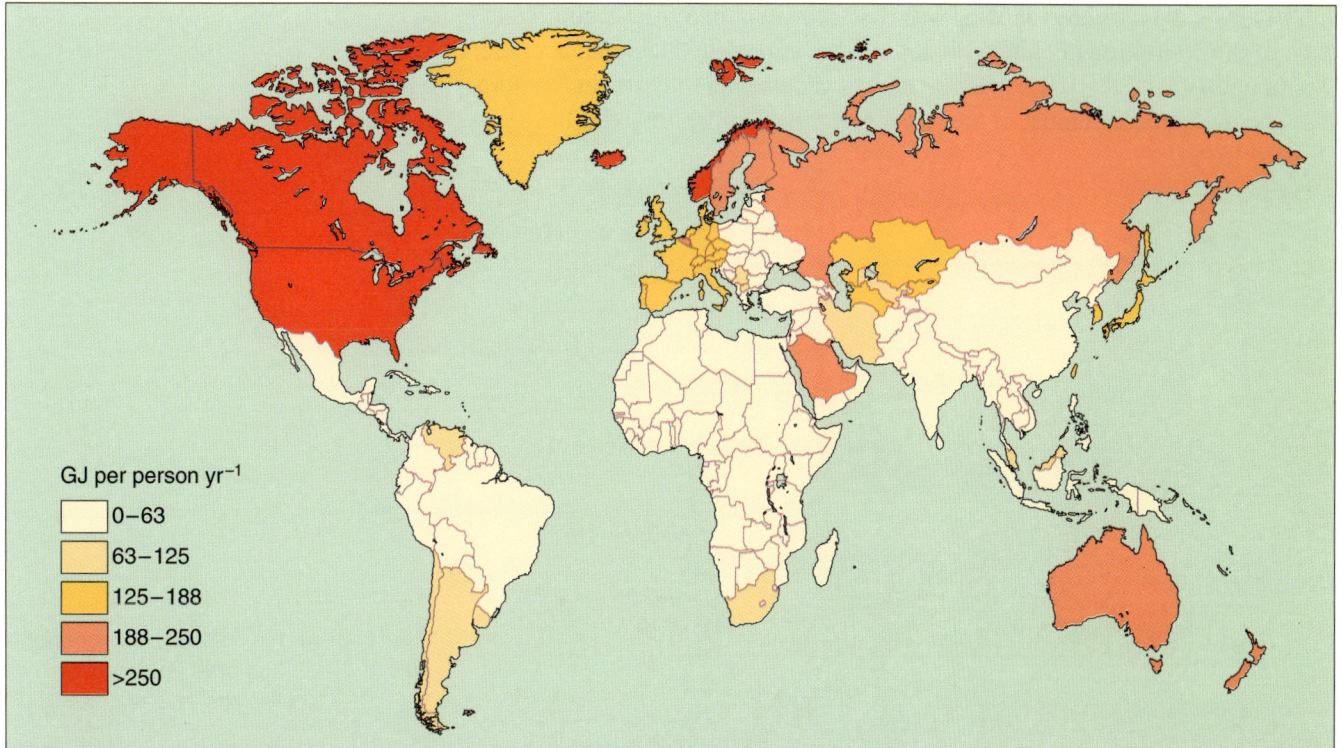

(a)

(b)

Figure 1.5 (a) Energy consumption by region, in GJ per person per year, 2004.
(b) Composite satellite image of the Earth at night. The white areas show outside lighting.

In 2002, UK primary energy used was the equivalent of 9.7 EJ: about 164 GJ for each of the 59 million people in the UK, just over double the global average. About one-fifth of the UK's primary energy requirement is used in the home, 30% lost in conversion and most of the rest for services, transport and industry (Figure 1.6).

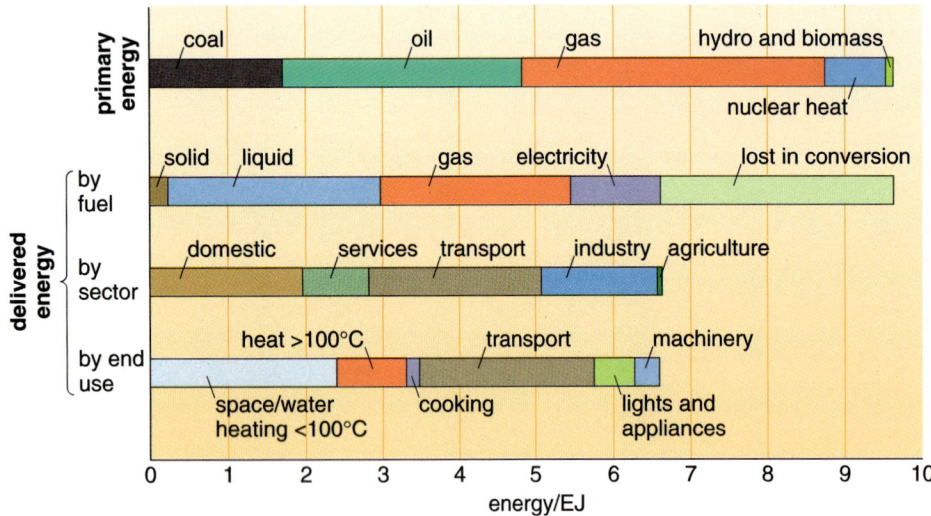

Figure 1.6 UK primary energy by fuel, and delivered energy by fuel, sector and end use 2000.

1.2.3 Global power demand

In Section 1.2.2 we calculated a value of 14.3 TW for the average global requirement for primary power in 2002.

● For a global population at this time of 6.2 billion, how much primary power was needed to support the activities of each person in the world, on average?

○ Dividing 14.3 TW (14.3×10^{12} W) by the world population of 6.2×10^9 gives an *average* primary power requirement of 2.3 kW per person.

The average global figure of 2.3 kW per person is about five times less than that needed to enable each North American citizen to sustain the lifestyle to which he or she has grown accustomed. If every individual in the world were to demand as much energy as the average person uses in North America, the global energy supply industries would require a fivefold increase in their use of primary energy sources. Even more daunting is the prospect of continued growth of both world population and per capita energy demand. We examine the likelihood, consequences and implications of these prospects in Chapter 6.

1.3 Sources of energy from the natural environment

The natural environment itself is bathed in energy from other sources. Standing on a cliff top on a bright spring day you can feel the warmth of the Sun and the freshness of the breeze and hear the crashing of breaking waves below. All these energetic processes can be compared in terms of energy and power.

In order to put the total global energy supply and demand into proper perspective we need to know the contribution made by different *natural* energy sources to the Earth's energy supply. By far the most important source on Earth is the Sun, but some energy comes from the gravitational attraction between Sun, Moon and Earth, and some from the Earth's own internal heat. Figure 1.7 gives all the data on natural energy sources you will need in working through this section.

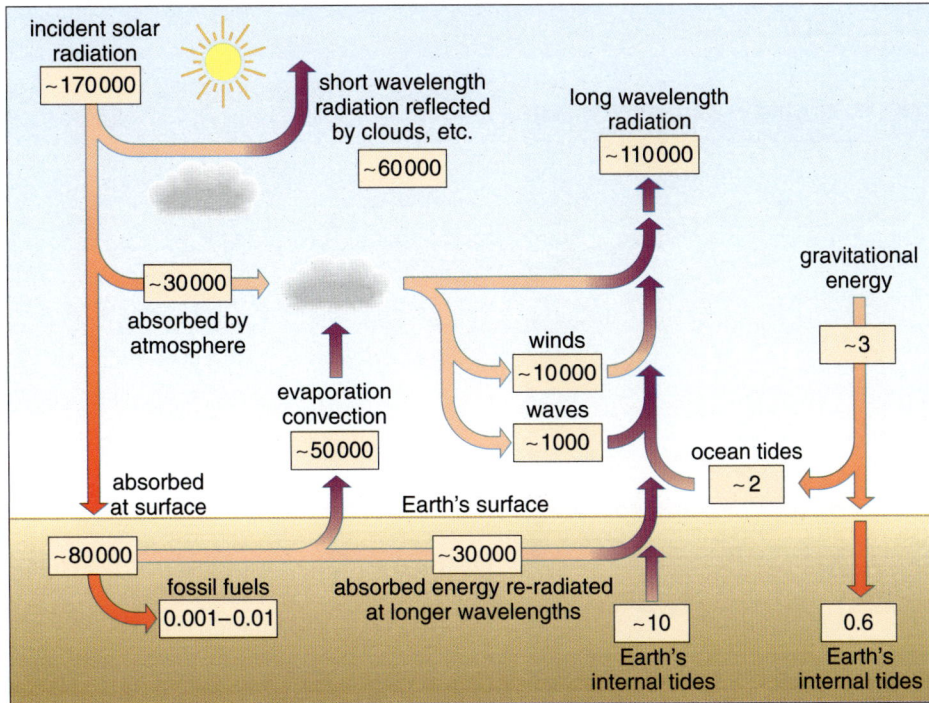

Figure 1.7 The exchange of energy between the Earth's natural energy systems per second — the power of these systems measured in TW.

1.3.1 Solar radiation

Over 99.9% of the energy available at the Earth's surface comes from the Sun. Solar energy emanates from a vast nuclear powerhouse producing heat, light and other types of electromagnetic radiation released by nuclear reactions. The Sun's power output is enormous, some 10^{14} TW, but only a tiny proportion, about 1.7×10^5 TW, reaches the Earth. About a third of this is reflected by clouds and the Earth's surface directly back into space (Figure 1.7).

Question 1.1

How much *energy* reaches the Earth (surface and atmosphere together) from the Sun each year? (Don't forget the numbers in Figure 1.7 are in terms of power.)

If all the solar energy that reaches the Earth could be harnessed, current human needs would be supplied thousands of times over. The reason it cannot is explained later.

Solar radiation is potentially available as an energy resource either as *direct* **solar energy** using solar cells or heating devices (Figure 1.8 and Chapter 5), or naturally through plant and animal growth. Some 50 000 TW of absorbed solar radiation is transferred back into the atmosphere through evaporation and convection (Figure 1.7). Some of this energy reappears in a usable form when water eventually returns to the surface as precipitation. This energy is largely dissipated as frictional heat and sound during the return flow of the water to the oceans, but it can also be harnessed as **hydropower**, an *indirect* form of solar energy.

Figure 1.8 The roof of this petrol station in London has a large array of solar cells that convert solar radiation directly into electricity. This provides enough electricity to power the petrol pumps and lights.

However, if the electrical power requirement of an average British household — around 3 kW — came from solar energy alone, even if every day were sunny each dwelling in Britain would need about 100 m² of solar panels, on a very large south-facing roof. But is solar energy the answer to energy supply in sunnier parts of the world?

- California has an area of about 4.1×10^{11} m². Even if the Sun shines there for 12 hours each and every day of the year, what percentage of California's land surface area would need to be covered by solar panels to supply the energy demands of the whole of the USA? The USA used 96.5 EJ of energy in 2003. A 10% efficient, metre-square solar panel would supply 31 J every second.

- Assuming that solar panels operated for 1.6×10^7 seconds a year (12 hours a day), one such panel would supply about 5×10^8 J. The USA would therefore need around 2.0×10^{11} such panels (i.e. 96.5×10^{18} J divided by 5×10^8 J). Since the area of California is 4.1×10^{11} m², about 50% of it would have to be covered by solar panels to meet US energy demand. (The assumptions we have made are so unrealistic that the figure would be much higher, were a 'solar energy only' scheme to be implemented.)

What happens to the solar energy that reaches the Earth's surface? Because the Earth is a sphere, the heating effect of the Sun is greater in equatorial latitudes than at the poles. Coupled with the Earth's rotation, this produces winds which blow between belts of high and low atmospheric pressure. About 10% of the Earth's *absorbed* solar power (10 000 TW; Figure 1.7) gives rise to winds. Much of the energy of winds is dissipated as heat, but some 10% of this wind power (i.e. 1000 TW) is transferred to waves through frictional effects at the sea surface.

● Could the world be supplied from wind and wave power one day?

○ Theoretically, yes, but practically, no. Total global wind power amounts to 10 000 TW, i.e. around one thousand times human power demands. However, tapping wind power over one-thousandth of the world's surface (some 510 000 km^2 or over twice the size of Britain) would be a colossal undertaking and the process would have to be fully efficient. Similar arguments apply to wave power, except that here a maximum of 1000 TW is theoretically available (100 times annual demand), requiring twice the surface area of the Mediterranean Sea to be tapped at 100% efficiency.

1.3.2 Tides

Tides are caused by the gravitational pull of the Moon and to a lesser extent the Sun. Although tides affect all fluid bodies on Earth in some measure, including some parts of the solid Earth itself, their main effect is on the seas and oceans. Ultimately the kinetic energy of tides is converted into heat, mainly through friction between water and the sea bed. Tides can be exploited as an energy resource, and the total amount of power available can be calculated from knowledge of the gravitational effects of the Earth–Moon–Sun system. At about 2 TW, it is many orders of magnitude less than the power of solar radiation (Figure 1.7) and less than 20% of the current power demand for human activities.

1.3.3 The Earth's internal heat

The occurrence of both volcanoes and hot springs shows that the Earth's interior is hot, producing molten rock at temperatures up to 1250 °C, and also superheated steam. However, these phenomena are mainly confined to several narrow zones along the world's active plate boundaries. Many measurements have now been made of the amount of heat flowing from the Earth's interior. Outside the distinctive zones mentioned above, heat flow varies from 40–120 milliwatts per square metre (mW m^{-2}), largely generated by the decay of long-lived radioactive isotopes within the Earth. The total power output from the Earth's interior, estimated at some 10 TW, is many orders of magnitude less than the total incident solar power (Figure 1.7).

● Could the global energy requirement come entirely from geothermal sources eventually?

○ No. At 10 TW, this source is roughly only the same as current power demand for human activities. All of it would need to be used (including that from the ocean floor) at an impossible 100% efficiency.

1.4   Fossil fuels

Part of the incoming solar energy becomes stored in **fossil fuels** (oil, gas and coal; Figure 1.7). To understand why fossil fuels exist you need first to know where the major stores of carbon are on the planet and how, through organic activity, this carbon becomes fixed in rocks and thus liable to be stored for geologically long periods.

1.4.1   Natural stores of carbon

The major natural stores of carbon (called 'reservoirs') are shown in Figure 1.9.

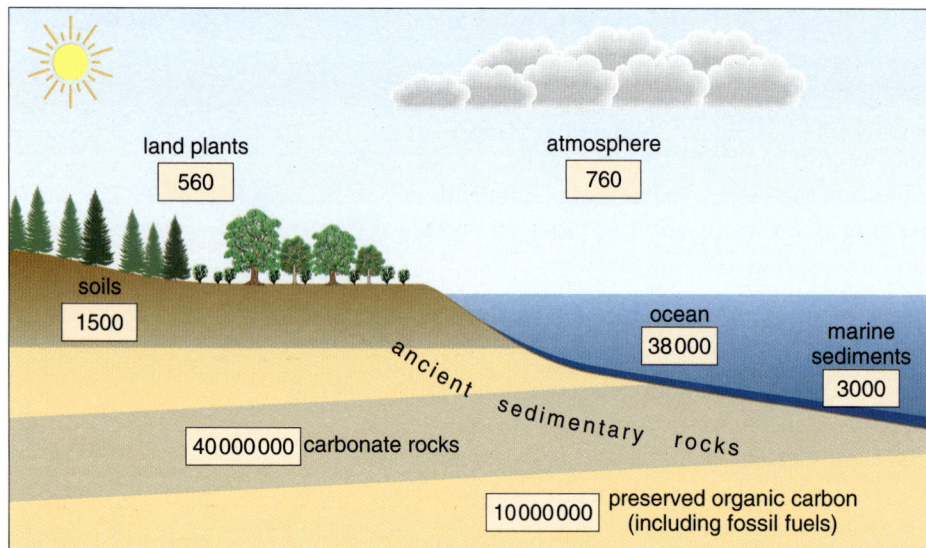

Figure 1.9 The seven major global reservoirs of carbon. Amounts of estimated carbon are in 10^{12} kg. Only about 0.1% of the preserved organic carbon is in the form of fossil fuels, the rest being finely divided at low concentrations in sedimentary rocks.

Which two reservoirs contain most of the carbon?

Carbonate rocks and preserved organic carbon.

Carbon is being *exchanged* continually between the principal reservoirs shown in Figure 1.9. However, since most stored carbon is held in carbonate rocks and preserved organic carbon (POC), the principal carbon exchange over geological timescales (millions of years) is from the surface reservoirs into limestones and POC.

For our purposes two exchange systems can be distinguished (somewhat artificially because in practice the two are intimately linked):

1   the land-based or *terrestrial* system in which carbon is exchanged between land plants and both the soil and the atmosphere;

2   the *marine* system which exchanges carbon within the oceans and between the oceans and the atmosphere.

Together they form the natural **carbon cycle**.

1.4.2 The terrestrial carbon cycle

Figure 1.10 shows the *rates* of natural carbon exchange between the terrestrial system and the atmosphere.

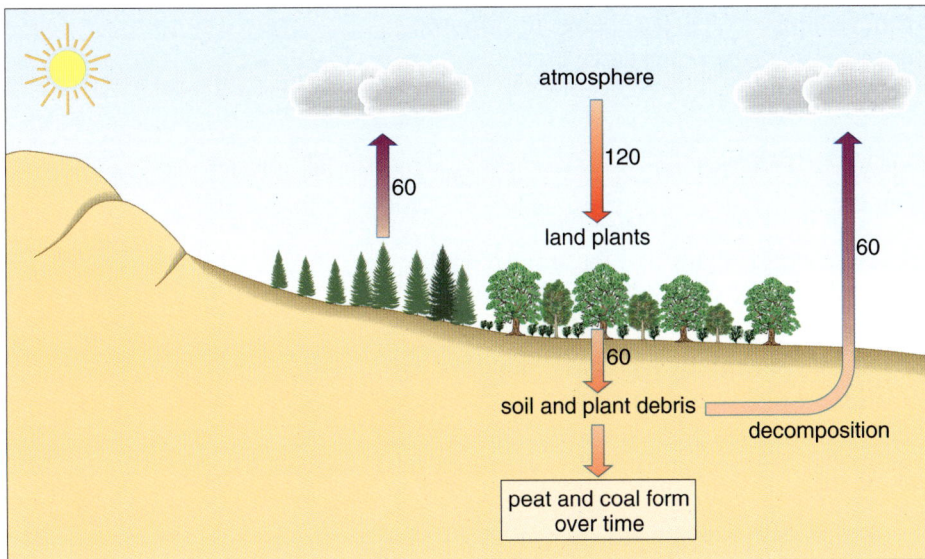

Figure 1.10 The terrestrial part of the natural carbon cycle, showing the rate of exchange of carbon between the land environments and the other carbon reservoirs. Plants absorb carbon dioxide from the atmosphere by photosynthesis and release it back into the atmosphere by respiration. Rates of carbon exchange are expressed as 10^{12} kg yr^{-1}. Only a tiny proportion of carbon enters geological storage.

● Using the data in Figure 1.10, calculate whether there is a net movement of carbon into or out of the atmosphere, as far as the terrestrial carbon cycle is concerned.

○ Figure 1.10 shows that 120×10^{12} kg yr^{-1} flows from land plants and soil and plant detritus into the atmosphere, and 120×10^{12} kg yr^{-1} moves from the atmosphere into land plants. Therefore, as far as the terrestrial carbon cycle is concerned, there is *on average* a net balance of carbon flow. (We show later how special conditions allow accumulations of plant material.)

The average time that carbon stays in a reservoir before moving to another reservoir is known as the **residence time**, and is measured by the amount of carbon in the reservoir divided by the transfer rate of carbon between it and other reservoirs.

● What is the average residence time of land-derived carbon in the atmosphere, which contains about 760×10^{12} kg of carbon?

○ The residence time of terrestrial carbon in the atmosphere, as defined above, is 760×10^{12} kg$/120 \times 10^{12}$ kg yr^{-1}, or just over 6 years, on average.

As Figure 1.10 shows, each year land plants take 120×10^{12} kg of carbon from the atmosphere. In the next section we consider how exactly carbon is exchanged between the atmosphere and plants.

1.4.3 Photosynthesis, respiration and decay

Green plants absorb solar radiation and use its energy to fuel **photosynthesis** — a chemical reaction in which carbon dioxide (CO_2) from the atmosphere is combined with water (H_2O) to form **carbohydrates** with the general formula $C_nH_{2n}O_n$. One of the simplest carbohydrates, glucose, has the chemical formula $C_6H_{12}O_6$, so in its simplest form photosynthesis can be represented by the balanced chemical equation:

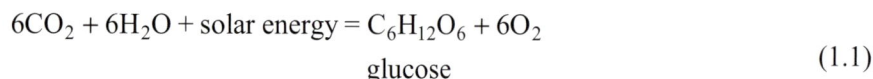

$$6CO_2 + 6H_2O + \text{solar energy} = C_6H_{12}O_6 + 6O_2$$
$$\text{glucose}$$

(1.1)

The oxygen produced by this reaction is released by plants into the atmosphere. Carbohydrates act as a store of energy for plants and also for other organisms that eat them. Such organisms use oxygen from the air to react with the carbohydrates (and other substances) to liberate energy by a process called **respiration**. During respiration, carbon dioxide and water are returned to the atmosphere. Expressed in the simplest chemical terms, the balanced reaction is:

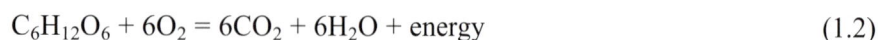

$$C_6H_{12}O_6 + 6O_2 = 6CO_2 + 6H_2O + \text{energy}$$

(1.2)

Equation 1.2 is the exact reverse of the photosynthesis reaction in Equation 1.1.

Carbon exchanges or *fluxes* link the chemistry of the atmosphere with plant and animal chemistry. The carbon taken from the atmosphere (*fixed*) by plants enables them to grow, but in addition much of it enters the food chain as either living or dead material. Living plants are eaten by herbivores which themselves may become food for carnivores. The dead material provides food for the decomposers (bacteria and fungi) that live in plant detritus, in the soil, and on the rotting remains of dead animals. Almost all organisms return some carbon to the atmosphere through respiration, but by far the greatest contribution comes from the activities of the decomposers. The timescale by which this takes place is measured in months and years, so plant and animal material is not normally available to be preserved as fossil fuels.

However, if organic matter decays in an environment where the oxygen supply is limited, carbohydrates cannot be broken down completely to form water and carbon dioxide. In this special oxygen-poor (**anoxic**) environment, a carbohydrate comparatively enriched in carbon may be produced. For example, within the waterlogged environment of a swamp (mire), cellulose (a common constituent of plants) can be broken down according to the following reaction:

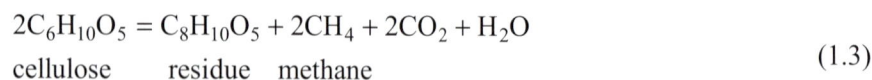

$$2C_6H_{10}O_5 = C_8H_{10}O_5 + 2CH_4 + 2CO_2 + H_2O$$
$$\text{cellulose} \qquad \text{residue} \quad \text{methane}$$

(1.3)

The residue produced, $C_8H_{10}O_5$ is relatively enriched in carbon compared with the original cellulose ($C_6H_{10}O_5$). This breakdown reaction releases **methane** (CH_4), as well as carbon dioxide and water. Methane is an organic compound containing carbon and hydrogen but no oxygen; one of a family of organic compounds known as **hydrocarbons**. (Hydrocarbons are covered in more detail in Section 3.1.) So anoxic environments prevent some fixed carbon returning to the atmosphere as CO_2, and these hydrocarbons together with carbon-rich residues represent a chemical half-way-house within the carbon cycle, which make carbon available to form the basis for fossil fuels.

Although significant layers of decaying plant debris are found on the floor of modern forests, these are often oxygen-rich environments thanks to the constant reworking of decaying material by plants and animals, fungi and bacteria. However, one modern environment with which you are probably familiar does contain plant material decaying in anoxic conditions — the peat bog. As you will see in Chapter 2, this is where useful preservation of terrestrial carbon occurs.

1.4.4   The marine carbon cycle

The ocean stores much more carbon than the terrestrial system (Figure 1.9). How is this marine carbon fixed into organic carbon within the sediments, and what are the main reasons for marine carbon fluxes? Figure 1.11 shows the rates of natural carbon exchange within the ocean and between the ocean and the atmosphere.

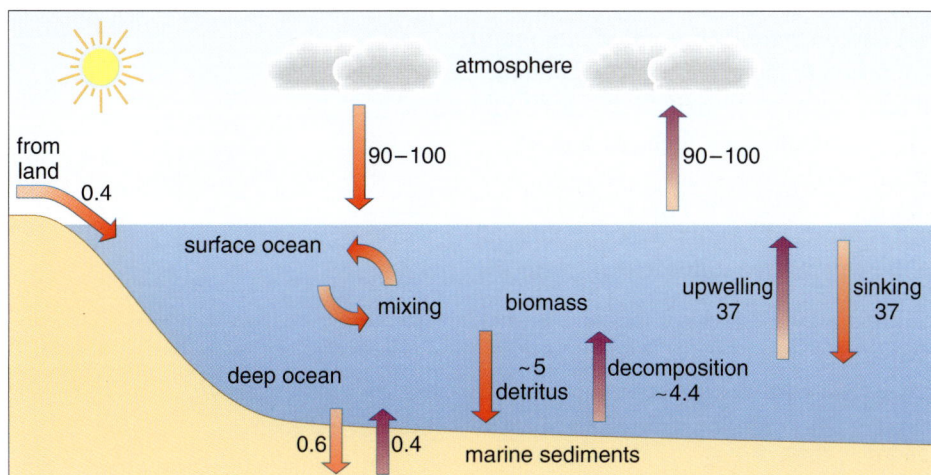

Figure 1.11 A schematic representation of the rate of exchange of carbon between stores in the marine environment and between the sea and the atmosphere. Rates of carbon exchange are expressed as 10^{12} kg yr^{-1}.

● From the data presented in Figure 1.11, is there a net movement of carbon into or out of the atmosphere from the ocean?

○ Figure 1.11 shows that $90–100 \times 10^{12}$ kg of carbon flows each year from ocean surface waters into the atmosphere, while roughly the same amount of carbon moves from the atmosphere into the ocean: there is no net movement.

Carbon dioxide is constantly being exchanged between the atmosphere and the upper levels of the oceans, by physical and by chemical processes. At the base of the food chain that produces organic material in the oceans are the marine phytoplankton (microscopic water-borne plant life) which require dissolved CO_2 for photosynthesis. The process also requires solar energy, so phytoplankton can live only in the sunlit upper parts of the ocean. Their photosynthesis releases oxygen, which dissolves in seawater. The productivity of phytoplankton depends on sunlight, temperature and supply of nutrients, and therefore varies geographically.

Carbon dioxide is more soluble in cold water than it is in warm water, so the concentration of dissolved CO_2 tends to be higher in cold polar waters than in warm tropical waters. Cold, dense polar water sinks and flows under the influence of gravity along the ocean floor towards the Equator. It returns to the surface by *upwellings* at various places in the oceans, to supply nutrients and promote unusually high phytoplankton productivity there.

Zooplankton (water-borne animal life, mostly microscopic) and higher marine organisms consume these phytoplankton. The dead remains of phytoplankton, zooplankton and larger organisms sink through the water column, transferring carbon from the upper few hundred metres towards the ocean depths. However, little of this organic matter gets a chance to accumulate on the ocean floor. It provides food for filter feeders in deep water and on the ocean floor, and through them for predatory animals, and ultimately feeds the ubiquitous decomposers. All these organisms release CO_2 back into solution through respiration. As Figure 1.11 shows, only a small proportion of the marine carbon cycle, 0.2×10^{12} kg yr^{-1} of carbon, is incorporated into marine sediments.

- Assume that 0.2×10^{12} kg yr^{-1} of carbon were incorporated into marine carbonates and fossil fuels in proportion to their present-day amounts (Figure 1.9). How long would it have taken to deposit all the carbon found in the current global store of POC, of which fossil fuels are a part?

- We know from Figure 1.9 that the global carbonate store is some 4×10^{19} kg of carbon, and that the POC store is some 10^{19} kg of carbon, a total of 5×10^{19} kg of carbon. If carbon were deposited into each reservoir in proportion, the rate of deposition of carbon into POC would be:

$$\left(\frac{1 \times 10^{19} \text{ kg}}{5 \times 10^{19} \text{ kg}}\right) \times 0.2 \times 10^{12} \text{ kg yr}^{-1} = 0.04 \times 10^{12} \text{ kg yr}^{-1}$$

At this rate, it would have taken 1×10^{19} kg$/0.04 \times 10^{12}$ kg yr^{-1}, which is 2.5×10^8 years, i.e. 250 million years to deposit enough carbon to form the global POC store, which includes fossil fuels.

However, not all fossil fuel formed in the marine carbon cycle; much of it formed on land (Section 1.4.3). The preservation of organic material in sediments depends not only on the supply of dead organisms, but also on anoxic chemical conditions where they accumulate. This will be discussed further in Chapters 2 and 3.

1.4.5   Generating carbon — the legacy of volcanoes

What is the origin of the carbon within the carbon cycle? Figure 1.9 showed that the greatest proportion of the global carbon store is locked into carbonate rocks. Over the 4.5 billion years of the Earth's history, carbon must have moved from the atmosphere into the oceans and thence into carbonates. How did the atmospheric carbon originate?

The Earth's atmosphere as a whole was derived mainly from gases brought to the surface from the Earth's interior. For example, the 1991 eruption of Mount

Etna in Sicily released an estimated 0.01×10^{12} kg of carbon in the form of CO_2 (Figure 1.12). Most volcanic carbon comes from the steady degassing of lava flows rather than from volcanic vents.

In the short term, volcanic sources release insignificant volumes of CO_2 compared with other fluxes of carbon, but over geological time, degassing of the Earth's interior can reasonably account for all the carbon in the natural surface system.

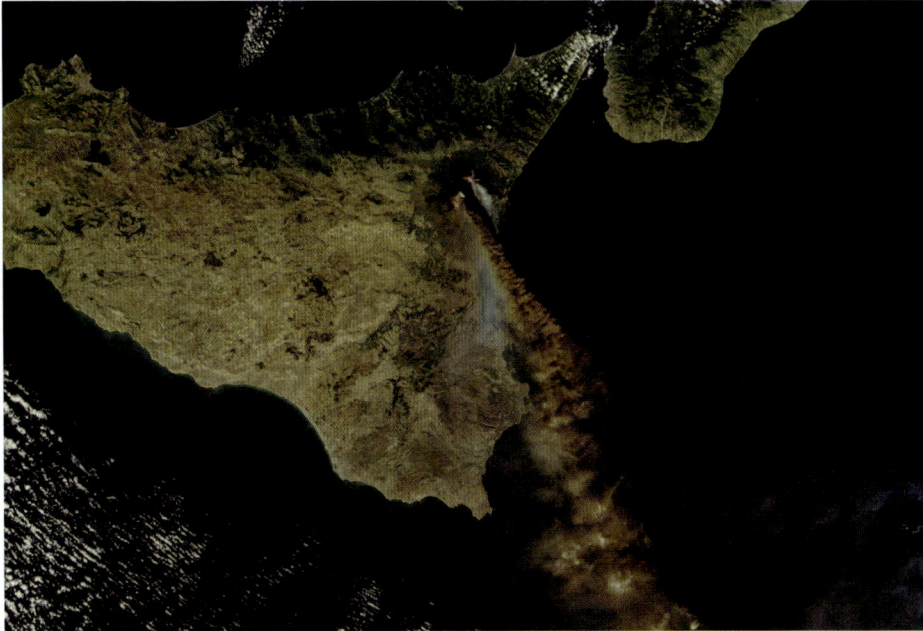

Figure 1.12 The 2002 eruption of the volcano Mount Etna, which added carbon to the global carbon cycle. The plume of gas and ash that shows up was blown towards the south-south-east from the volcano's vent.

1.4.6 The fossil fuel 'bank'

During the period of accumulation of most coal and petroleum, the past few hundred million years, the equivalent of around 10^{23} J of chemical energy have been 'banked' by Earth processes. As you have seen in Sections 1.4.3 and 1.4.4 carbon is added to these reservoirs continually, at a rate today that is equivalent to about 5×10^{17} J yr^{-1}. That rate of growth represents roughly a mere thousandth of present world energy demand.

Question 1.2

If the consumption of energy from all primary sources in 2002 were 451 EJ, how long would the fossil fuel energy 'bank' last at that rate of consumption?

In fact, the 'bank' contains only about 0.04% of the amount of organic carbon preserved in sedimentary rocks. However, that total store of buried organic carbon is finely dispersed, at an average concentration of about 0.4% in sedimentary rocks. At that level, it can never be exploited, either economically or with an efficiency that yields more energy than it consumes.

Although buried organic carbon is widely distributed, concentrations sufficient to constitute resources are rare and very restricted in space and geological time relative to small amounts of organic carbon that are present in many sediments.

1.5   Nuclear energy

Einstein's famous equation $E = mc^2$ shows that mass (m) and energy (E) are proportional to one another. The constant c^2 linking the two is the square of the speed of light c (3×10^8 m s^{-1}). Implicit in the equation is that mass can be converted into energy, and vice versa, although the conversion of energy into mass occurs only in very powerful particle accelerators. The conversion of matter into energy is the basis of nuclear energy (Chapter 4). When unstable nuclei split (nuclear fission) the sum of the masses of the isotopes that are produced is slightly less than that of the original fissile isotope. That tiny mass deficit is converted into a relatively huge amount of energy, because c^2 is a very large number (9×10^{16} m^2 s^{-2}). When isotopes fuse at immense temperatures in the interiors of stars, the product again has lower mass than the original isotopes. So both nuclear fission and fusion potentially produce energy that can be exploited, but in both cases the phenomena have to be artificially induced.

The 'fuel' for nuclear fission occurs naturally in the Earth's crust in the form of unstable isotopes of uranium and thorium, added to by other isotopes that are formed artificially inside nuclear reactors. Note that uranium and thorium isotopes naturally break down by emission of various particles (helium nuclei and electrons), as a result of which their 'daughter' isotopes have lower atomic masses and numbers. Some of the 'daughter' isotopes are also unstable and undergo radioactive decay, which eventually results in the formation of several stable isotopes of the element lead. Such radioactive decay is *not* the same as nuclear fission and occurs at a constant pace. Radioactive decay also releases energy, but at an amount far lower per decayed atom than in nuclear fission. That energy continually heats the Earth's interior (Section 1.3.3) and is the source of *geothermal* energy (Chapter 5).

Potential 'fuel' for nuclear fusion also occurs naturally in the form of an isotope of hydrogen that includes a neutron as well as a proton in its nucleus, instead of the single proton. This isotope (^{2}H or *deuterium*) occurs in tiny amounts in water, and was produced originally by processes early in the history of the Universe.

1.6   Concentrating, storing and transporting energy

The Earth is awash with energy from sources other than fossil fuels; thousands of times as much as humans use. Why then do we need to use any other energy supply?

1.6.1   Concentrating energy

As far as human needs are concerned, there is a marked difference between 'dilute' and 'concentrated' energy. Water vapour in the atmosphere, for example, has considerable potential energy since a huge mass globally (about 13×10^{15} kg — Smith, 2005) is held high above the Earth's surface. But this potential energy represents a very dilute form of energy; falling rain could not turn a water wheel. It is only when energy can be 'concentrated' that it can be put to good use — in this case by rainfall accumulating in streams and rivers, or being stored in reservoirs at high elevations. The concentration can be expressed colloquially in terms of **energy density**, which is the amount of energy stored by a resource divided by the volume of the space that it occupies.

Some forms of energy are relatively difficult to concentrate, so have a low energy density, whereas others are easier to concentrate. The energy contained in moving air is rather difficult to concentrate; windmills and wind farms have to be sited where natural factors enhance wind speed and constancy. Solar power has a low energy density, so requires large collecting devices. The potential energy of rain is naturally concentrated and held in mountain lakes; we concentrate this energy artificially when rainwater is stored in a reservoir. This emphasizes why fossil fuels are so valuable as they represent naturally concentrated forms of the solar energy that reached the Earth millions of years ago.

The ultimate form of concentrated energy is matter itself, in the form of nuclear energy.

1.6.2   Storing and transporting energy

To be useful to us, energy must be available where and when we want it, and in a form and in amounts we can handle. Severe weather systems concentrate natural energy wonderfully, but hurricanes associated with storm-force winds, driving heavy rain, thunder and lightning wreak havoc rather than top up our energy supplies.

Storing most forms of energy is very difficult. We have to re-heat our homes daily in the wintertime because they constantly lose heat, despite our attempts to insulate them. We cannot store light when the Sun goes down; we have to turn electricity into light until the Sun reappears. In fact, only two forms of energy are truly storable. Potential energy can be stored almost indefinitely by mechanical means, as in springs or lifted weights — the basis of clocks. Far more convenient is storage that exploits chemical energy — batteries, or even better, chemical fuel.

Fuels are compounds whose combustion liberates a large amount of energy per unit mass: they commonly have a high energy density. Wood was the major fuel before the Industrial Revolution, and remains the most important fuel for many non-industrial societies today. Wood, and other plants that can be used as fuels, produce **biomass** energy. As industry develops, energy demands grow and fuels with higher convertible energy content per unit mass are needed. Modern energy supply is centred on the fossil fuels: coal, oil and gas (Figure 1.4). Note that although the isotopes whose fission or fusion forms the basis for nuclear power are not burnt, they are generally known as nuclear *fuels* (Chapter 4).

A further advantage of fuels as energy sources is their transportability, so that conversion can take place on selected sites or in mobile units. Highly concentrated fuels require less energy to transport than those with a low energy density but since lots of energy can be released accidentally from badly handled concentrated energy sources, care has to be taken to ensure that transport is safe. For some applications, such as cars, generating energy from stored chemical energy has the advantage of the ease of transport of small amounts of fuel.

Fossil fuels therefore represent extremely useful energy sources because they have a high energy density, they store energy for very long periods to be used when needed, and they can be transported simply and relatively safely.

A very important form of energy transport in developed countries is the generation and transmission of electricity. Figure 1.6 shows that around one-sixth of energy used in the UK is distributed as electricity via the National Grid. Since every primary energy source is used to some extent in generating electricity, a brief introduction to generators is useful (Box 1.2).

Box 1.2   Electricity generation

By far the greatest contribution to electricity supplies is made by exploiting the way in which an electrical current is induced in a conductor when it is moved through a magnetic field. Most such generators are made up of a cylindrical coil of conducting wire that is rotated between extremely strong magnets, so that an electrical current is induced in the coil (Figure 1.13a). There are other means, such as fuel cells and photovoltaic devices, but generators are common to conversion of most kinds of primary energy. Driving the rotation depends on harnessing energy released from primary energy sources through a specific form of engine, the turbine.

A turbine is a rotary engine driven by a moving fluid. A propeller on a boat or an aircraft, or a jet turbine, provides a driving force because of the design of its blades. A turbine used in electricity generation simply exploits the reverse of this effect. Moving 'fluid' provides the force that spins the turbine. The rotation transmits energy, and for electricity generation it is connected to a rotary generator. The simplest turbines have one moving part, as in windmills or water wheels (Figure 1.13b).

Electricity generation from flowing water or wind exploits the kinetic energy of the fluid. However, from the 19th century to the present day, the most common turbines used to generate electricity have been driven by high-pressure steam created by boiling water using coal, oil, natural gas, nuclear or geothermal energy (Section 5.1). In some electricity generating stations the moving fluid is the gas produced by burning oil or natural gas, in the manner of a jet turbine, thereby missing out the intermediary of steam and the associated inefficiency. The principle of a steam turbine is somewhat different from those used to harness the energy of wind and flowing water, and so too is its design (Figure 1.13c). Steam contains energy in three forms: as

heat; as the energy needed to change water's state from liquid to gas (*latent heat of vaporization*); and as the energy bound up by the compression of a pressurized gas. The last is a form of potential energy, released when the pressure drops, as in the case of the air in a tyre when its valve is opened. Together, these three forms of contained energy are known as the **enthalpy**. The enthalpy of a gas is given by:

$$H = U + PV \qquad (1.4)$$

where H is enthalpy (in J), U is heat (including latent heat of vaporization; in J), P is pressure (in pascals or kg m^{-1} s^{-2}) and V is the volume of steam (in m^3). (Note that pressure times volume — in kg m^2 s^{-2} — gives units in joules.) It is the PV term that provides much of the driving force for a steam turbine, because the pressure gradient across the turbine leads to rapid expansion of the steam and therefore high speed in the flow. Many steam engines, including the original piston engines invented in the early Industrial Revolution, also exploit the latent heat of vaporization, by condensing the steam at the outlet of the turbine. The change of state from gas to liquid results in a near vacuum that increases the pressure gradient across the turbine — condensation adds 'suction' to the 'blowing' by steam. Latent heat is released and this can be 'recycled' in the generating system.

The late-20th century saw an increased reliance on electricity generation that uses natural gas as a fuel. Gas turbines are similar in design to those used in aircraft jet turbines. Turbines rotated by combustion gases have two important advantages over steam turbines: they are more efficient, and can be turned on and off very quickly, thereby suiting variable demand for electricity. Natural gas also contains a lower proportion of carbon than coal and oil, and so less carbon dioxide is emitted to the atmosphere.

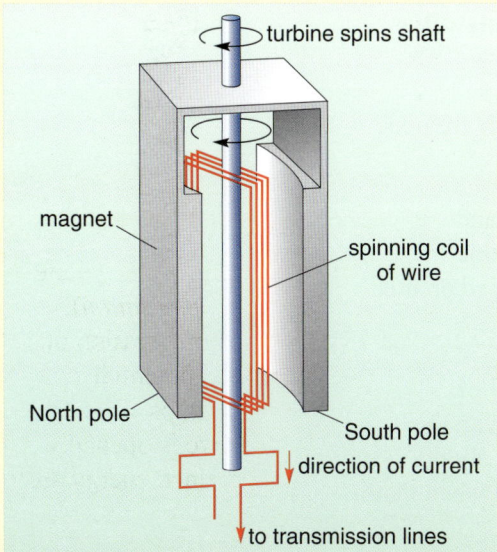

(a)

(b)

(c)

Figure 1.13 Basic principles of electricity generation. (a) A schematic generator. (b) A water wheel is a form of turbine, in this case used to lift water for irrigation. (c) The blades of a modern steam turbine with its casing removed.

1.7   Renewable and non-renewable energy supplies

Energy resources can be considered in a completely different way from their energy density — whether or not they are renewable. Some energy sources incorporate energy released comparatively recently from the Sun and are replenished naturally over a timescale of days to tens of years. Therefore solar, wind and wave energy resources, being continually available, are **renewable** energy supplies. Other examples of renewables are geothermal and tidal energy sources.

Other potential energy resources are legacies of solar power received and converted into stored energy in the geological past; coal is a good example. Coal seams are replenished naturally over a timescale of millions of years. Once coal began to be exploited faster than its rate of creation it became **non-renewable** and its use at current levels cannot be sustained.

- Which energy sources mentioned so far are renewable, and which are non-renewable on the scale of a human lifetime?

- Energy sources such as solar, biofuels, hydro, wind, waves, tides, and geothermal are continually replenished, so they are renewable on this timescale. The fossil fuels and nuclear energy are not being replaced on this timescale, so they are non-renewable.

All fossil fuels are slowly being renewed by the death, burial and decay of present plant and animal life, but at an extremely slow rate compared with the pace of human economic activity. The Earth's natural systems may eventually replace all the fossil fuels that humanity has already used, but how and when that might become possible is impossible to judge.

Likewise, only extremely slow geological processes renew fissionable nuclear fuels. The instability of naturally radioactive isotopes such as uranium and thorium, which formed in stars much larger than the Sun during the moment of their destruction as supernovae, results in them gradually diminishing with time. Replenishing the ores of uranium and thorium by geological processes takes hundreds of million years. So, nuclear fuels too are non-renewable. Even the hydrogen isotope deuterium, which is potentially a vast source of energy from thermonuclear fusion, is a 'one-off' legacy resource and is finite. But even so its potential vastly outstrips that of all other non-renewable energy sources.

The distinction between renewables and non-renewables is one of timescale and energy concentration, but it is critical for human society. Think of renewable energy resources as *income*, and non-renewable energy resources as *inheritance*. We 'spend' the Earth's energy resources constantly for cooking, travelling, heating or cooling buildings, manufacturing and in many other ways. At present, modern industrial societies generate energy mostly from fossil fuels, thereby depleting an inheritance accumulated from millions of years of 'banked' solar energy and internal heat. Much less energy is currently generated from day-to-day energy 'income', i.e. from renewables: the global energy 'accountancy' has become unbalanced and unsustainable.

Setting aside the environmentally damaging effects of burning fossil fuels, such as atmospheric pollution and global warming (discussed in later chapters, particularly Chapter 6), sooner or later present energy generating policy will deplete our stock of fossil fuel. To stay solvent in energy terms *over the long term* leaves no choice other than to transfer society's day-to-day energy supply to renewable sources, or return to a low-energy society. The other alternative is harnessing energy from nuclear fission or fusion, but that too cannot last indefinitely. This book aims to provide a scientific basis to understanding some of the decisions that will need to be made to enable humanity to stay 'solvent' in energy terms. Yet you will no doubt be well aware that such decisions are not those of individuals, but are presently dominated by economic and political factors, irrespective of the scientific facts.

1.8 Summary of Chapter 1

1 Energy is the basis of modern society. Other physical resources can only be effectively extracted, processed and transported if there is a ready supply of energy at the right price.

2 Energy is defined as the capacity to do work, while work is a force acting on an object that causes its displacement (i.e. force × distance). Both energy and work are measured in joules or, more fundamentally, in newton metres. Power (measured in watts, i.e. joules per second) is the rate of doing work or the rate at which energy changes from one form to another.

3 All conversions of energy are inefficient to varying degrees, so that primary energy consumption far exceeds the useful work that it makes possible.

4 The Sun is by far the most important source of natural energy on Earth. The solar radiation that reaches the Earth contributes to winds, waves, atmospheric water circulation, atmospheric heating and surface water evaporation, and to organic activity.

5 The gravitational attractions of the Sun and the Moon combine to produce tides, and rocks in the Earth's interior also generate heat by the decay of radioactive isotopes in them. These are small but potentially exploitable sources of energy.

6 Fossil fuels are ultimately derived from solar radiation, through photosynthesis and the carbon cycle.

7 Most of the world's carbon is locked within carbonate rocks. A large amount of carbon also exists as preserved organic carbon, which includes fossil fuels.

8 Green plants use solar radiation to build carbohydrates and plant tissue from carbon dioxide and water in the atmosphere and dissolved in the oceans, in a process known as photosynthesis. Photosynthesis releases oxygen into the atmosphere and oceans. When they respire, organisms use oxygen to generate energy from food, releasing carbon dioxide and water vapour back into the atmosphere and oceans. These respiratory reactions are the reverse of photosynthesis.

9 Concentrations of marine phytoplankton occur in the upper sunlit layers of the oceans where upwelling currents bring nutrients. These form the basis of the marine food chain. There is only a build-up of carbon within marine sediments where there is an adequate supply of organic material and where physical conditions are right for its preservation.

10 Nuclear energy is derived from the conversion of matter into energy and has a very high energy density.

11 The world is plentifully supplied with solar-derived energy, but most of it is not in a sufficiently concentrated form to be useful to modern industrial society.

12 Fuels are of immense value because they are concentrated sources of energy (they have a high energy density) that can easily be stored, transported and used at will. At the current level of technology, energy

transport has become an essential aspect of energy use, dominated by electricity. The most convenient way of electricity generation is through the conversion of mechanical motion — most usually produced today from thermal energy — using generators driven by turbines.

13 Energy sources can be subdivided into renewables, like solar, wind and wave power, and non-renewables, like peat, coal, oil and gas. Renewables are effectively everlasting, but non-renewables are finite.

COAL

There are many environmental reasons why coal is a rather undesirable source of energy. Burning it introduces large amounts of gases into the atmosphere that harm the environment in a variety of ways, as well as other, solid waste products. Coal extraction leads to spoil heaps and mines that scar the landscape, land subsidence that affects roads and buildings, and in some cases water pollution.

With apparently so little going for it, why do we rely so much on coal to meet our energy needs? In this chapter, it will become apparent that the most appealing quality of coal is that there is plenty of it. Coal is twice as important globally as any other fuel in generating electricity, and could remain so for the next 200 years. That is reassuring for a future where energy demands continue to increase and when the alternatives to coal are currently looking less dependable. The downside is that continued burning of coal could have dire consequences for the environment in the coming centuries (Chapter 6), unless 'cleaner' ways can be found to harness energy from it.

Before these issues are explored in more detail, it is necessary to explore the basics: what coal is, how and where it is found, and how it is extracted at a variety of depths below the surface. Another important theme concerns the distribution of coal reserves and resources, and the control exerted on them by both economics and politics.

2.1 The origins of coal

If you examine a piece of coal, at first sight it appears black and rather homogenous. However, closer inspection generally shows a series of parallel bands up to a few millimetres thick. Most obvious are shiny bands that break into angular pieces if struck. Between them are layers of dull, relatively hard coal and thin weak layers of charcoal-like carbon. Coal splits easily along these weak layers, which crumble to give coal its characteristic dusty black coating.

Microscopic examination of these bands shows that the bright coal represents single fragments of bark or wood, often showing a well-preserved cellular structure. The hard coal is an assemblage of crushed spores, fine wood debris and blobs of resin. The charcoal-like layers comprise charred fragments of bark and wood probably produced by oxygen-starved burning beneath the surface while dead vegetation was accumulating. These features suggest that **coal** forms from the highly compressed remains of land plants. Further evidence for this is provided by the preservation of individual plant fragments in sediments associated with beds of coal (Figure 2.1).

(a)

(b)

Figure 2.1 (a) A fossilized leaf. (b) Fossilized leafy branches and reproductive organs of a *Lepidodendron* tree.

2.1.1   Coal-forming environments today

Coal formation begins with preservation of waterlogged plant remains to produce **peat** and then slow compression as the peat is buried. About 10 m of peat will compress down to form about 1 m of coal; clearly large amounts of plant debris must be available for preservation. Even so, for a significant thickness of peat to accumulate there must be a balance between the growth of plants and the decay of underlying dead material to form peat (a process known as **humification**).

Such a balance occurs in areas of poorly drained land known as **mires** (swamps). Whilst there are different types of mire, they all require the water table to be at or above the land surface for most, if not all of the year. This waterlogging of mire soils restricts the supply of oxygen. Such anoxic conditions prevent the complete decay of plant matter to carbon dioxide and water by aerobic bacteria. Instead, anaerobic bacteria convert some into methane, thereby reducing the hydrogen content of the decayed matter and increasing its carbon content, which is essential for the formation of coal. Rapid burial ensures that plant material decays and compacts fast enough to accommodate new plant growth in the mire above. The process of humification is fast in hot, humid tropical areas, but peat also accumulates in cooler, higher latitudes providing a humid climate is maintained. In regions such as Siberia and Canada, mosses rather than trees are the primary source of plant material.

Currently, 3% of the Earth's surface is covered by peat. However, not all of this is likely to form coal in the future.

● Consider the extensive areas of peat which typify many upland districts of the UK and Ireland (where peat has been extracted for use as a fuel). Is this peat a good candidate for future coal formation?

○ Almost certainly not. Such peat was formed during a much wetter period of the geological past and in today's drier climate, large areas are above the water table undergoing oxidation. Therefore, this peat is likely to be eroded, rather than buried and turned to coal.

The next two sections look at different types of potentially coal-forming mires that occur today and are thought to have been significant in the geological record.

Peat formation in deltas and coastal barrier systems

Since mires require poor drainage, low-lying land close to coastal areas might provide the right conditions for peat to form. Most extensive areas of modern peat formation are indeed situated not far above sea-level, and as Figure 2.2 shows, they are commonly associated with river deltas and coastal barriers. Such environments would also have been significant areas of peat production in the geological past. However, the flooding of an area alone does not guarantee significant accumulations of peat; high productivity of organic matter is also required.

(a)

(b)

(c)

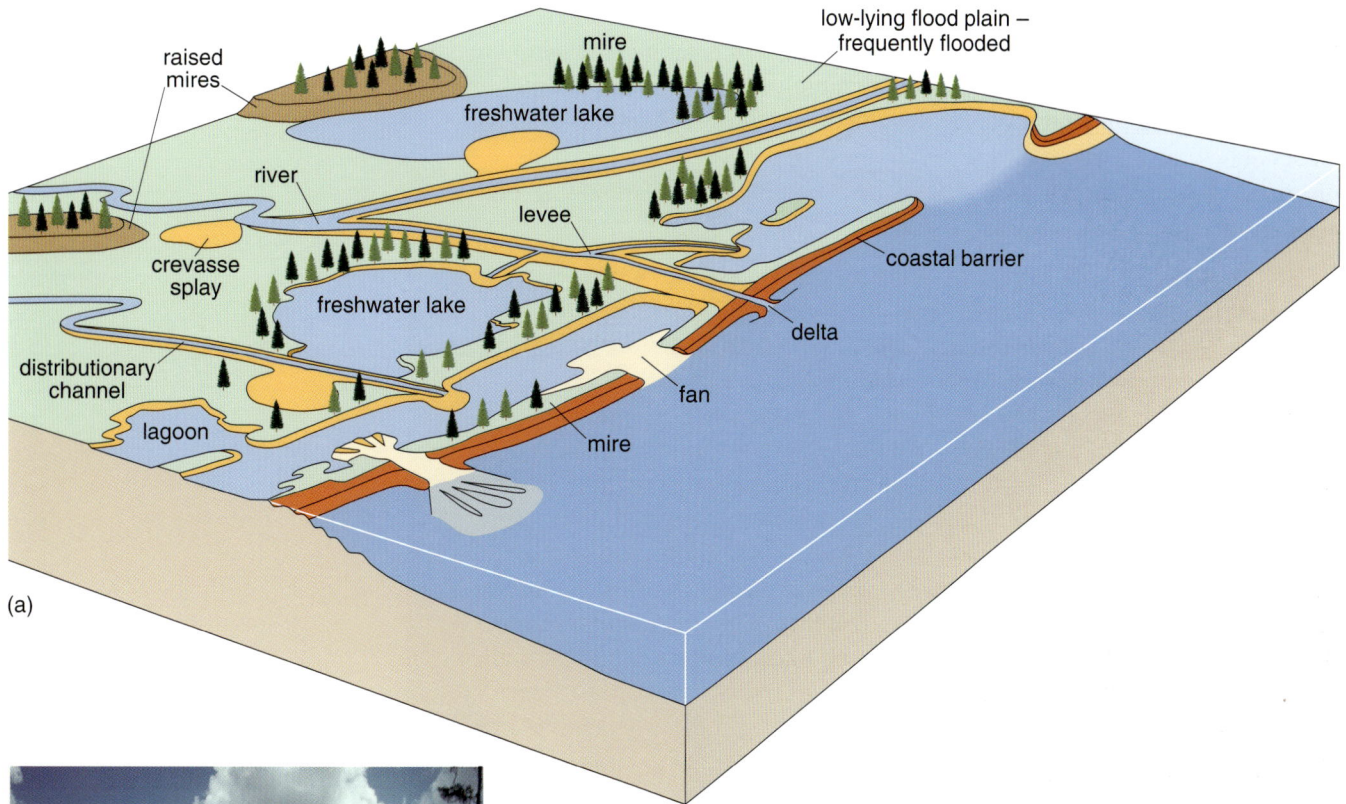

Figure 2.2 (a) The location of mires in deltas and coastal barrier systems.
(b) A modern-day example of a coal-forming mire from the Okeefenokee Swamp, Florida.
(c) Reconstruction of a late Carboniferous low-latitude swamp forest.

Deltas act as conduits bringing sediment via distributary river channels out into an open body of water (often the sea). In between these river channels are areas that receive less sediment, and it is here that high rates of subsidence and organic matter production may promote the development of mires (Figure 2.3).

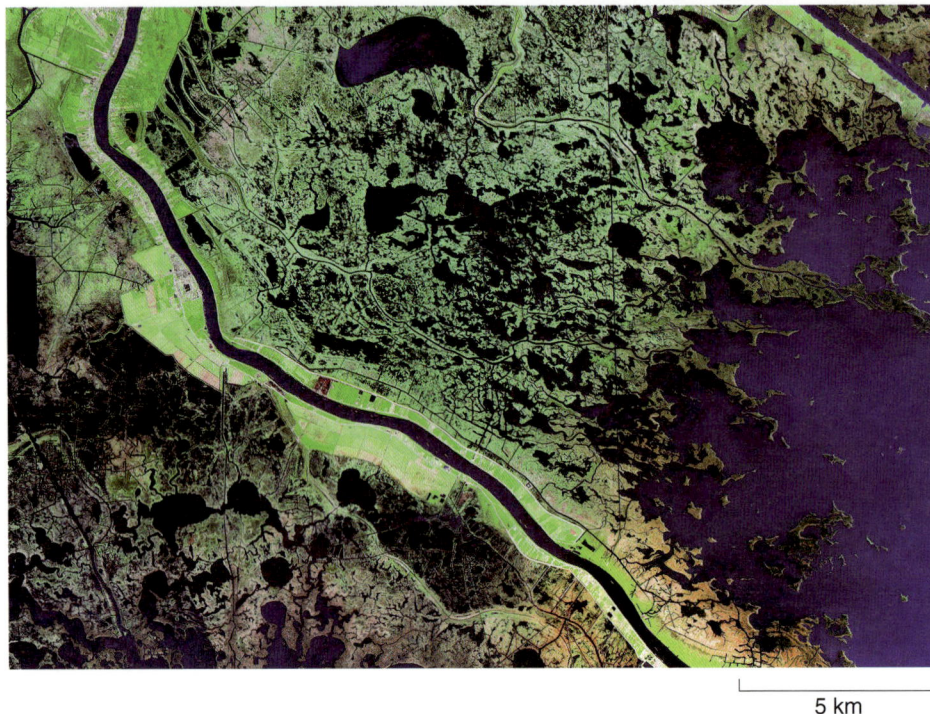

5 km

Figure 2.3 Satellite image showing dark-coloured mires on the Mississippi delta south of New Orleans. These mires are forming around small lakes, and both are surrounded by a network of stream channels. The main river channel is approximately 500 m wide, and is flanked by natural levees that have been strengthened to reduce flooding. As a result the mires grow unchecked by annual deposition of sands and silts during floods. (This image was taken in 2001 — well before the major floods in this area caused by Hurricane Katrina in 2005.)

● What is likely to happen to these areas when adjacent rivers are in flood?

○ Floodwaters will inundate the area and deposit a layer of sand or mud, which will contaminate the peat. This will (temporarily at least) halt the development of peat and, as we shall see later, contribute to the impurity of coal so formed.

As Figure 2.2a shows, a coastal barrier is a ribbon-like beach that protects a lake (or lagoon) on the landward side of the barrier from the sea. In this protected environment mires can develop along the fringes of the lagoon, or adjacent to tidal channels that cross the area.

● Considering the proximity to the sea, what process is most likely to contaminate the mire with sediment?

○ Storms will tend to wash sand off the coastal barrier and redeposit it on the mire.

Peat formation in raised mires

Mires can also form inland within low-lying depressions, provided the rate of precipitation exceeds the rate of evaporation (Figure 2.4a). Peat is impermeable and so its accumulation progressively impedes drainage. This attribute gives mires the ability to maintain a water table independent of the area surrounding them. Therefore, the water table *rises* as the peat layer increases in thickness, thus elevating the surface level of the mire (Figure 2.4b–d). Such raised mires are now thought to have been very significant environments for coal formation in the geological record.

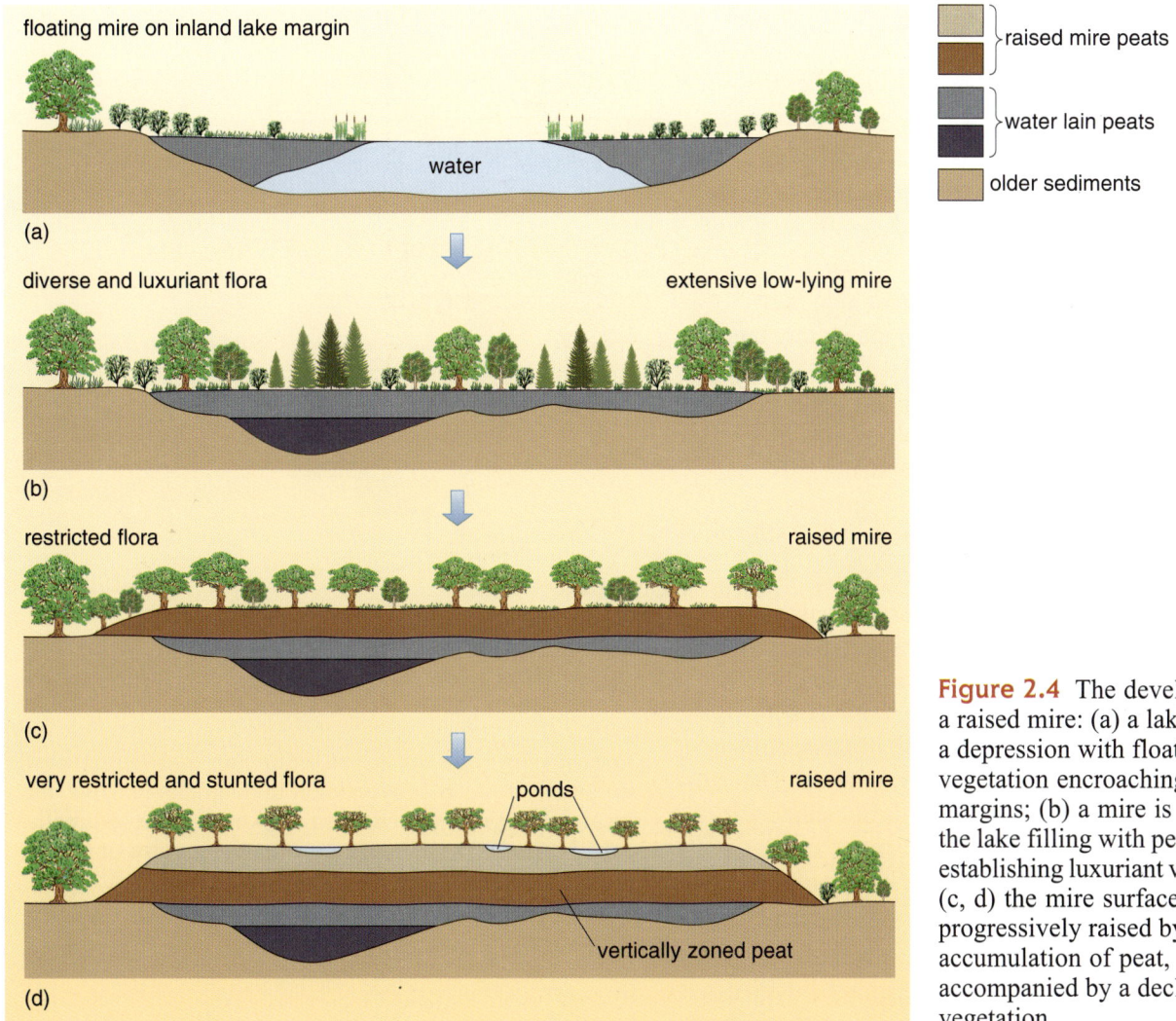

Figure 2.4 The development of a raised mire: (a) a lake forms in a depression with floating dead vegetation encroaching from the margins; (b) a mire is formed by the lake filling with peat and establishing luxuriant vegetation: (c, d) the mire surface is progressively raised by accumulation of peat, accompanied by a decline in the vegetation.

Having identified the modern environments in which the coals of the future may currently be forming, the next section will look at the evidence from ancient coal-bearing rock sequences, to see whether they formed in environments similar to those in which peats accumulate today.

2.1.2 Coal-forming environments in the geological record

Figure 2.5 simplifies a typical vertical succession of sedimentary rocks found in many coalfields. The sequence from the base of the section upwards reveals the following:

1. When a mire starts to form, the first plants take root in underlying clays or sands that form the soil. Their rootlets are preserved beneath the coal seam as black carbonaceous markings, and the fossil soil in which they are found is called a **seatearth** (Figure 2.6). The presence of rootlets shows that the peat formed *in situ*, rather than being transported into the area by water currents.

2. The coal seam itself consists largely of plant material with small but variable amounts of mud. The seam itself can vary in thickness from a few millimetres to tens of metres.

3. Immediately above the coal seam there may occasionally be a mudstone containing rare but distinctive marine fossils (for example, brachiopods and cephalopods). This is unusual because most of the other fossil remains associated with coal-bearing sequences are normally of freshwater or land-based species. Where present, marine beds suggest that peat formation ceased as the sea flooded the area.

4. The muddy sediments overlying the coal seam pass upwards first into siltstones and then into sandstones. This sequence of rocks usually gets steadily coarser upwards, but variations are common. Fossils within this part of the sequence invariably indicate non-marine conditions. The sandstones and siltstones may show sedimentary features that indicate action of waves and currents in relatively shallow water. Some sandstones show evidence of river channels.

5. The sandstones often pass upwards into seatearths and another coal layer.

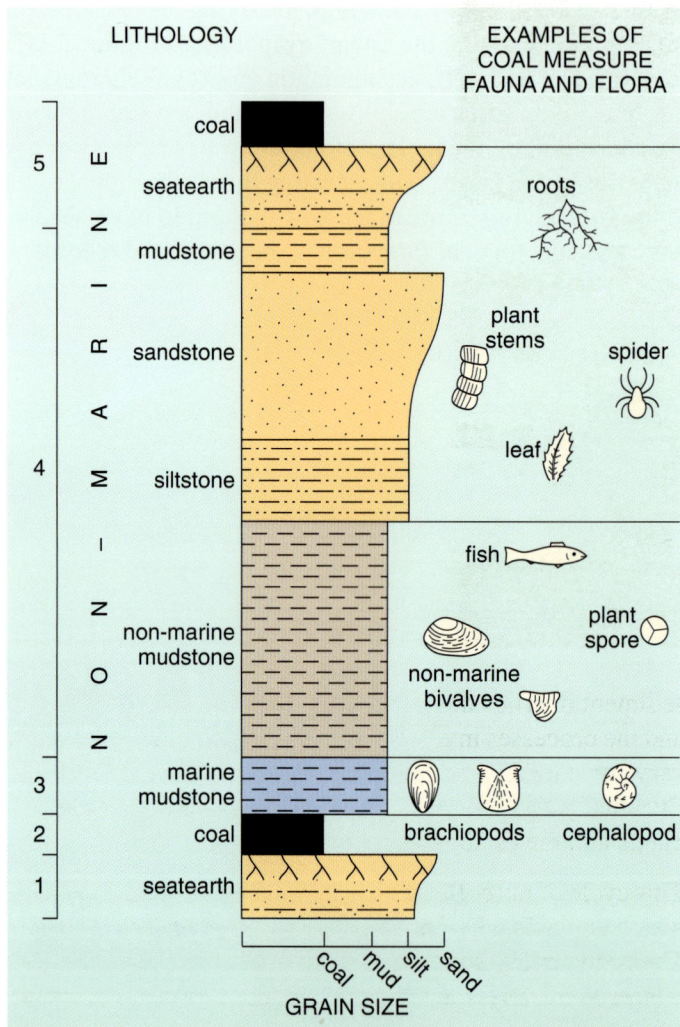

Figure 2.5 A vertical section through a typical coal-bearing sequence, showing a variety of sediments, and some typical fauna and flora. As is conventional in diagrams like this, the vertical axis represents sediment thickness (here schematic) and the horizontal scale represents maximum size of grains within the sediment. The numbers 1 to 5 on the left of the figure refer to sediment layers 1 to 5 listed in the accompanying text.

These vertical successions are thought to typify sediment deposition in certain areas of deltaic and coastal barrier environments similar to those shown in Figure 2.2. They are interpreted as representing initially flat low-lying sandy (or muddy) areas covered by vast freshwater lakes containing a variety of land plants growing in mires. Mire formation is then terminated by flooding of these areas, either by adjacent rivers bursting their banks, or by the sea flooding into the area. In both cases, these submerged areas would have filled up with mud, silt, and sand or, as depicted in Figure 2.5, mud grading into sand. Exactly what type of

Figure 2.6 A seatearth showing fossil rootlets as black carbonaceous lines towards the top.

sediment overlies the mire depends on the source of the sediment (i.e. river or sea) and the processes involved in depositing it. Individual sedimentary successions can vary considerably, both laterally and vertically from one succession to the next. Eventually, this sediment pile will fill the submerged area enabling recolonization by plants and the establishment of a new mire.

This cycle of mire–flooding–sediment infill–mire, is repeated time and time again, which explains why there may be many seams stacked vertically in a coalfield. The sedimentary succession in coalfields can reach hundreds or even thousands of metres in thickness, even though all the sediments were deposited in shallow water.

- What does this suggest about the stability of the land surface on which such sediments accumulated?

- Great thicknesses of coal-bearing rocks are clear evidence that the sedimentary basins in which they formed were subsiding. The compaction of peat and mud (sand is relatively less compactable) would have been contributory factors, but cannot alone account for such sediment thicknesses.

Question 2.1

Estimates of the current rate of subsidence for the Ganges and Nile deltas vary between 1 and 5 mm yr^{-1}. Using a rate of 1 mm yr^{-1}, calculate the time needed to deposit sufficient peat to form a 3 m coal seam. What assumptions does this calculation make?

Subsidence would not always have been uniform, so whilst mires existed in one part of the delta, sands, silts or muds were burying mires elsewhere. This variability results in seams that split laterally into two or more beds separated by bands of sandstone or carbonaceous shale, or that converge with an adjacent seam to become a single, thick one.

How do you think the vertical succession observed within raised mires might differ from the succession in Figure 2.5?

Most importantly, as raised mires can form far inland, flooding is less likely and so the cyclicity of mire–flooding–sediment infill–mire will not be seen. Instead, unimpeded development of the mire will mean that individual seams of greater thickness will develop. With flooding less likely, these coal seams will be less contaminated by sediment in comparison with deltaic and coastal barrier-formed coals.

The downside to coals formed in inland raised mires is that subsidence rates are not likely to be as high, and so total coalfield thicknesses are in general unlikely to be as great.

2.1.3   The physics and chemistry of coal formation

Coal is a type of sediment made up mainly of lithified plant remains. But how does spongy, rotting plant debris become a hard seam of coal? As discussed earlier, plant material growing in mires dies, and then rots under anoxic conditions to form peat (by the process of *humification*). With time, the mire becomes covered with layers of sediment, the weight of which squeezes water and gas out of the pore spaces and compacts the vegetation. As subsidence allows deposition of further mire–sediment cycles, the process of **compaction** continues. The vegetation matter interbedded with sand, silt and mud progressively increases in density to become indistinguishable from coal.

The first stage in the chemistry of turning plant material into coal is one of biochemical decomposition. Bacterial breakdown of the more soluble components, principally the cellulose, results in enrichment of the more resistant, waxy leaf coatings, spores, pollen, fruit and algal remains. Decomposition also expels some gases originally contained in the rotting matter — chiefly water, carbon dioxide and methane — leaving organic residues rich in carbon.

The second phase starts when the plant deposits are progressively buried beneath substantial amounts of mud, sand and silt. As depth of burial increases so too does pressure. Because of the Earth's internal heat flow, temperature also increases with depth. **Coalification** of the deposit involves progressive physical and chemical changes brought about by the increased temperature and pressure. The degrees of change result in distinguishable stages of coal quality, or **rank**, which reflect the maturity of the coal. The different rank stages are listed in Table 2.1, together with some of the parameters used to define them. Changes in rank are gradual and so the boundaries of the rank categories are somewhat arbitrary.

Compaction under pressure progressively increases the hardness of the coal. The continuous variation of chemical composition through the rank series is shown in Figure 2.7. Low-rank coals (b, c) contain more volatiles than do

Table 2.1 Changes in the chemical composition of buried plant remains with increasing rank.

Rank		Description	Typical content of volatile matter/weight %	Typical element analysis/weight %		
				Carbon	Hydrogen	Oxygen
a	peat	decomposed fibrous material	>50	50–55	5–10	40
b	lignite (soft brown coal)	still contains woody material; hard blocky appearance	50	60	6	30
c	sub-bituminous coal (hard brown coal)	dark material developing slight lustre	45	75	5	20
d	bituminous coal	black, shiny and strongly banded appearance	35	85	5	10
e	anthracite	metallic lustre; bands absent	<10	90	4	5
f	graphite*	malleable, with a silvery lustre	<5	>95	<4	—

* Graphite is a silvery-grey mineral formed entirely of carbon produced by metamorphism of originally organic materials at temperatures over 300 °C and pressures equivalent to burial of several kilometres. It is, of course, the 'lead' used in pencils, and does not readily combust.

high-rank coals (d, e), reflected by the variation in their oxygen and hydrogen content in Table 2.1. **Anthracites**, for example, usually contain less than 10% volatile matter. To form, they require high pressures associated with tectonic deformation or high temperatures near to igneous intrusions. Metamorphism at very high temperatures and pressure transforms coal to graphite deposits, which are not energy resources, although valuable in their own right.

Question 2.2

Using Table 2.1 and Figure 2.7, describe how the chemistry of coal changes with increasing rank.

Changes in rank involve the expulsion of water, carbon dioxide and methane (CH_4). Only small amounts of methane are liberated during the early stages of coalification, but during the transition from **bituminous coal** to anthracite (particularly over the range 85–92% carbon), expulsion of methane and other hydrocarbons removes hydrogen, whereas the emission of carbon dioxide declines. As the complex organic compounds in buried vegetation are slowly transformed to simpler, more carbon-rich compounds, coal changes colour from brown to black.

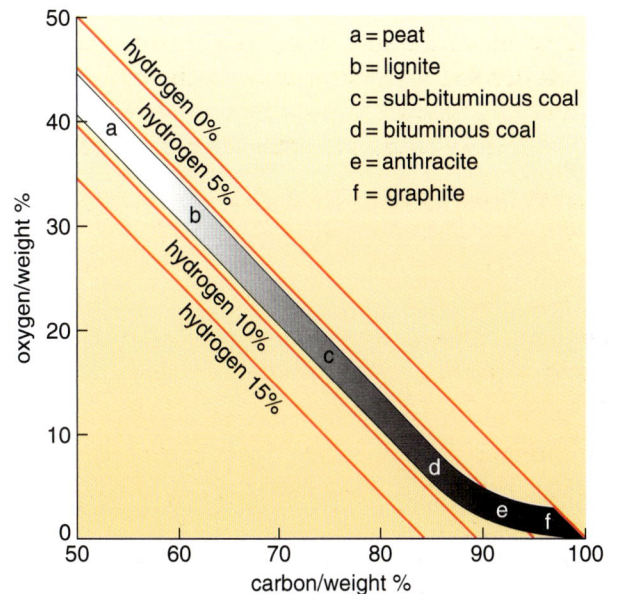

Figure 2.7 The relationship between carbon, oxygen and hydrogen contents for the six stages in the rank series in Table 2.1. (Rank increases from a–f.)

● Would you expect the density of coal to increase or decrease with increase in rank?

○ As rank increases, the porosity and water content of the coal decrease, so density increases.

The density of coal is reflected in its structure, hence the arbitrary distinction between 'hard' and 'soft' coals used in Table 2.1. In fact all coal is soft compared with most rock forming minerals, and the terms refer to how coal breaks: *soft coal* is crumbly, whereas *hard coal* breaks in a brittle fashion, and remains in lumps when transported.

The relative proportions of carbon and volatiles in coals affect their physical properties and their uses.

● Low-rank coals rich in volatile matter (more than 30%) are easy to ignite and burn freely but with a smoky flame. Low-volatile (high-rank) coals are more difficult to ignite, but they burn with a smoke-free flame; they are natural smokeless fuels.

● For industrial use, coal is usually heated to expel remaining volatiles. The carbonized residue that remains after the volatile matter has been driven off in the absence of air is called **coke**. High-rank bituminous coals become partly fluid on heating and swell up to form a porous coke, especially valuable for the iron and steel industries. Coke makes a useful artificial smokeless fuel because it is free of volatiles.

● The **calorific value** of coal (i.e. the amount of heat liberated under controlled conditions) generally increases with rank. Nevertheless, coals with a high volatile content (>30%) are usually burned in power stations for the ease with which they burn, even though they give out less heat than higher ranked coals.

2.1.4   Impurities in coal

Coal rank reflects the maturity of a coal, but another variable is the ratio of combustible organic matter to inorganic impurities found within the coal. As discussed earlier, impurities result mainly from clay minerals washed into the mire prior to its eventual burial. In addition, some impurities are formed from the plant material itself during coalification.

These inorganic impurities are non-combustible and therefore leave an inert residue or *ash* after coal combustion. High-ash contents increase the volume of particulate matter ejected into the atmosphere following combustion. So coals used as boiler fuels or for coking require less than 10% ash content.

A second important group of impurities are carbonate minerals. During the early stages of coalification, iron carbonate is precipitated either as concretions (hard oval nodules up to tens of centimetres in size) or as infillings of fissures in the coal. Interestingly, it was probably the breakdown of such iron carbonates to molten iron when coal that contained them was burned, which led to the accidental discovery of iron smelting.

Other impurities are nitrogen and sulphur that are chemically reduced during coalification to the gases ammonia (NH_4) and hydrogen sulphide (H_2S), which become trapped within the coal. However, most sulphur is present as the mineral *pyrite* (FeS$_2$), which may account for up to a few per cent of the coal volume. Burning coal oxidizes these compounds, releasing oxides of nitrogen (N$_2$O, NO, NO$_2$, etc.) and sulphur dioxide (SO$_2$), notorious contributors to *acid rain*.

In addition, relatively high concentrations of sodium chloride (NaCl), deposited by seawater floods into coal-forming mires, make the coal virtually unusable in power plants because salt causes severe boiler corrosion. Lastly, trace elements (including germanium, arsenic and uranium) are significantly enriched in coal and are released by burning it, contributing to atmospheric pollution.

2.1.5   How old is coal?

Not surprisingly, the distribution of coal deposits through time corresponds closely to the origin and distribution of land plants. (This is discussed further in Section 2.4.4.) Coals are commonly found in rocks from Carboniferous times onwards, Devonian coals are rare, and pre-Silurian true coals are never found. This coincides with evidence for the evolution of land plants, which first appeared in Silurian times about 400 Ma (million years) ago, colonized the land surface rapidly through the Devonian and became abundant by Carboniferous times. Not all post-Carboniferous terrestrial sediments contain coals, however. Coal-bearing rocks tend to be concentrated in rocks of specific ages in certain locations, reflecting the development of ancient mires under specific climatic conditions.

Carboniferous mires

During the late Carboniferous, mires developed over vast areas of the UK. Much of today's land area was an extensive, low-lying plain bordering a sea to the south (a sea that was soon to be the site of a mountain-building episode). Any mountains that existed lay hundreds of kilometres to the north. Large river systems meandered southwards across these plains.

At that time, the UK lay in tropical latitudes, almost on the Equator (see Figure 2.32). The low plains were covered by extensive forests: the Carboniferous equivalent of the present-day tropical rainforests (see Figure 2.2c). However, most of the 'trees' were hollow, not solid, and more closely resembled modern horsetails than modern trees. No flowers or birds existed, but insect and reptile life was abundant in the forests.

Tropical storms were probably as common then as they are today. Such storms would devastate vast areas of forest, reducing trees and plants to jumbled leaves, branches and crushed hollow logs. The same storms would also have caused extensive flooding. After this devastation, the forest would quickly re-establish itself, only to be devastated again by subsequent storms. The forest floor was probably often metres deep in rotting vegetation destined to become peat and, much later, coal.

In the late Carboniferous, cycles of global sea-level rise and fall resulted from the melting and re-growth of continental ice sheets in the Southern Hemisphere. Consequently, numerous mire–flooding–sediment infill–mire episodes occurred on low-lying coastal plains, which led to thick sedimentary successions with numerous coal seams.

Question 2.3

Use Figures 2.2a and 2.5 to help you match the following sediments in a coal-bearing sequence A–E with their likely environment of deposition (a)–(e).

A   Mudstone with freshwater shells

B   Siltstone

C   Coarse sandstone

D   Seatearth

E   Coal

(a)   Peat accumulations on the swampy areas of a delta plain

(b)   Distributary stream channels cutting through the delta plain

(c)   Fossil soil beneath the mire

(d)   Deposits laid on delta plain in times of flooding

(e)   Shallow lakes and lagoons on the delta plain

2.2   Finding and extracting coal

Coal is often regarded as the principal fossil fuel, and with good reason. There is almost three times more energy available from the global proven coal reserves as there is from proven oil and gas reserves taken together. Therefore, it is unsurprising that even today much time and effort is spent locating it.

This section considers the techniques used in coal exploration and how coal is produced from surface and underground mines. But first, a brief look at a few of the historical aspects of coal mining will put subsequent developments into context.

2.2.1   Winning coal in former times

Coal was probably first used as a fuel by early Chinese civilizations, and there is evidence for coal working in the UK since Roman times. However, early approaches to mining were limited by the available technology, and left much of the coal behind.

At first, coal was dug from seams exposed at the surface in shallow excavations into valley sides that followed the coal seam. The amount of coal that could be extracted from these **drifts** and **adits** was small, even when wooden props were used to stop the overhanging roof from collapsing. The first true underground mines, which signified some knowledge of hidden coal seams, were **bell pits**. Miners would dig a vertical shaft down to a coal seam less than 10 m below the surface. Once the underground seam was reached, the coal was worked outwards in all directions from the bottom of the shaft. When the bell pit became unsafe, the shaft was simply abandoned and a new one started nearby (usually 20–30 m distant).

Digging vertical shafts through overlying strata over 10 m thick is somewhat unproductive, and so **pillar-and-stall** working was later adopted. As Figure 2.8 shows, a rectangular grid of tunnels was driven horizontally into the coal from the base of a vertical shaft. Later, the wide *pillars* of coal within the grid were systematically removed from the furthest limits of the mine back towards the shaft, allowing the roof to collapse (Figure 2.8b).

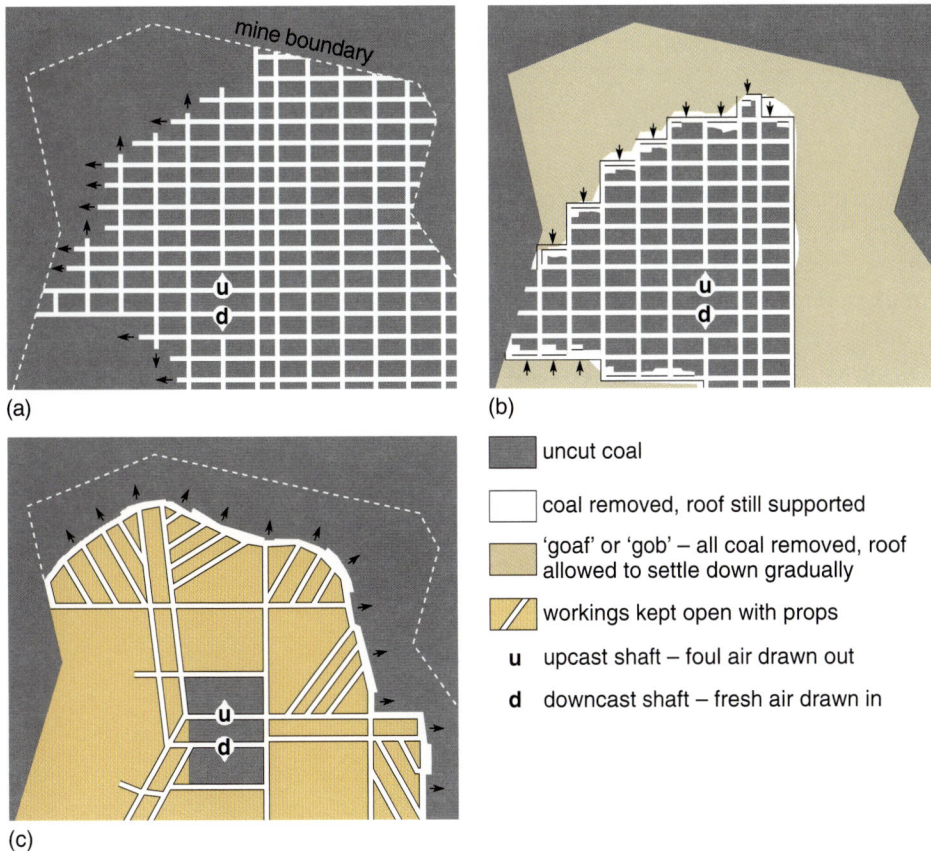

(a)

(b)

▨	uncut coal
☐	coal removed, roof still supported
▨	'goaf' or 'gob' – all coal removed, roof allowed to settle down gradually
▨	workings kept open with props
u	upcast shaft – foul air drawn out
d	downcast shaft – fresh air drawn in

(c)

Figure 2.8 (a) and (b) Pillar-and-stall workings involved driving a right-angle grid throughout the whole area to be worked, leaving support pillars. In (b) the roof has subsided after removal of the pillars. (c) Longwall workings that removed all the coal as working advanced became common in the 19th century.

For coal seams at depths greater than 300 m, the support pillars were often crushed by the weight of overlying rock and so **longwall mining** was widely used after 1850. This technique involved a 'wall', usually about 30 m long, being cut into the coal seam in one continuous process, advancing away from the shaft (Figure 2.8c). Any stone available (for example, sandstone within the seam) was built into support walls 2–5 m wide running at right angles to the working face. These walls and wooden props were used to support the roof after the coal was cut and to prevent crushing of the coal at the working face.

Poor ventilation and drainage limited the size of these early mines. As soon as shafts were sunk beneath the water table, mines began to flood and had to be abandoned if they flooded faster than they could be drained. Flooding ceased to be a major problem following the introduction of steam pumps as early as 1712 in UK mines.

As mines extended further underground, ventilation became a significant problem. Fires were lit at the foot of an 'upcast' shaft — one of a pair of shafts close together. The upward movement of hot gases drew a corresponding draught of fresh air into the mine through the nearby 'downcast' shaft. This method continued to be used until the advent of mechanical air pumps in the 1830s.

Coal extraction became more efficient in the mid-19th century with the invention of rotary cutters powered by compressed air. However, even in the mid-20th century miners still relied on picks, shovels and crowbars to win coal from seams inaccessible to machines. By the early 20th century, electric conveyor belts were used to transport the coal away from the coalface. Even though coal could be cut and loaded onto conveyor belts in one continuous automated operation, the manual loading of coal was commonplace until the 1950s and 1960s.

2.2.2 Exploring for coal

Early miners would have found it easy to trace the distinctive black colour of coal along an outcrop (for example, a coastline or river valley), and surface trenches were used to locate less obvious outcrops. However, tracing an outcrop underground was problematical as the only means of exploration was by digging costly trial shafts. The development of exploratory steam-powered drilling in the early 19th century improved matters, but it was not until the mid- to late-20th century that more advanced techniques made it possible to significantly reduce the uncertainties associated with estimating the size of a coalfield.

Modern exploration techniques are aimed at accurately assessing the location, quality and quantity of coal in a coalfield. In order to achieve this there are three broad categories of tools available to geologists: mapping, geophysical methods and drilling. They are considered here in the order in which they are likely to be employed.

Geological mapping of coalfields

Coalfields can be divided into two categories: **exposed coalfields**, where the coal-bearing strata outcrop at the surface, and **concealed coalfields**, where they are hidden beneath younger rocks. Exposed coalfields can be defined with considerable precision by surface geological investigations; indeed geologists recording field data still represent the cheapest exploration 'tool' available to the coal industry.

In populated regions, the locations of coal outcrops are well known and mapped. However, in remote areas new finds are still possible. In such areas data acquired from satellites or aircraft are assessed before geologists start exploring on the ground. The Global Positioning System (GPS), now increasingly used for navigation, uses signals from satellites to pinpoint locations precisely, so that geologists can more easily create accurate maps in the field. Field data are increasingly processed using spatial analysis software to create digital maps of coal outcrops and to model the likely extension of the coal beneath the surface.

Geophysical methods — seismic surveying

Geophysical survey methods use measurements made at or near the Earth's surface to investigate the subsurface geology. The most widely used geophysical method is seismic reflection surveying; a rapid and highly cost-effective way of gathering data.

A seismic source (produced either by the explosive release of compressed air in a shallow borehole, or a heavy pad vibrated hydraulically at the surface) generates seismic waves that travel through the ground (Figure 2.9). These are reflected at buried geological boundaries and return to the surface where their amplitude and time of arrival are recorded by an array of detectors (Figure 2.9). Data produced by moving the positions of both source and detectors along a surface traverse is processed by computer to produce a **seismic section** through the Earth along the line of the survey, which take the form of a vertical cross-section. Further details of seismic surveying follow in Section 3.3.3, where its principal use in petroleum exploration is discussed.

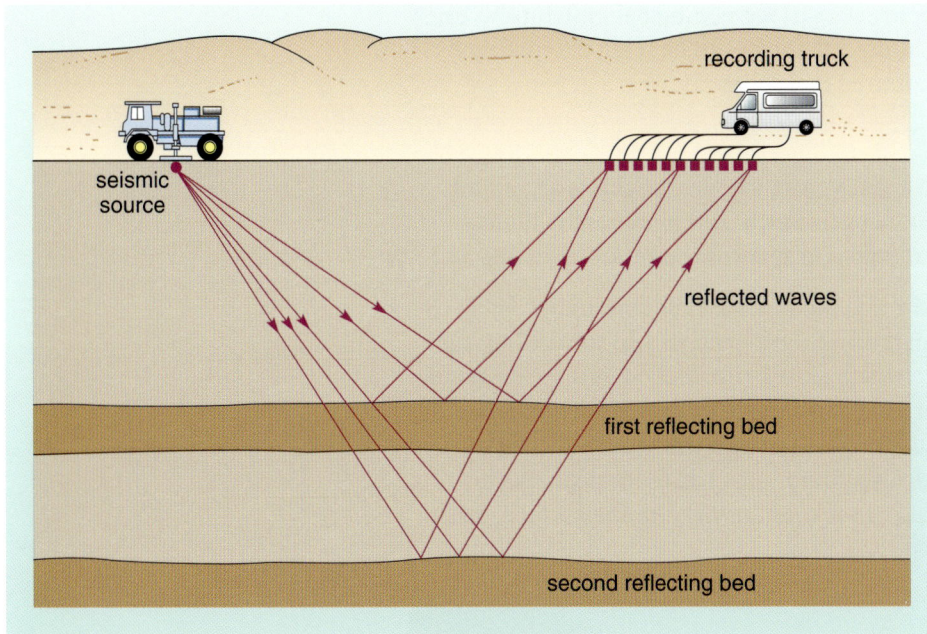

Figure 2.9 The layout of a seismic survey, showing the reflection of seismic waves by two reflecting layers.

Figure 2.10 shows an example of a seismic section. A series of light and dark lines (or *reflectors*) approximately show the orientation of the sedimentary strata at depth. Coal has very different seismic characteristics from those of most common rocks. This results in large variations across coal–sediment boundaries that produce strong reflections. Seismic sections can also indicate geological problems that might be encountered in a coalfield. In particular, the displacement of a number of reflectors might indicate the presence of a fault, of which there are several in Figure 2.10.

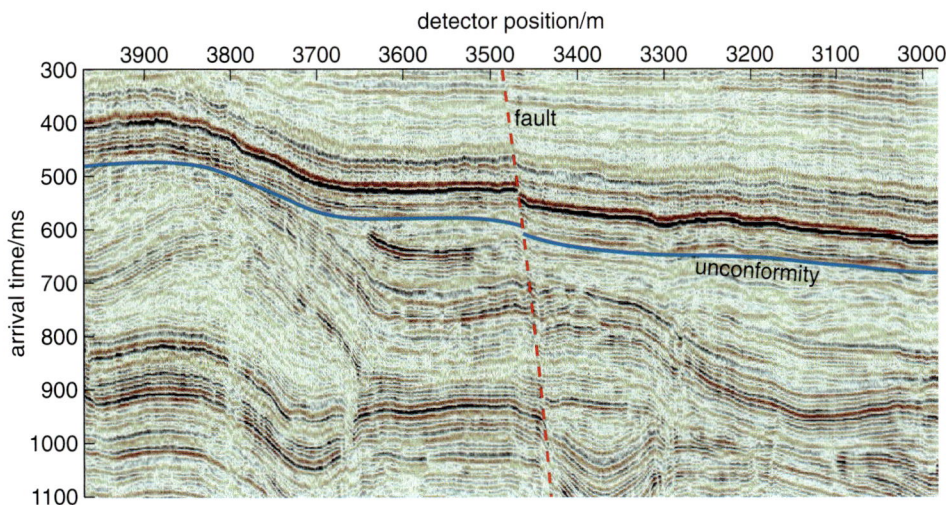

Figure 2.10 An example of a seismic section. The (vertical) arrival time axis in milliseconds (ms) is roughly equivalent to increasing depth. Towards the top of the section a pair of dark lines indicate major coal seams. They are displaced by a fault near the centre of the traverse (marked by the dashed red line). Many other features show up, including greater complexity in the deeper part of the section, and towards the left of the section deep, more steeply dipping reflectors are truncated by the simpler ones at shallower depths: this is an unconformity (solid blue line).

The data in a seismic section enables the subsurface geological structure to be visualized. However, eventually, the interpretation has to be related to reality by drilling vertical boreholes to sample inferred rocks directly.

Drilling

Drilling is expensive, so this next phase of exploration only begins when all the data have been gathered from pre-existing geological and topographic maps, aerial/satellite photographs, geological mapping and from seismic surveying.

The thickness and quality of a coal seam in an area are first determined by drilling boreholes a few kilometres apart using a grid pattern. Mobile drilling rigs (Figure 2.11) use a powerful motor to rotate a drill bit attached to a series of steel rods within the hole. The bit, made of tungsten carbide or studded with diamonds, grinds away the rock, cutting a cylindrical hole through the rock sequence as pressure is applied to it. Specialized drilling fluids are used to lubricate the bit. The same fluids bring small fragments of rock, or *cuttings*, to the surface, where they can be examined by the geologist.

Figure 2.11 A mobile drilling rig in operation.

If substantial samples are required when a coal seam is penetrated, a cutting barrel can be used in place of the solid drill bit to drill out a cylinder of rock, called a **core** (Figure 2.12). Coring sequences of strata is slow and expensive, and is only undertaken when details from cores are essential.

Question 2.4

Given that drilling a borehole currently costs around £190 per metre, and that coring is two and a half times more expensive than this, calculate the cost of drilling a borehole 800 m deep, which includes 130 m of core.

Although expensive, cores do make it possible to constrain the thickness and depth of the coal seam. Furthermore, detailed analysis of other sediments in the cores can reveal the environment in which the rocks were originally deposited. Recalling Figure 2.5, this may permit geologists to predict parts of the rock sequence where coal is either more or less likely to be found. Coal samples are usually recovered from the core for chemical analysis, to measure its carbon (i.e. rank), sulphur and ash content.

Geophysical methods — borehole logging

If a core is not recovered from a borehole, another way to assess the types of rock that it penetrates is to measure their physical properties. Mounting a string of electronic instruments behind the drill bit most conveniently does this: it allows the properties of the rock to be monitored as the borehole is drilled. An alternative is to lower instruments down the completed borehole by cable; hence the name *wireline logging* (discussed further in Section 3.3.4).

Figure 2.12 Core samples being removed from a cutting barrel in the field.

Such logging measures several physical properties of the rocks surrounding the borehole. These include the velocity at which sound travels through each rock type, their density and their emission of natural gamma radiation (from unstable isotopes of uranium, thorium and potassium that vary a great deal between different rock types). Because a single physical property cannot define a rock type, the logging geologist must compare a range of properties at each depth to interpret the rock type present.

As the location, quality and quantity of coal in a coalfield will affect the profitability of a mine, mining will not commence until the mining company and its financial partners are satisfied that these crucial parameters have been sufficiently constrained. Even after mining has commenced, further exploration (including the drilling of more boreholes) periodically allows initial estimates to be refined.

2.2.3   Modern mine planning

Once the geological data gathered during the exploration phase has been evaluated, geologists will estimate the quality and quantity of coal present. Coal *reserves* (in tonnes) are calculated from volume × density (Section 2.4.5). The volume of coal is controlled by seam area and seam thickness. Hence:

tonnage = seam area × seam thickness × coal density

Seam area is not the same as surface land area, as the coal seam may dip. However, the surface land area will define the land area required for mining. Other considerations are the purity and rank of the coal, the likelihood of encountering any geological problems (which will be discussed a little later), and whether the seam has been worked in the past.

The first decision to be made is whether to open a **surface mine**, or an underground mine. In surface mines, exposed and/or shallow coal seams are accessed by removal of waste rock or **overburden** from the surface (Argles, 2005). They are therefore rarely deeper than 100 m below ground level.

● What do you think controls the maximum depth to which coal seams can be worked economically using surface mining methods?

○ There is no profit to be made in removing overburden, so the maximum economic depth for surface mining depends on the value of coal produced compared with the cost of removing the waste material. In addition, surface mining machinery is unable to operate below a certain depth.

The ratio of the amount of overburden to the total amount of workable coal is therefore of critical importance. Within the coal industry, this is known as the **stripping ratio** (or *overburden ratio* in Argles, 2005). Stripping ratios can be calculated either in tonnes or in thickness ratios (commonly used in coal operations). The maximum economic stripping ratio for surface mining has steadily increased over the years to around 20:1, helped by improvements in the productivity and life of the plant and equipment.

Question 2.5

Figure 2.13 shows a cross-section through coal-bearing rocks. By counting the squares in the overburden and coal, and by assuming a maximum stripping ratio of 20:1, determine whether the coal seam could be extracted profitably using surface mining techniques.

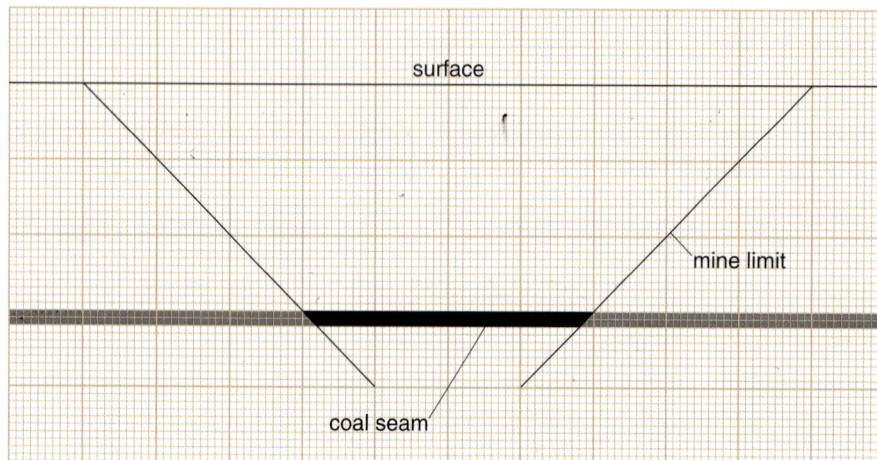

Figure 2.13 A cross-section through coal bearing rocks. For use with Question 2.5. Note that areas on this cross-section are proportional to volumes of rock. The lines showing the mine limit are at angles where bare rock surfaces are stable, so that the mine walls do not collapse during operations.

Mines working high-rank coal, which is more valuable, can operate higher stripping ratios than those working lower-rank coal; the surface anthracite workings in south Wales for example, operate a stripping ratio of 35:1. Note also that because coal is a moderately high place-value resource (Sheldon 2005), it is economic for some present-day surface mines in the UK to extract coal at greater depths than underground mines in other parts of the world that are remote from points of sale.

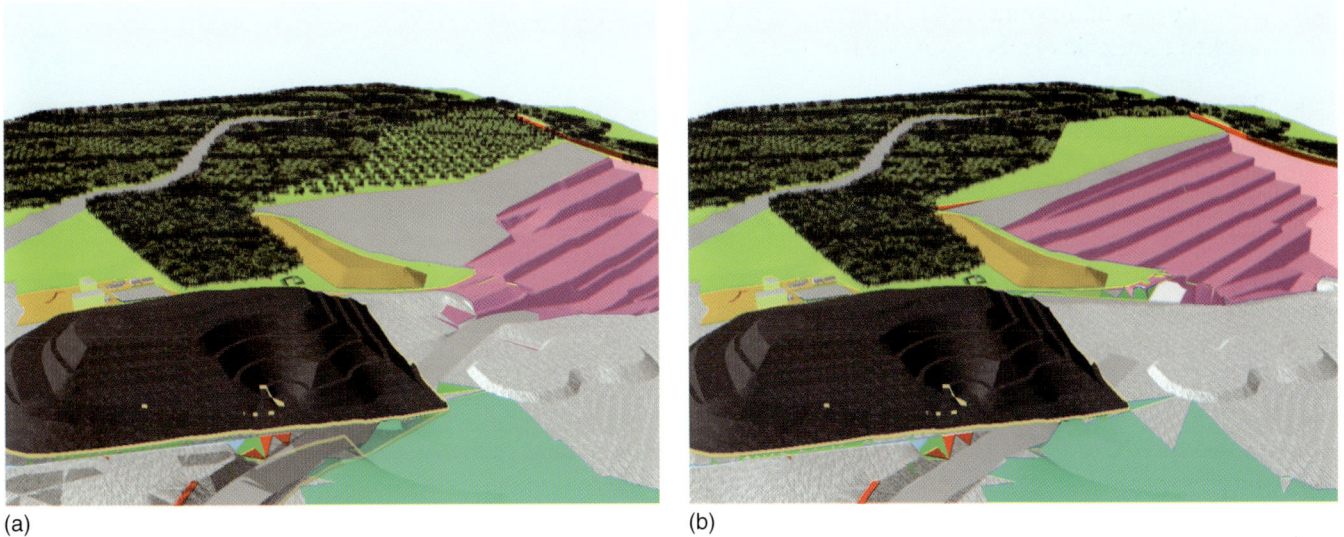

(a)

(b)

Once it has been decided whether to use either a surface or underground mine, engineers will begin planning the optimal mine layout and the processes that will be used to extract and remove the coal from the coalface. As Figure 2.14 shows, computers are often used to model the continually changing layout of the mine with time.

Figure 2.14 Three-dimensional computer model showing two stages in the development of a surface mine. The pink area shows the projected development of the coalface through time.

2.2.4   Surface mining

In surface mining (sometimes called 'opencast' mining in the UK), the coal seam is accessed by removing the rock overburden; a process that benefits from economies of scale by using some of the world's largest machines (Figure 2.15).

Figure 2.15 A dragline operating to remove overburden above a shallow coal seam being cut by a mechanical shovel.

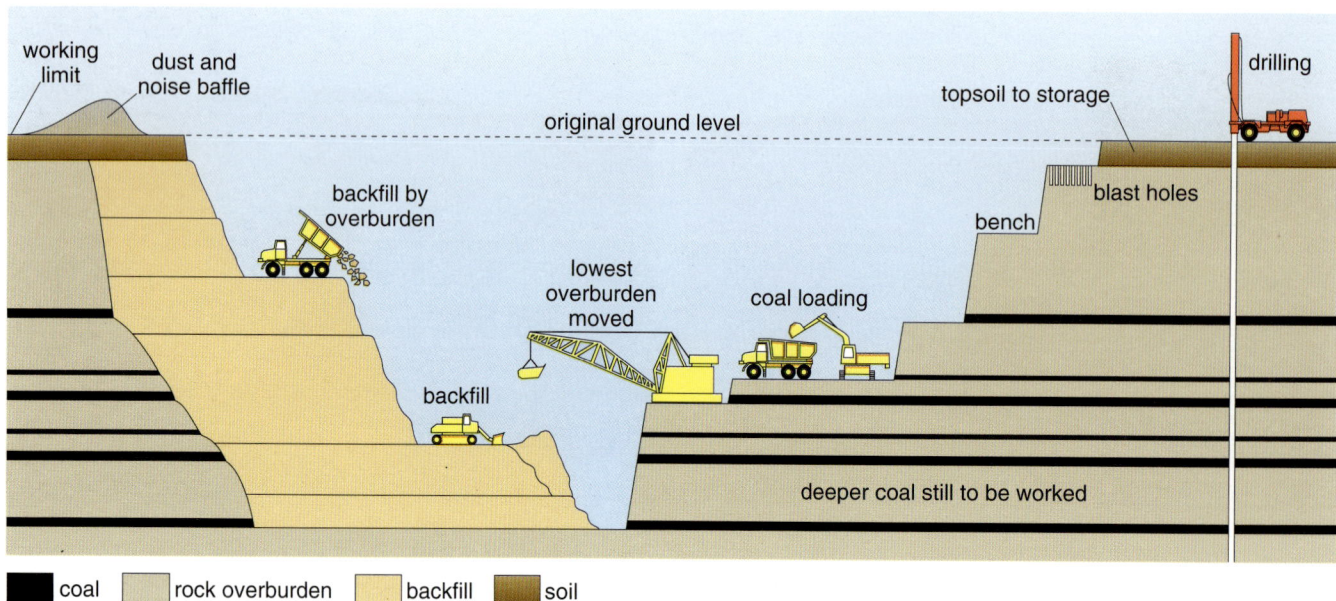

working limit · dust and noise baffle · original ground level · topsoil to storage · drilling · backfill by overburden · blast holes · bench · lowest overburden moved · coal loading · backfill · deeper coal still to be worked

■ coal · rock overburden · backfill · soil

Figure 2.16 shows schematically how surface mines are organized. Topsoil is removed and stored or used immediately for land restoration elsewhere. Shallow or soft overburden is removed by draglines, hydraulic shovels and dump trucks, but at deeper levels with harder layers of rock explosives may have to be used. When the first coal is reached, seams are usually worked by **bench mining** methods (Figure 2.16). The top surface of coal exposed on each bench is carefully cleaned to remove any adhering waste rock. This careful exclusion of non-coal material enables high-quality coal to be produced consistently by surface mining. The clean coal is then broken up by diggers or with the aid of explosives, and loaded into trucks using mechanical shovels. Bench methods allow every seam to be worked, not just the thickest. Seams as thin as 0.1 m can be worked where they form part of the overburden to a lower, profitable seam.

The precise sequence of overburden and coal removal in a surface mine depends on the area covered by the mine, and the thickness and vertical spacing of the coal seams. In some cases, a single large pit is opened and the overburden is then moved around within it to gain access to different parts of the coal seam (Figure 2.17). In mountainous areas, benches may be cut into the slopes and the removal of coal and overburden may result in the mountain being levelled.

The critical factor is the angle at which the seams dip beneath the surface. In the case of horizontal strata (Figure 2.13) and an almost flat land surface, wherever the coal seam is present beneath the surface, the overburden is roughly the same. Consequently, the potential size of a surface operation to mine a horizontal coal seam is limited only by environmental legislation, ownership of the land and any areas of rising ground that would increase the overburden ratio. Finding seams that remain horizontal over vast areas is uncommon, however. Where the strata dip beneath the surface, overburden increases as the seams dip to deeper levels, thereby posing a limit on mine size.

Figure 2.16 A diagrammatic section through the surface mining process showing bench mining methods. The ultimate depth of excavation can exceed 100 m. Benches are cut where coal is exposed, but also at other levels to allow access. The benches are often left in place to ensure stability of the mine walls.

Figure 2.17 Surface mining using bench mining methods at Orgreave, Staffordshire. Nine seams, ranging in thickness from 0.1 to 1.35 m, were worked in this surface mine.

Question 2.6

Figure 2.18 shows a cross-section through coal-bearing rocks that dip at 45° to the horizontal. By counting the squares in the overburden, and by assuming a maximum stripping ratio of 20:1, determine whether it would be economic to strip mine the coal seam down to depth A and additionally down to depth B. You can assume that the coal seam covers 125 mm^2 squares down to depth A and 150 squares to depth B.

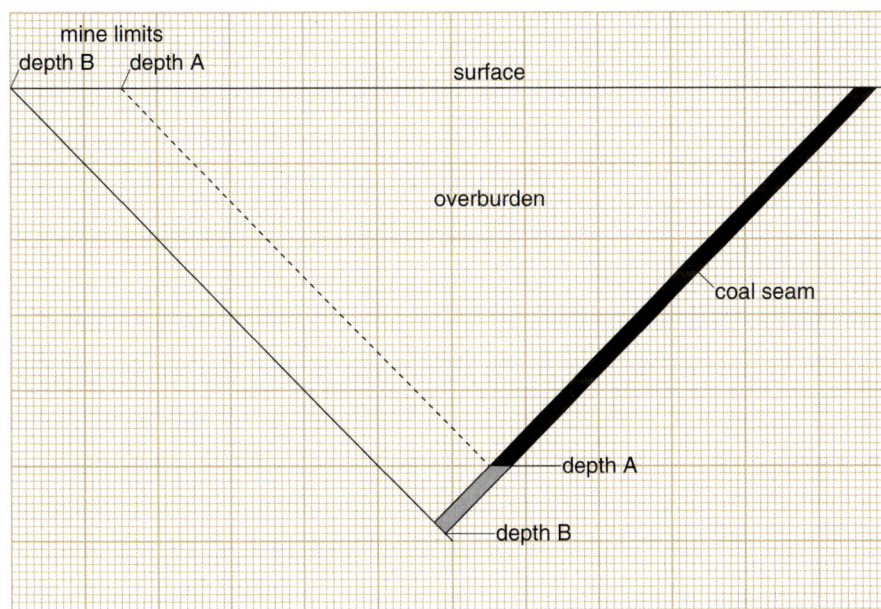

Figure 2.18 A cross-section through coal bearing rocks. For use with Question 2.6. Note that areas on this cross-section are proportional to volumes of rock.

Your answer to Question 2.6 shows how the extent of surface mining is limited laterally in the direction of dip, the limits of a mine becoming narrower as the dip increases, because the stripping ratio increases more with depth of excavation. A dip of 45° is at or above that where unconfined rock becomes unstable and begins to slip: surface mining of coal with steeper dips is too dangerous to be undertaken, and, if at all, the seams must be won by underground operations. At lesser dips, surface mining takes the form of operations that are along and parallel to the outcrop of coal seams: this is termed **strip mining**. Such operations are widespread in the western USA (Box 2.1) and are particularly appropriate for single, horizontal coal seams at shallow depths beneath flat topography, as in some Australian coalfields and the low-grade brown coals of Germany. Only a narrow strip is 'sterilized' from agricultural use for a year or so as the strip mine progresses across country. Bench mining, as shown in Figure 2.16, is favoured where there are a number of seams within 100 m of the surface.

In the UK, surface mines are now located in areas of shallow coal where underground mining has ceased, and include districts where the old mining methods had extracted only a proportion of the coal. Typically, surface bench mines are favoured in the UK where access to large areas is highly restricted. They usually operate for between three to five years, cover about 2 km², and contain about two million tonnes of coal in a dozen or so seams. In 2004, just under half of UK coal production came from 42 surface mines (Figure 2.19). This is half the number of operating surface and underground mines just eight years before.

The UK has approximately 43×10^6 t of coal reserves in operational and non-operational surface mines. (75% of this coal is in Scotland, the rest being shared equally by England and Wales.) It is becoming increasingly difficult to estimate

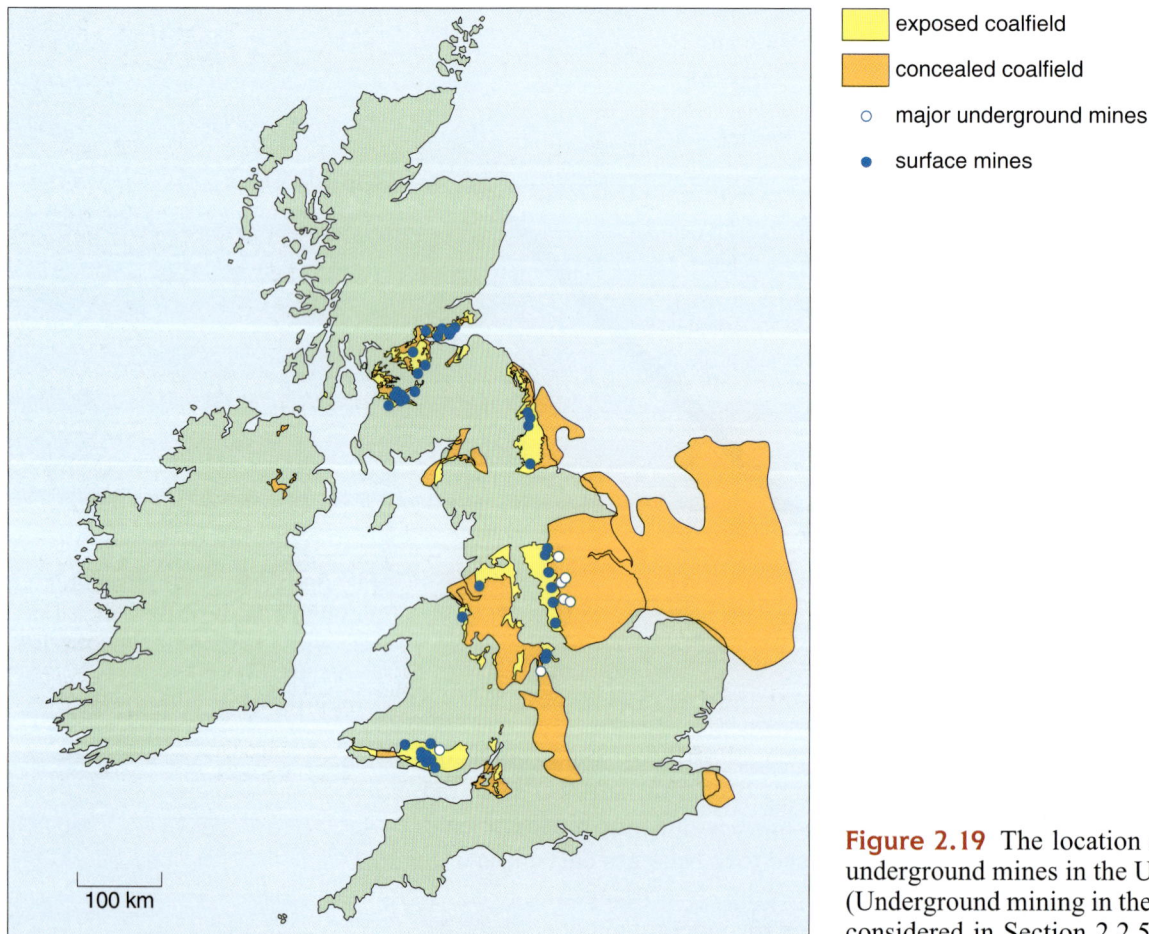

Figure 2.19 The location of surface and underground mines in the UK in 2004. (Underground mining in the UK is considered in Section 2.2.5.)

how long this coal will supply the UK demand, as cheaper imports of coal from around the world have forced the European coal industry, both in surface and underground mines, into rapid decline since the mid-1990s. As a direct consequence, 53% of the coal that is consumed in the UK is now imported and this figure is likely to increase in the future.

The decline in the production of surface mined coal in the UK is not reflected elsewhere in the world where in general, production from surface mines has increased in recent years. Despite the fact that most coal reserves are only exploitable through underground mining, globally some 40% of coal is now surface mined. The USA produces 50% of its coal from such mines (see Box 2.1); Australia, 70%; and Venezuela and Columbia close to 100%. However, China — by far the largest producer of coal — is dependent on deep coal seams and only mines 7% of its coal at the surface.

Box 2.1   Surface coal mining at the Black Thunder Coal Mine, Powder River, Wyoming

In the USA, three states (Montana, Illinois and Wyoming) possess over 56% of the country's coal reserves (Figure 2.20a). Until recently, Wyoming laid claim to the largest single coal mine in the USA; the Black Thunder surface mine. Black Thunder, located in the Powder River Basin (Figure 2.20b), opened in 1977. In 2003, the mine produced 51.5×10^6 t of coal.

To place this in perspective, the whole of the UK produced just over half this amount of coal in the same year.

The coal at Black Thunder is of Early Tertiary age (56.5–65 Ma), so it is significantly younger than the Carboniferous coal of the UK. (The ages of the

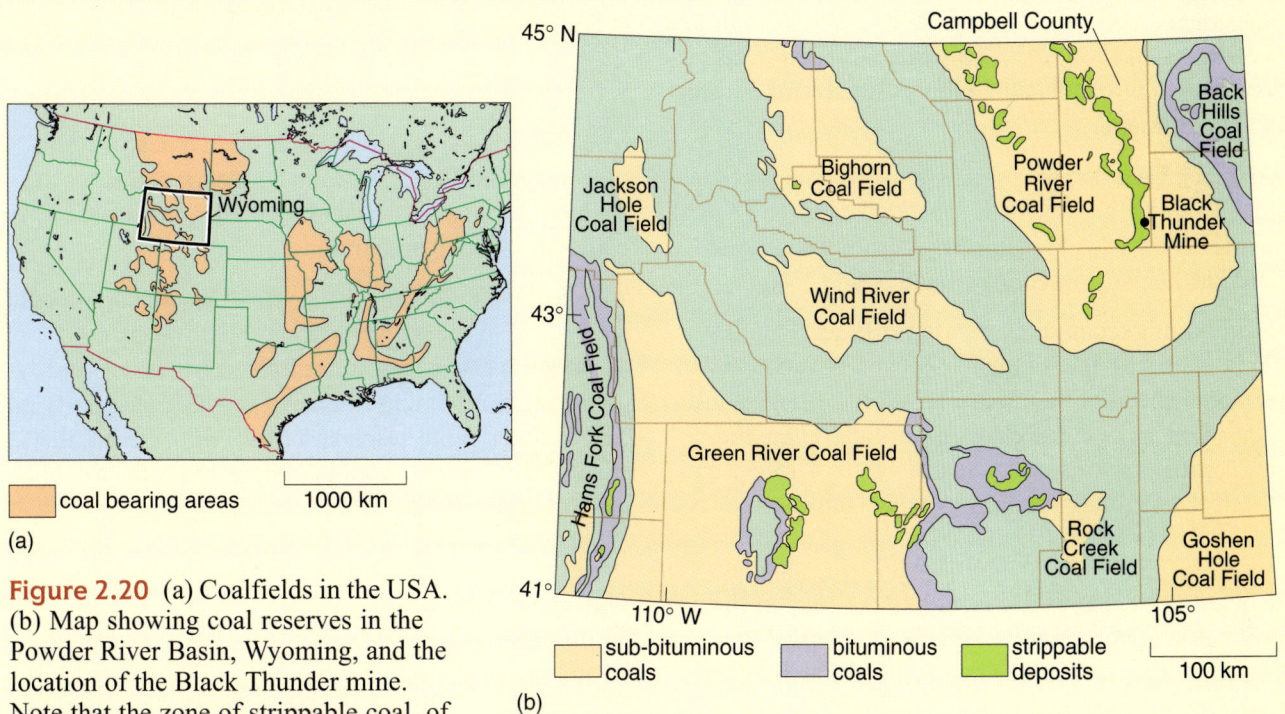

Figure 2.20 (a) Coalfields in the USA. (b) Map showing coal reserves in the Powder River Basin, Wyoming, and the location of the Black Thunder mine. Note that the zone of strippable coal, of which the mine is a part, extends for almost 2° of latitude and 2° of longitude in Campbell County alone.

(a)

(b)

Figure 2.21 (a) The Wyodak seam, which at the Black Thunder mine is 22 m thick (Photo © Arthur Meyerson Photography). (b) The dragline 'Ursa Major' at the Black Thunder mine. Note the trucks for scale.

various coal deposits are discussed further in Section 2.4.4.) The principal Wyodak seam is an impressive 22 m thick (Figure 2.21a). This single coal seam covers some 35 000 km² of the Powder River Basin, making it the most important in the USA. Its great thickness and the fact that it is composed of the remains of woody material suggest that the coal formed in a stagnant tropical mire where the rate of subsidence was sufficient to allow peat to accumulate over thousands of years. This extensive mire was associated with a delta that moved westward during the Palaeocene.

At Black Thunder, the seam is gently dipping, and locally splits into two beds separated by up to 18 m of waste. The coal is low-sulphur and sub-bituminous and has an ash content of around 5%.

Strip mining techniques are used at Black Thunder. They start with the removal of topsoil by mechanized scrapers ahead of the pit and this soil is trucked and placed on areas undergoing restoration. Then, the overburden is stripped away to a depth of 15–75 m to reach the Wyodak seam in a two-stage process.

Around 20 to 30% of the overburden is removed by blasting it directly into areas where the seam has already been mined. Moving such large volumes of rock in this way requires an enormous amount of explosive. The remaining overburden is removed by four large draglines. The largest, called 'Ursa Major', has a 110 m-long boom and carries a 122 m³ bucket, which can clear 150 t of overburden in one scoop (Figure 2.21b).

Once the coal seam is exposed, it too is blasted to ease its removal. Electric mining shovels then load the coal onto trucks, which transport it to a nearby crushing plant. A 3.5 km-long conveyor belt takes the coal to a further crusher and then into silos from which it is loaded onto trains.

With a low-sulphur and ash content, Wyoming's coal is well suited to fuel power stations without any preparation except crushing. Indeed, 96.7% of coal produced in Wyoming is used to generate electricity in 22 US states. With reserves of over 900×10^6 t, the Black Thunder mine could continue to produce coal at its current rate until 2022.

The quantity of coal extracted per worker-day in surface mining can be many times that in underground mines — a factor that is reflected in the generally lower cost of surface mined coal (helped also by lower capital and operating costs). Other advantages of surface mining are that geological problems are more easily resolved and the working environment is safer for mining personnel.

One disadvantage for mine operators is that some major coal-producing countries, including the UK and US, have legislation requiring rehabilitation of the mined area, or laws prohibiting surface mining on land where rehabilitation would not be possible. In such cases, and where the coal is deeper than the economic stripping ratio, underground mining is the only option.

2.2.5 Underground mining

Coal extraction is of course less straightforward using **underground mining** techniques. The associated costs are higher, and these begin with the sinking of two shafts, an 'upcast' and a 'downcast' shaft for ventilation (Figure 2.22). Sinking these to a depth of a kilometre may take a few years and during this time, no coal is extracted. However, not all underground mines involve deep, vertical shafts. Coal seams less than 350 m deep may be reached by inclined shafts, or drifts. In hillsides, horizontal adits may be opened up directly into the coal seam, producing an earlier return on investment.

The coal seams exploited in most European mines are typically about 1.5 m thick, but can vary from 0.5 m to 3 m. Modern underground mines use longwall extraction methods (Figure 2.8c), relying on highly mechanized extraction techniques. A cutting machine works its way along the start of the planned extraction to develop a roughly 250 m long coal face (Figure 2.23). Temporary hydraulic jacks support the roof at the face. Cut coal falls onto a conveyor belt laid out parallel to the coal face, and is carried away.

Figure 2.22 Winding houses above coal mine shafts at Harworth Colliery in Nottinghamshire, UK (now mothballed). The inclined structures are covered conveyor belts.

Figure 2.23 A cutting machine working its way along a coal face. A large rotating drum armed with steel picks cuts coal from the seam to fall on the conveyor belt, which moves to the left in this case.

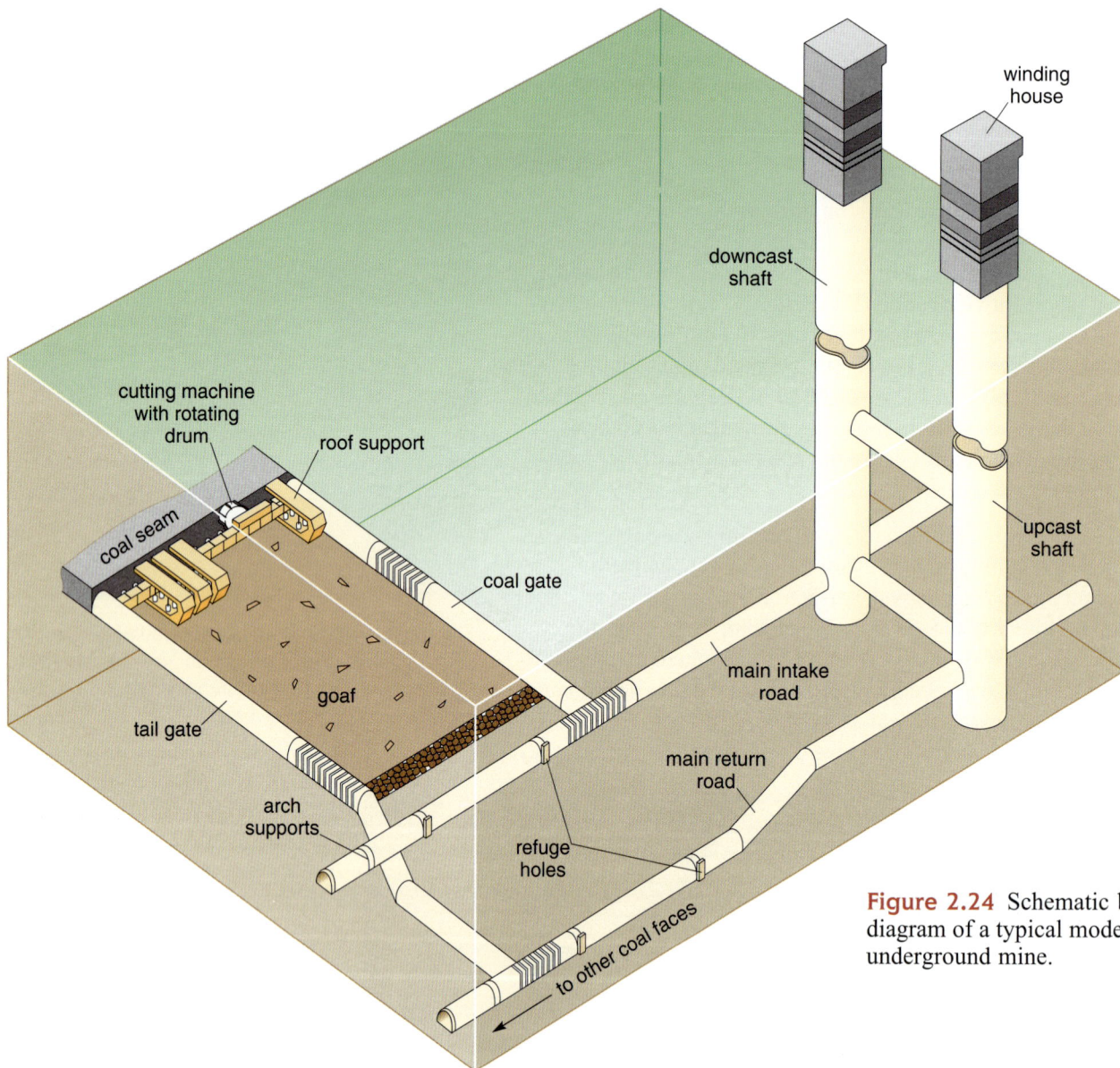

Figure 2.24 Schematic block diagram of a typical modern underground mine.

In the most recent installations the conveyor is articulated, and advances automatically, together with the roof supports, as the cutter moves along. By the end of each traverse, the face is ready for the next run in the opposite direction. As the face advances, supports are removed and the roof is allowed to fall into the cavity (*goaf*) left when the coal is removed (Figure 2.24).

Tunnels or *gates* at either end of the face link it to the colliery's permanent 'road' system. These roads are used for access of miners and machinery, to ventilate the face and remove the coal. In both new and established mines, roads have to be driven to new areas of coal working. Cutting gates and roads is not a profitable operation in itself, but it does provide valuable geotechnical information to assist the mining operation. Like faces, roads and gates require support, using concrete linings, steel arches or bolts secured in the rock with resin. Wire mesh, secured by the latter, prevents minor rock falls.

The road network links the faces (most modern mines have several working at any one time) to the two shafts (Figure 2.22 and 2.24) that allow coal and stale air to be removed, and fresh air, workers and supplies to enter the mine.

Figure 2.25 illustrates the two common types of underground mine layout used in the UK. At an **advance face**, the coalface is advanced into a block of coal. The access gates at either end of the face are also advanced to keep pace with the face. At a **retreat face**, the two access gates are first driven to the far boundary of the block of coal before the coalface is opened at their extremities. The face is then worked back towards the main roadways and the roof collapses in the same direction.

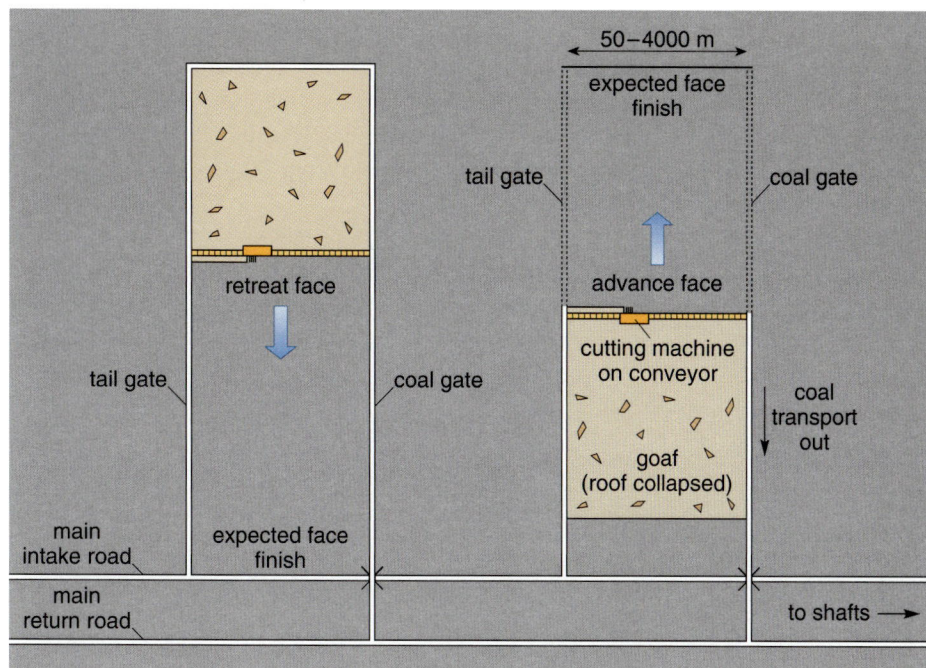

Figure 2.25 Two types of modern coal face layout in plan view. In the advance face, the coalface and gates advance into a block of coal. In the retreat face, the access gates are first driven to the boundary of the block of coal. The coalface is then opened at their extremity and worked back towards the road network.

Until the late 20th century, most underground mines in the UK were separate operations, dating from the period before nationalization in 1945, although some adjacent mines were linked. The last major underground coal development in the UK focused on newly proven major reserves in North Yorkshire, well away from the existing coalfield of South Yorkshire. It involved mine planning on a huge scale, involving several linked mines (Box 2.2).

Box 2.2 The rise and fall of an underground coalfield: the Selby complex in North Yorkshire

The UK's most recently developed underground coalfield, below Selby in North Yorkshire, was initiated between 1972 and the first production of coal in 1983. The Selby complex provides not only a good illustration of the development of a modern coalfield, but also how economic forces can dictate the future of a prospect.

Coal seams in the Yorkshire coalfield dip eastwards under a cover of younger Permian rocks. As demand for coal increased and the exposed coalfield became exhausted, the working area extended progressively eastwards to the deeper levels of the concealed coalfield.

In the 1960s, exploration drilling showed that the principal seam (the Barnsley seam) is between 1.9 m and 3.25 m thick under the Vale of York, north of the town of Selby (Figure 2.26). This represented an exciting prospect of workable coal covering an area of 260 km^2. A major drilling programme commenced in 1972, and from 68 deep boreholes, drilled approximately 1.5–2.5 km apart, reserves of 600×10^6 t were calculated in the Barnsley seam. The coal was found to be very clean, with 2–8% ash content (ideal for the power station customers who specified no more than 18% ash).

Stripping ratios of up to 550:1 at North Selby were far too high for surface mining, so the coal could only ever be extracted by underground methods. The mining complex consisted of two parallel 12.2 km long tunnels that provided the main coal outlet to the railway at Gascoigne Wood. Other tunnels connected working areas in different parts of the coalfield to this main outlet. Shafts served each working area, by providing ventilation and allowing access for workers and machinery (Figures 2.26 and 2.27).

The drilling programme had proved nine other seams of workable quality lying at depths between 300 m and 1100 m. However, the land in the Vale of York is only some 7 m above mean sea-level, so surface subsidence with a consequential risk of flooding was an important consideration. Because of this, planning permission limited extraction to just the Barnsley seam, thereby limiting the surface subsidence to less than a metre, with no extraction around the towns of Selby and Cawood (Figure 2.26) and some industrial areas. Leaving nine seams behind, as well as a substantial part of the Barnsley seam itself reduced the total amount of recoverable coal from 2000×10^6 t to just 224×10^6 t.

● What proportion of the coal in the Selby coalfield was recoverable? How long would it take the miners in the Black Thunder strip mine to extract a similar volume of coal?

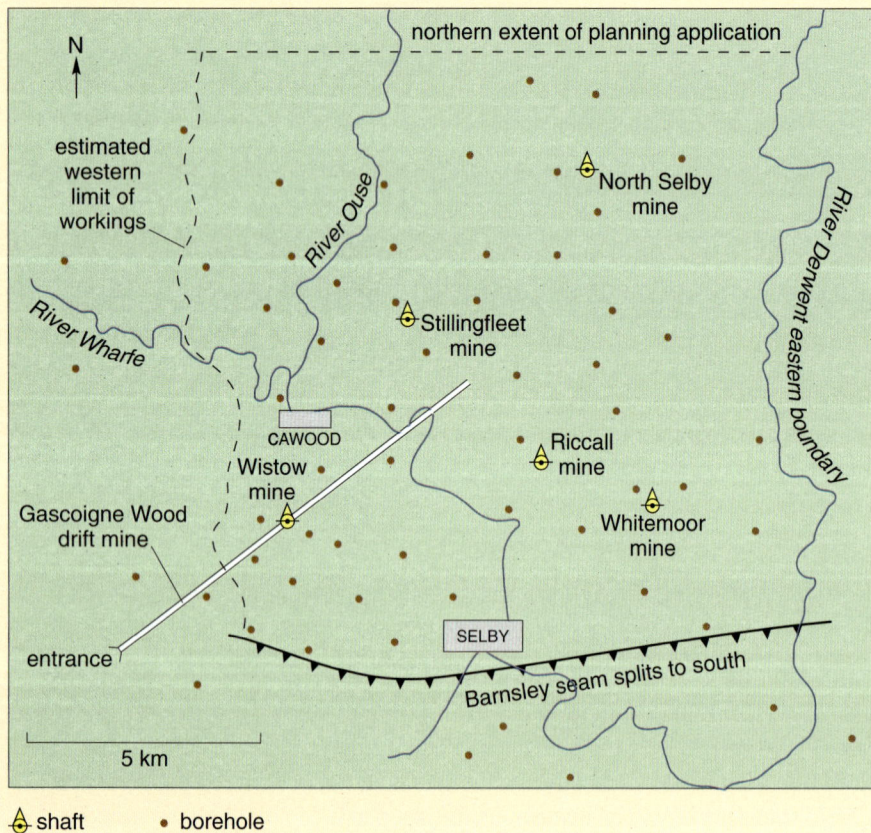

Figure 2.26 Exploration of the Selby coalfield. The planning limits of the Selby project are shown, together with the locations of the Gascoigne Wood mine entrance and underground drift, and the locations of the Wistow, Stillingfleet, North Selby, Riccall and Whitemoor mines. The exploration boreholes are also shown as dots.

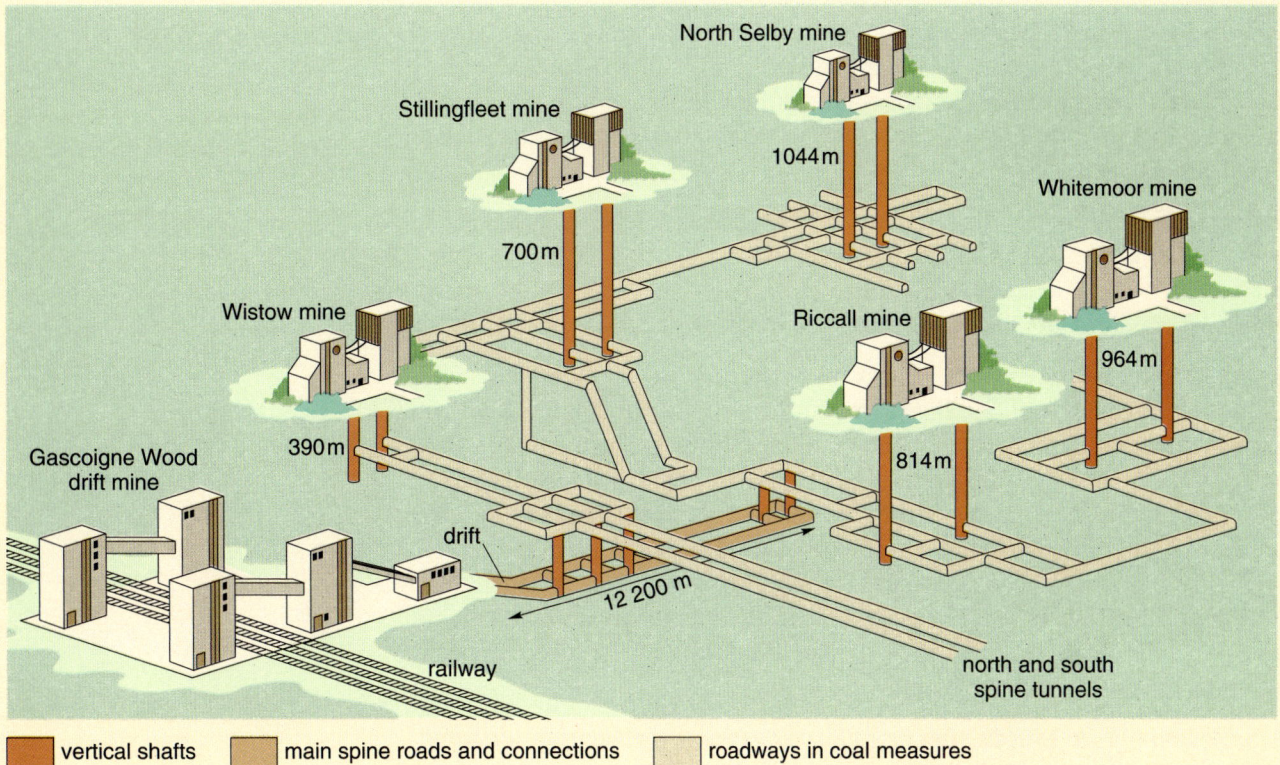

Figure 2.27 The Selby coalfield in 1992, showing the mine shaft, roadway and tunnel layout. The depth from the surface to the Barnsley seam is shown at each mine.

The proportion of recoverable coal at Selby was

$$\frac{224 \times 10^6 \text{ t}}{2000 \times 10^6 \text{ t}} \times 100 = 11.2\%$$

produced compared with the cost of removing the waste material. In addition, surface mining machinery is unable to operate below a certain depth.

The Black Thunder mine produced 51.5×10^6 t of coal in 2003. Therefore, to produce 224×10^6 t would take

$$\frac{224 \times 10^6 \text{ t}}{51.5 \times 10^6 \text{ t yr}^{-1}} = 4.3 \text{ yr}$$

The £400 million colliery complex started production under very strict environmental controls in 1983 and output peaked in the early 1990s at 10×10^6 t yr^{-1}. At the height of production, coal from ten retreat faces was transported along the network of conveyor belts to the coal-handling point at Gascoigne Wood. There the coal was washed and blended (by mixing coals of different ash contents from different mines) before being transported by rail to the nearby power stations at Ferrybridge, Eggborough and Drax A and B. Although Selby coal then represented about 15% of all coal used for electricity generation in the UK, it also supplied the domestic 'house coal' market too.

Output at the Selby complex fell to 6×10^6 t yr^{-1} by 1999. This decline in production resulted from an over-reliance on the power stations as customers, with contracts agreed at a time of low global coal prices. In the end, Selby was unable to maximize its potential, despite long-term contracts with the power stations, the expansion of their market beyond power station fuel, and despite repeated initiatives to maximize production and cut costs. The last mine at Riccall closed late in 2004, some five years short of that originally envisaged for the complex.

2.2.6 Geological problems in coal mines

A modern coalface is a very complex operation that represents a large investment in terms of capital, labour and planning. Cutting machines and lengthy conveyors are inflexible and require uniform geological conditions to maximize output. What then are the effects of geological variations on such a mining system?

Geological factors control the selection of working areas. The two principal geological conditions that affect mining operations are, first, the nature of the coal-bearing rocks and lateral variations in rock type. The second concerns geological structure: the dip of the strata and the presence of faults.

Variations in thickness and type of rock may occur both within the seam and in the strata forming its roof and floor. The thickness of a seam has a considerable bearing on the profitability of mining operations. Seams 1–2 m thick are particularly suited to mechanized longwall mining, whereas seams less than a metre thick suffer from the law of diminishing returns, since to obtain the same volume of output from a thin seam as a thick one, a greater area of extraction (and hence more rapid face advance) is required. The presence of layers of mudstone within a seam reduces its value, because the quality and marketable value of coal is determined by its ash content. Such layers may also indicate the beginning of a split within the seam.

Of equal importance in achieving rapid face advance is the nature of the roof and floor strata. Mudstones and siltstones usually provide good working roofs for a coalface, but a soft seatearth below the seam will cause heavy equipment to gouge out several centimetres of floor material, reducing the quality of the coal. To counteract this, coal may have to be left at the bottom of the seam.

The presence of sandstones can also result in operational problems at the coalface. The most serious of these arise where **channel-fill deposits** are encountered in the mudstones above a coal seam. These structures formed when erosive drainage channels on the original delta plain cut down into the underlying sediments and filled with coarse sand. If such channels cut down into the mudstones above a seam, they usually result in unstable roof conditions, and serious roof falls can occur. Sometimes the channels cut down into the coal seam itself and extreme examples may locally have eroded the seam away completely (Figure 2.28). Where these **washouts** occur they will bring a working face to a standstill.

The stripping ratio for the two seams in Figure 2.28 is less than 7:1 for most of the area to the NW of the fault: the two seams there are economic for surface working and can be considered as reserves. If only the thin seam were present, that would be economic only in the western part of the area. To the SE of the fault, the stripping ratio rises to 23:1, due to the displacement and the angle of dip. This is uneconomic for surface mining and the downfaulted block cannot be considered a reserve for a surface operation. The coal seams in that area would possibly be a prospect for underground mining. However, the features revealed to the west of the fault — washouts, old workings and seam splitting — that pose no great problem for surface mining, would considerably reduce the reserves, were they to be encountered underground. The split might make

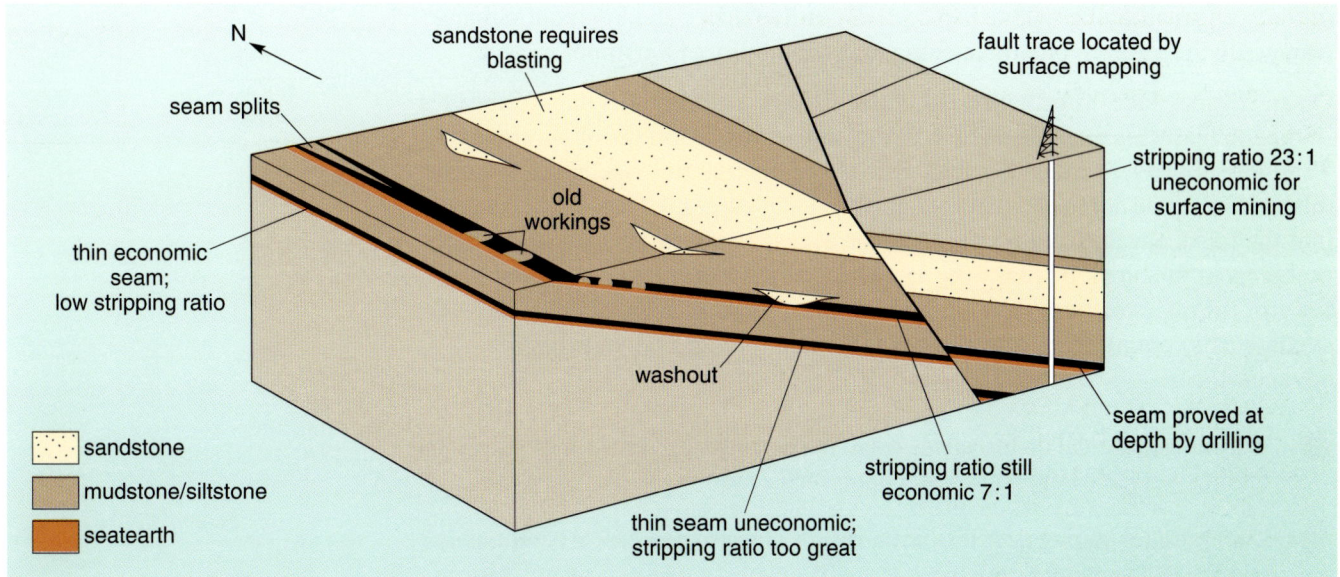

Figure 2.28 Some of the geological factors that influence coal-mine planning and reserve assessment.

both branches of the thicker seam too thin to work. Old workings and washouts present hazards from flooding as well as the absence of coal. Large volumes near them would not be workable. Consequently the reserves underground would be considerably reduced.

Whereas the sedimentary setting of a seam and its associated strata determines the profitability of the mining operations, the structural setting imposes physical constraints on the layout of mine workings. Faults form major problems for mechanized faces, as any displacement greater than the seam thickness may bring production to a halt with consequent loss of output (Figure 2.28). Substantial displacements may also result in the stripping ratio beyond the fault becoming too great for the coal to be economically recoverable by surface mining. Such faults may therefore mark the boundary of a prospect's reserves.

In addition, in some areas of the world igneous rocks intruded into coal-bearing strata will adversely affect the quality of the coal adjacent to the intrusion.

Recognizing geological problems

The most profitable coal mines are those that possess unbroken, horizontal seams of constant thickness and quality. In mines where this is not so, profit levels will depend on the ability of the mine geologist to predict changes in the seam before they are encountered at the face.

Geological problems fall into two categories — gradual changes and sudden changes. Where a change is gradual, such as a seam thinning or splitting, data from boreholes in advance of the workings, supplemented by observations on working faces and development roads can be used to identify those areas where a seam becomes too thin or split to be worked.

Where the changes are sudden, for example with washouts or faults, other methods have to be employed. This may involve the detailed study of the strata exposed in roadways and in borehole sections in order to plot the paths of distributary channels. Forecasting fault positions is more difficult. In particular, unforeseen small faults with a displacement of less than 5 m may require major changes to the mine layout, threatening the profitability of the mine.

On an advance face, information about the geological conditions that lie ahead is limited to that which can be deduced from any nearby workings or boreholes. Therefore, an advance face is a high-risk layout, though this risk has now been mitigated to some degree by using geophysical techniques to 'see' horizontally into the rocks ahead. On a retreat face, the access roads expose any geological problems in the block of coal to be extracted, and the face can be worked back towards the main roadways without risking interruptions. Consequently, the average daily output from retreat faces is some 25% higher than that from advance faces.

- Why do geological disturbances result in so many problems for underground mining, yet have very little effect upon surface mining?

- Underground mining is inflexible and needs uniform geological conditions to maximize its potential. Such mines are incapable of negotiating any serious variations in the thickness of the seam. Surface mining is extremely flexible and can strip away all the non-coal rocks regardless of geological variations, leaving the coal to be extracted easily.

For sound geological, economic and safety reasons, surface working will always be preferred to underground mining, so long as the stripping ratio is favourable. Over 95% of all near-surface coal can be recovered in this way, and it is relatively straightforward to exclude all the valueless roof and floor sediments during production. By comparison, underground mining rarely recovers more than 70% of coal in thick seams to ensure exclusion of roof and floor rock. Thin seams, even within the mined sequence, cannot be economically extracted. Areas where the seam is disturbed geologically may not be workable, and support pillars may need to be left to prevent roof collapse and subsequent damage to underground roadways or surface buildings.

Question 2.7

When a coalface first enters a new area of reserves, it may encounter a variety of geological hazards, which may halt production. Which hazards may be described as gradual changes, and which are sudden changes? What would be the effect on production of each of these two classes of hazards?

2.3   Environmental aspects of coal mining

Coal produced by both types of mining is used either to fuel electricity generation or for industrial and domestic heating, both of which result in atmospheric pollution (Chapter 6), but here we are concerned with direct environmental impact on the land. Surface and underground mining operations cause significantly different environmental problems. Those that surround surface mining are common to any large quarrying operation: sterilization of the land and restoration of quarry sites; dust; and noise while operating. Mining waste is not a problem since it helps to fill the hole created. Underground mines produce less noise and dust at the surface, but cause land subsidence and generate long-lasting waste tips, even though some waste is used as backfill. As you will see, both operations can potentially cause water pollution.

2.3.1 Surface mining

Many environmental issues arise when surface mining is considered, and such mines regularly arouse local opposition. By their very nature, surface mines have a major impact on the landscape, involving the digging of enormous pits with accompanying noise, dust and traffic movements, and destruction of mature landscape. Increasingly, in recent years the environmentally conscious public has used the planning processes to oppose and sometimes prevent mining on sites where the environmental impact would be severe.

Many steps can be taken to minimize the nuisance of surface mines. Topsoil is commonly stored in graded embankments around the boundaries of mines as a baffle against visual intrusion, noise and dust. On-site vehicles can be fitted with effective silencers. To prevent dust being raised on site, water bowsers spray haulage roads. Lorries leaving the site pass through wheel washers and their loads are often covered. Furthermore, individual mines have lives limited to 5–10 years and operators are required to restore former sites to productive farmland, forestry or recreational use by re-spreading the original topsoil.

2.3.2 Underground mining

Underground mining operations have four significant environmental impacts — spoil heaps, methane build-up, subsidence and water pollution. Spoil heaps have always been the principal surface feature of underground mining operations. However, legislation and technical advances have brought improvements in modern mines, and the closure of many of the UK's older mines has often been followed by successful rehabilitation of mine sites and spoil heaps by landscaping and tree planting.

Coal seams naturally generate methane, and as this is an explosive gas, build-up in old mine workings has to be prevented by venting to the atmosphere. Such methane has been successfully used to generate saleable electricity or diverted into household gas supplies, which lessen its effect as a powerful greenhouse gas (Chapter 6).

Mining subsidence

Subsidence is an inevitable hazard wherever underground mining is carried out. The major factors affecting the extent of subsidence are *seam thickness* and its *depth beneath the surface*.

The amount of subsidence can be calculated roughly by using the formula:

$$s = \frac{4t}{\sqrt{d} + 4} \tag{2.1}$$

where s is the amount of surface subsidence (in m), t is the thickness of the worked seam (in m), and d is the depth to worked seam (in m).

Activity 2.1

(a) Plot a graph on Figure 2.29 to investigate the subsidence produced at depth by mining a seam of 3 m thickness under the Selby area.

Hint: You will need a calculator to do this. You will have to work out, using Equation 2.1, the amount of subsidence s for a seam thickness of 3 m for a range of working depths. Use a range from 50 m to 750 m, at 100 m intervals, and then record your calculated values of s and d in a table. Then plot your values of s and d on Figure 2.29, and draw a curve through them.

(b) From your graph, what is the minimum depth at which a 3 m seam can be worked to produce less than the 0.99 m of surface subsidence demanded at Selby?

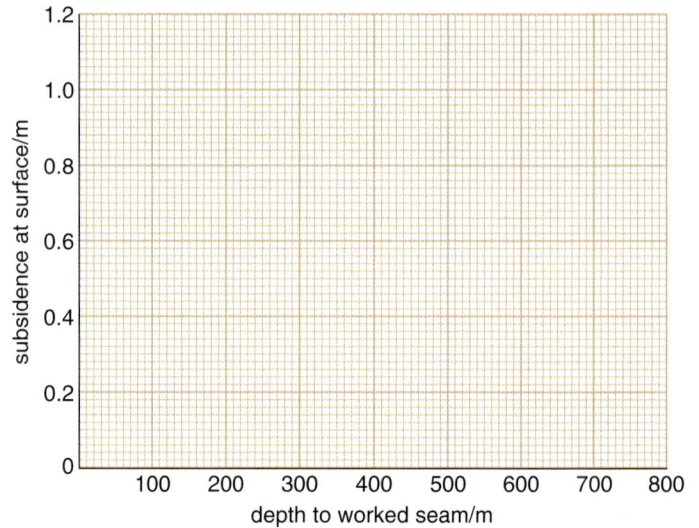

Figure 2.29 For use with Activity 2.1.

Roof collapse will often start within 24 hours of coal extraction, but the full effects are transmitted rather slowly upwards, eventually resulting in subsidence at the surface (Figure 2.30). It may be over 10 years before the surface is completely stable again. Vulnerable structures, such as conurbations, dams, viaducts, and historical buildings are protected by leaving coal unworked beneath them, but such protection may be extremely costly where it significantly affects the layout of the mine. The Selby mine highlighted a particular problem of subsidence by affecting the land surface in a low-lying area. At Selby, government-imposed restrictions minimized both the visual impact of mine site buildings and the effects of ground subsidence in areas particularly liable to flooding or subsidence damage. The mine operator also had to finance flood protection measures and the relocation of the main east coast railway line.

Figure 2.30 Surface subsidence effects above an abandoned coal mine in Sheridan, Wyoming, USA. The mine was in operation between 1904 and 1921. The indistinct, rectangular-shaped depressions occur where the coal has been removed; their arrangement suggests that pillar-and-stall mining was used (see Figures 2.8a and 2.8b). The old workings are approximately 10 to 15 m below the surface. Circular pits at the margins of the rectangular depressions are caused by localized collapses into open stalls.

Water pollution from coal mines

Most underground and some surface mines lie well below the water table. Both therefore have the potential to pollute any groundwater that flows through them. The root cause of the problem is the action of anaerobic bacteria on pyrite (FeS_2) within the coal sequence. This process releases metal and sulphate ions into solution, which in turn causes the acidity of the mine water to increase:

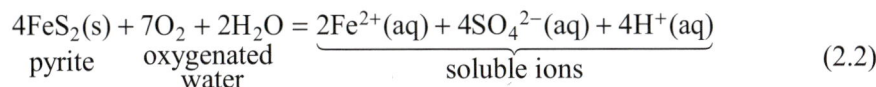

$$4FeS_2(s) + 7O_2 + 2H_2O = 2Fe^{2+}(aq) + 4SO_4^{2-}(aq) + 4H^+(aq) \qquad (2.2)$$

pyrite   oxygenated water   soluble ions

Before mining, groundwater flow through coal sequences is usually sluggish and in general chemically slightly reducing. During mining, most of the original water is extracted by pumping to keep the mine dry. Pumping exposes the coal-bearing rocks to moist air, leading to oxidation of pyrite within the sequence, helped by the catalytic action of bacteria. When mines close and pumping stops, the water table can rise again. Soluble products of pyrite oxidation will pollute groundwater with sulphuric acid, iron and manganese cations, sulphate anions, and sometimes highly dangerous arsenic ions. The roadways of abandoned coal mines form a very effective artificial network of underground watercourses, which can channel polluted water to the surface. Such **acid mine drainage** (**AMD**; see Smith, 2005) is highly damaging to most forms of aquatic life and to humans who come into contact with it. Where this water emerges, surface streams show the telltale sign of AMD: ochreous slimes (Figure 2.31).

Figure 2.31 Polluted waters from acid mine drainage.

Surface mines represent only a transient problem. Water is pumped out of them during working into settling ponds, which ensure that only clean water leaves the site. The mined area is backfilled after use, and the water table generally equilibrates to that within the surrounding area. Preventive measures usually centre around excluding oxygen, water or bacteria from underground exposures to prevent pyrite oxidation. In old mines, products of pyrite oxidation have already accumulated over centuries of exposure, and sealing old mine workings to reduce water flow is impractical.

A solution to AMD is not easy to find. Continued pumping after a pit has closed down is expensive, and routing polluted mine waters into special treatment areas is impractical because it is almost impossible to predict the movement of groundwater in anything other than general terms. Environmental scientists have looked at other ways of improving the quality of mine waters, such as the use of wetlands in which some plants can permanently absorb pollutants, and neutralizing mine waters by adding alkaline calcium carbonate or sodium hydroxide. Neutralization also decreases the solubility of dissolved metal ions, so that they precipitate and settle out of solution.

2.4 Global coal reserves

Having looked in detail at how coal is mined, this section focuses on those areas of the world that produce it. It begins by looking at how and why reserves of coal are distributed throughout the UK and Europe, before reviewing the current global reserves of coal.

2.4.1 Coal distribution in the UK and Europe

The UK and Europe were fortunate in having extensive coalfields that powered the Industrial Revolution. Figure 2.32 shows the distribution of the major Carboniferous mires which became coal-bearing rocks across Europe, either outcropping at the surface or buried beneath younger rocks. The first thing that is evident from this map is that not all countries shared the same good fortune; either such rocks were deposited, but were subsequently eroded away, or they were never deposited in the first place.

In deciding which of these two possibilities is correct you need to look more closely at Figure 2.32. It shows the geography of Europe as it might have appeared in Late Carboniferous times, about 300 Ma ago, when the entire region lay close to the Equator. It is clear that the Carboniferous landmass bears little resemblance to the present-day coastlines, which are superimposed on the map for reference.

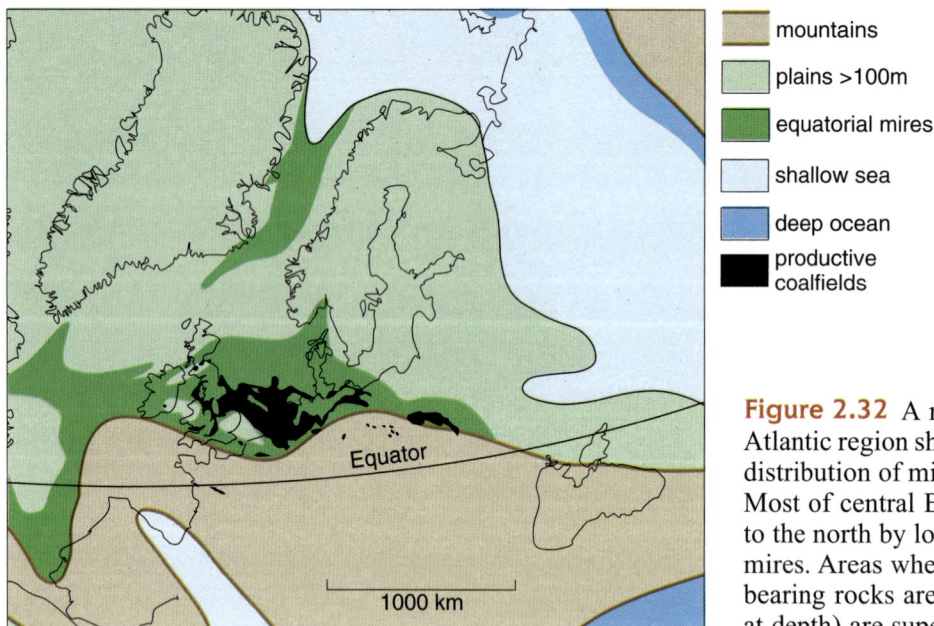

Legend:
- mountains
- plains >100m
- equatorial mires
- shallow sea
- deep ocean
- productive coalfields

Equator

1000 km

Figure 2.32 A map of the present-day North Atlantic region showing the palaeogeographical distribution of mires in Late Carboniferous times. Most of central Europe was mountainous, bounded to the north by low-lying plains and coal-forming mires. Areas where the major Carboniferous coal-bearing rocks are preserved (both at the surface and at depth) are superimposed in black.

● Why are the outlines of Greenland and Norway so close together in Figure 2.32?

○ The North Atlantic Ocean did not exist in Carboniferous times.

Carboniferous Britain and north Europe formed a low-lying plain backed by newly formed mountains to the south and a shallow sea to the north, beyond present-day Scandinavia. Tropical waterlogged mires developed across Britain and Ireland, through the southern North Sea into Belgium, the Netherlands and northern Germany and east into Poland, as well as between Greenland and Scandinavia. It is likely that coal formed across the whole of this area.

So, why aren't Carboniferous coal measures found over *all* the area that was land during that period? Consider the following.

- Is it likely that coal-bearing rocks would have formed in the mountainous areas?

- Coals form from mires found mostly (though not exclusively) in low-lying areas (see Section 2.1.2). Therefore, it is unlikely that Carboniferous coal would have formed in the mountainous areas marked on Figure 2.32.

- Why are no Carboniferous coals found over the plains area marked in light green on Figure 2.32?

- Although relatively low-lying, these areas (including Scandinavia, the Baltic States, south-eastern England and west Wales) were hilly enough to prevent the development of mires.

- Why aren't Carboniferous coals preserved over all the area once covered by mires?

- In many areas that have undergone uplift due to tectonic activity, erosion may have removed the coal-bearing sequence. (This happened in the Pennine area of the UK.)

Erosion also helps to *expose* Carboniferous coals by removing overlying, younger rocks. As a result, surface mining is common in the Ruhr area of Germany, in Poland, in the Midland Valley of Scotland, and in parts of Lancashire and Yorkshire.

2.4.2   The UK's coal reserves

Production of large quantities of coal in the UK during the 19th and 20th centuries led to the progressive depletion of reserves. In 2005 underground mining was limited to the Carboniferous coalfields of Yorkshire and the East Midlands, with only one underground mine operating in South Wales. However, surface mining sites still work coal in most of the coalfields (Figure 2.19).

As Section 2.5 will show, what is considered to be a *reserve* (i.e. the amount that is thought to be *recoverable in the future under existing economic conditions* — Sheldon, 2005) changes with time. The coalfields of northern England and the Midland Valley of Scotland that extend eastwards under the North Sea are examples of this. Although coal has been worked there in the past, it is not currently recoverable at a profit.

In 2004, the UK's coal *reserves* were estimated to be 1.0×10^9 t of anthracite and bituminous coal, and 0.5×10^9 t of sub-bituminous coal and lignite. Together, these two figures represent 0.2% of total global coal reserves.

2.4.3   Coal in the European Union

The EU's coal reserves in 2004, after enlargement to 25 member states, stood at 100×10^9 t. Table 2.2 shows the eight European Union Member States with the most significant reserves ranked in order of greatest tonnage. With a little over 100×10^9 t of coal of all ranks, the EU possesses approximately 10.2% of total global coal reserves. Germany has by far the largest reserves (dominated by 'hard' brown coal and lignite), rivalled only by Poland for coals of higher rank. By contrast, the UK's reserves make a minor contribution to the European total.

Of course, for a more complete picture, the coal *resources* (i.e. the *amount currently (reserves) and potentially profitable to recover in the future, given reasonably foreseeable changes in economic and technical conditions* — Sheldon, 2005) of these countries (Table 2.2) should also be considered. The resource figures (calculated four years before those of the reserves) should be treated with some caution as countries often adopt different criteria (in terms of seam thickness and depth) when calculating them. However, a comparison of the last two columns in Table 2.2 shows that there are huge quantities of coal currently considered uneconomic to mine. This is especially true in the UK, an issue that will be considered in more detail later on.

Table 2.2 also shows that many EU countries produce low-rank brown coal in large quantities. Germany, the Czech Republic, Greece and Hungary have greater reserves of brown coal than higher rank coals. Much of this coal formed during Tertiary times, about 20 Ma ago. Tertiary lignites in the UK are limited to small areas around Bovey Tracey in Devon and Lough Neagh in Northern Ireland.

Figure 2.33 illustrates the production of coal in Europe since 1981 for those countries listed in Table 2.2. With the exception of Greece, production has declined, especially among the major producers. This trend reflects the ending of generous EU-supported government subsidies, which had allowed otherwise loss-making mines to remain in operation. As European coal is around three times as

Table 2.2   European Union countries with the largest proven reserves (in 2003) and resources (1999).

Country	Coal reserves at end of 2003			Coal resources at end 1999
	Hard coal (anthracite and bituminous coal)/10^9 t	Brown (sub-bituminous) coal and lignite/10^9 t	Total tonnage from all ranks/10^9 t	Total tonnage from all ranks/10^9 t
Germany	23.0	43.0	66.0	122.0
Poland	20.3	1.9	22.2	64.5
Czech Republic	2.1	3.6	5.7	21.3
Greece	none	2.9	2.9	not known
UK	1.0	0.5	1.5	190.0*
Hungary	none	1.1	1.1	6.8
Spain	0.2	0.5	0.7	2.9
France	0.02	0.01	0.03	0.3
Total			>100	not known

*Privatization of the the UK's coal industry in 1994 makes it difficult to quantify resources. The figure quoted is from 1992 and includes 45×10^9 t which 'could be extracted with current technology'.

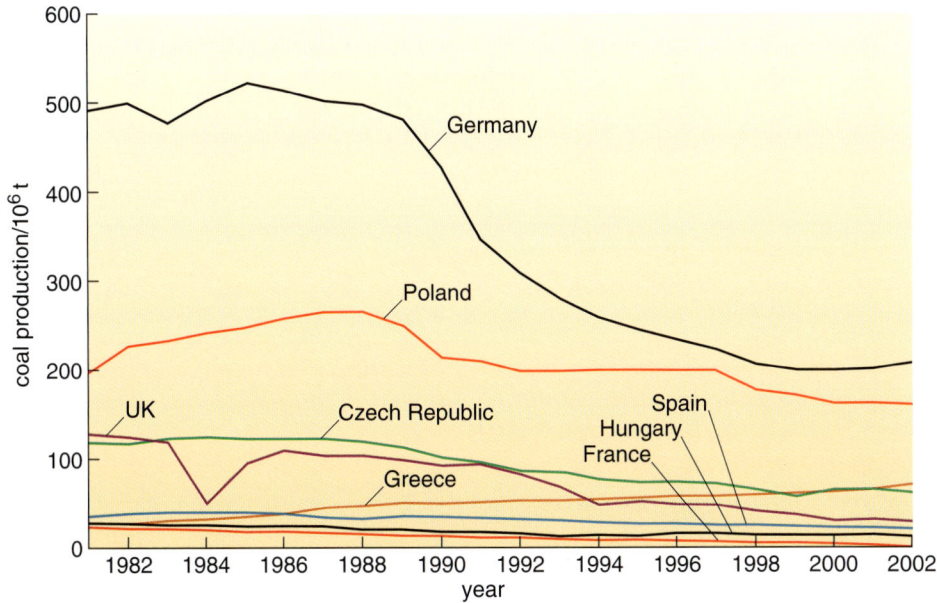

Figure 2.33 Coal production in Europe during the period 1981 to 2002. (The miners' strike accounted for the drop in UK production in 1984.)

expensive as imported coal, it cannot compete in the global market without such subsidies. By switching to fuels other than coal, the European domestic market is unlikely to come to the aid of its ailing coal industry.

Even so, in 2004, the EU produced approximately 593×10^6 t of coal. By contrast, it consumed 749×10^6 t yr^{-1}, making it a net importer of coal. Some countries (e.g. France and Spain) rely more heavily on imports than others.

2.4.4 Global distribution of coal

Figure 2.34 shows the global distribution of coal deposits. The major areas are principally in the Northern Hemisphere; with the exception of Australia, the southern continents are relatively deficient in coal deposits.

Figure 2.34 A map showing the global distribution of known coal deposits of all ages.

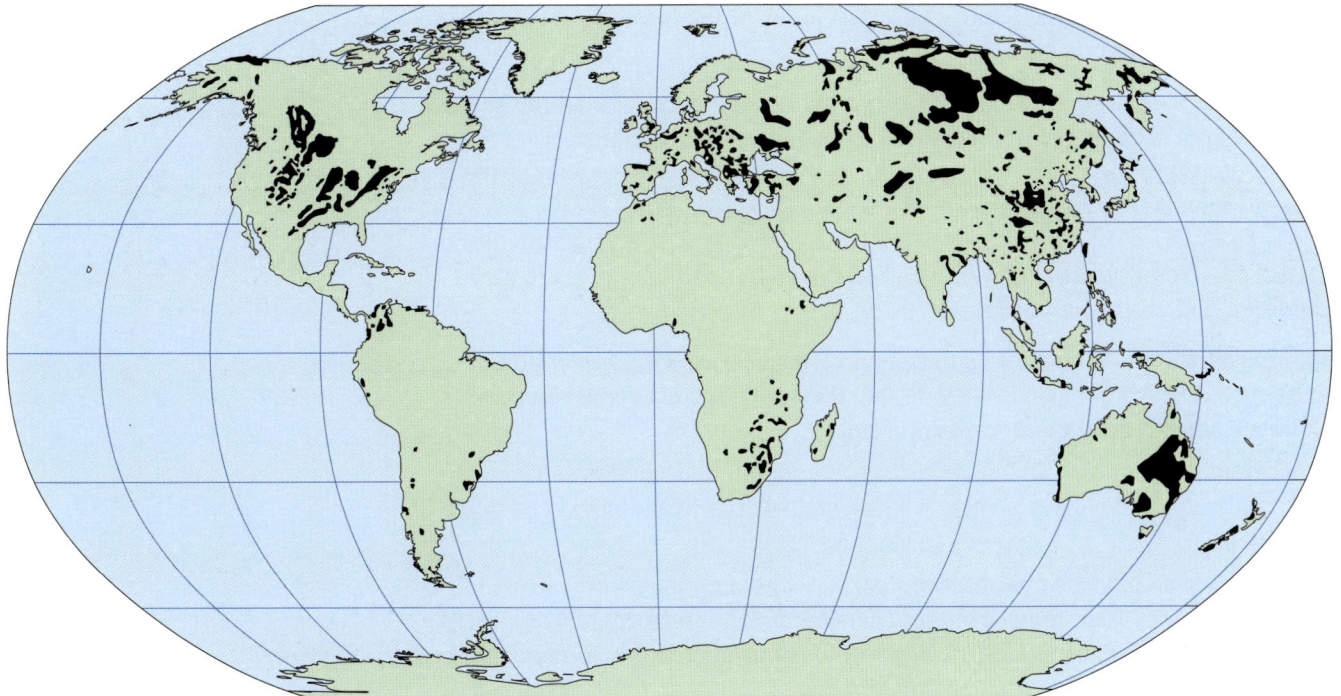

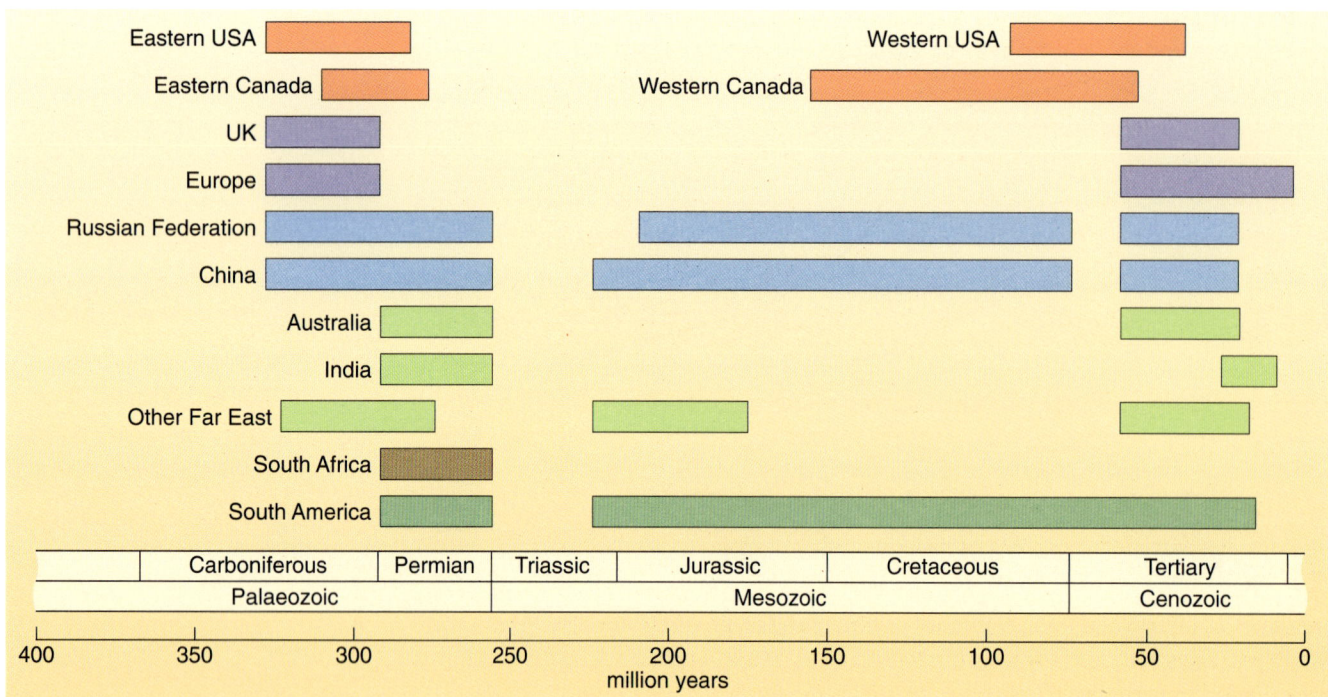

This relatively uneven distribution is the result of peat formation at different times in the geological past in predominantly tropical latitudes, and the subsequent drift of the continents to their present-day positions. As Figure 2.35 shows, the oldest coals of any economic significance date from the Middle Carboniferous Period — the earliest geological strata in which coal has been identified are of Devonian age but they are of little economic significance. With the exception of parts of the Triassic Period, major coal deposits have been forming somewhere in the world throughout the last 320 Ma. Sedimentary sequences of the last 2–3 Ma do not contain coals, simply because there has been insufficient time for them to develop from plant debris. Possible future coals exist today in the form of peats that are accumulating in modern mires.

A broad chain of large coalfields of Carboniferous age extends from the eastern USA, through Europe, the Russian Federation and south into China. A second chain of Permian coalfields is found in the southern continents — South America, India, southern Africa, Australia and Antarctica (the latter is not shown on Figure 2.34). The vast coalfields of the western USA (see Box 2.1) and Canada are of Cretaceous–Tertiary age (Figure 2.34). Mesozoic–Cenozoic lignites are also important global sources of coal.

Figure 2.34 shows that not all regions have coal reserves. They either do not possess coal-bearing rocks, or they do, but they are not considered reserves. There are several geological reasons for this:

- Regions where post-Devonian sediments were not deposited or have been completely eroded have no coal whatsoever, such as Scandinavia and much of Africa.
- Regions may have no coal-bearing rocks — although they were once deposited they have now been eroded away. Ireland once contained a substantial sequence of Upper Carboniferous coals; probably as much coal as the UK, but only a tiny fraction now remains.

Figure 2.35 Global distribution of major coal deposits by geological age. Some coals are known from the Devonian Period, but have minor economic significance.

- Some regions have coal-bearing rocks that are buried so deeply under younger sedimentary sequences that they cannot be worked economically. The Amazon basin is a good example.

- Few preserved sedimentary rock sequences formed under terrestrial conditions with which mires were associated. Some terrestrial sediments formed at high palaeolatitudes and others in arid climatic belts, neither of which are conducive to peat formation, whilst other sediments formed on the sea floor. The Carboniferous sediments of South Africa and the Permian of the UK (dominated by terrestrial glacial and desert deposits respectively) are good examples, as neither is coal bearing, whereas Permian terrestrial sediments of South Africa and those of the Carboniferous of the UK both contain substantial coal sequences formed in a humid climate.

2.4.5   Global coal reserves and their life expectancy

In 2003, global proven coal reserves were estimated at 984.5×10^9 t, of which slightly over half (52.7%) was anthracite and bituminous coal and the rest (47.3%) was sub-bituminous coal and lignite.

Figure 2.36 shows the breakdown of global reserves by continental regions. North America has 26% of total global coal reserves, South Asia and the Pacific (mainly Australia) have 30%, Europe 20%, and the Russian Federation 16%. The relatively sparse distribution of coalfields elsewhere (Figure 2.34) is borne out by far less significant reserves in Africa and the Middle East (6%), and in South and Central America (2%).

Table 2.3 shows known global reserves as of 2003, ranked in order of the countries with the highest annual production. However, current reserves represent a snapshot in time; what is considered a reserve depends on current extraction technology, and reserves are depleted as coal is mined. Moreover, coal is a high-volume, low-value resource, and so reserves are highly sensitive to economics, the more so because at the start of the 21st century there is a glut of easily mined coal worldwide.

Figure 2.36 Global coal reserves proven by the end of 2003. Amounts are in 10^9 t, with the total of anthracite and bituminous hard coal in brackets.

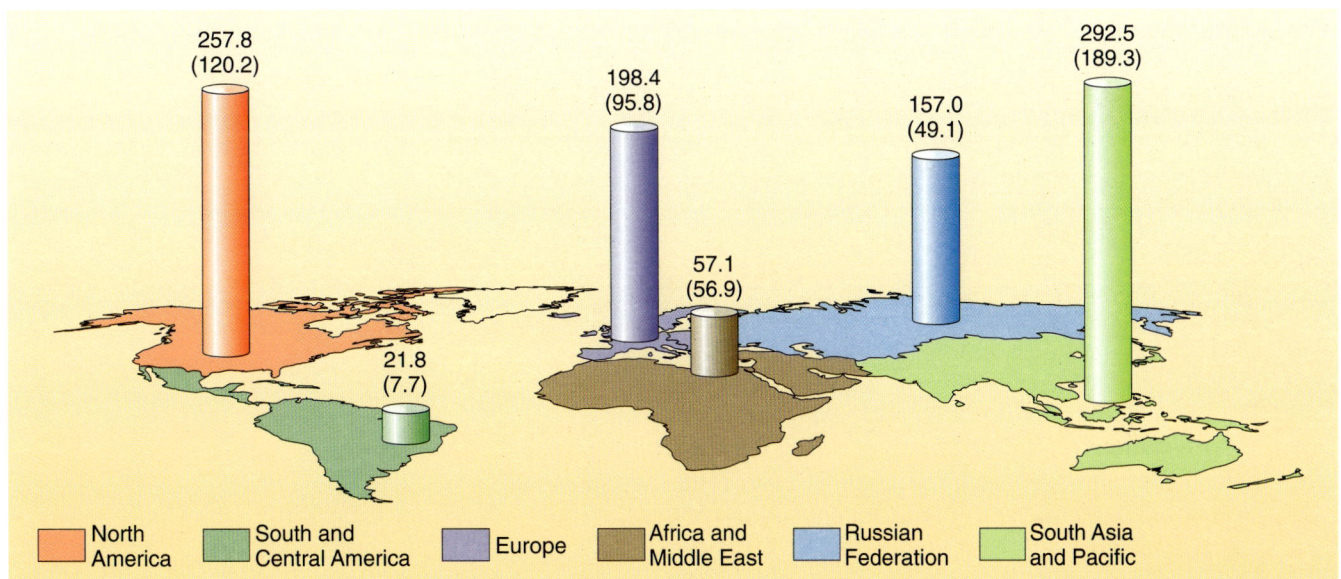

Table 2.3 Coal production, consumption, reserves, reserves/production (R/P) ratios and resources for the eleven globally dominant coal producers, and the UK at the end of 2003 (resource data from 1999). The UK is currently the world's 15th largest producer. Coal accounts for about a quarter of global primary energy, as it has done for the last 30 years.

Country	Annual production /10^6 t	Per cent of global production/%	Annual consumption /Mtoe	Reserves /10^9 t	Per cent of global reserves/%	R/P ratio /yr	Resources* /10^9 t
China	1667.0	33.5	799.7	114.5	11.6	69	not known
USA	970.0	21.9	573.9	250.0	25.4	258	457.5
India	367.3	6.8	185.3	84.4	8.6	230	not known
Australia	347.2	7.5	50.2	82.1	8.3	236	106.8
Russian Federation	274.8	5.0	111.3	157.0	15.9	>500	not known
South Africa	238.8	5.3	88.9	49.5	5.0	207	115.5
Germany	205.0	2.1	87.1	66.0	6.7	322	122.0
Poland	162.8	2.8	58.8	22.2	2.3	136	64.5
Indonesia	114.6	2.8	18.9	5.3	0.5	47	not known
Kazakhstan	84.7	1.7	26.9	34.0	3.5	401	not known
Ukraine	80.3	1.7	39.0	34.2	3.5	426	45.5
UK	28.2	0.7	39.1	1.5	0.2	53	190.0
Global total	5118.8	100	2578.4	984.5	100	192	not known

Note: The data for coal consumption in this table are given in million tonnes of oil equivalent. This takes into account the differences in heat value of various ranks of coal.

* Resource figures are not available for all countries.

Resource figures are also given in Table 2.3, however such data are not available for all countries, including many of the world's major producers.

● Compare the USA and China in terms of reserves versus coal production in 2003.

○ China has less than half the reserves of the USA, and yet annually it produces 1.5 times more coal. Clearly if this rate is maintained, China's coal will be exhausted before that of the USA.

This is in fact a recent phenomenon. Since 2000, China has increased production of coal by over 40%, overtaking the USA, which was the world's largest producer between 1999 and 2001.

Comparisons between the sustainability of coal production in different countries are made easier in Table 2.3 by calculating *the ratio of proved reserves in the ground (R) to the annual production (P)*. This **R/P ratio** has dimensions of tonnes divided by tonnes per year, which is equal to *years* (t/(t yr^{-1}) = 1/yr^{-1} = yr). The R/P ratio is therefore a measure of how long the coal reserve (or that of any other fossil fuel or non-renewable resource) will last, if the current annual production figure is maintained into the future.

● Would you expect the R/P ratio to remain reasonably constant for decades?

The pace of exploration and evaluation continually adds to reserves, so they are often maintained at a constant level for many years. However, economics also plays a role — if costs of mining go up or the price of coal goes down some reserves will be lost from the inventory (Sheldon, 2005).

Annual consumption figures and R/P ratios for 2003 are also given in Table 2.3. Some interesting facts underlie the data given in this table.

- Despite producing 21.9% of the world's coal, the USA has to import coal to satisfy its demand, even though coal supplies less than a quarter of its primary energy requirement.
- By contrast, China has four times the population of the USA, yet coal supplies two-thirds of its primary energy requirement.
- The very high R/P ratio of the Russian Federation (>500 years) is a consequence of the recent fall in industrial coal consumption following the break-up of the Soviet Union.
- India has the world's fourth largest reserves, but needs to import about a quarter of its coking coal for the country's steel industry.
- Australia is by far the world's largest exporter (as you can judge from its high production yet low consumption), selling coal to over 30 countries. About half the exports go to Japan (the world's largest coal importer).
- The figures for the UK reflect the widespread closures of mines in the 1990s that effectively reduced reserves enormously (Section 2.5).

In years to come, as reserves begin to run low, global coal prices may increase. More coal resources would then become economic to mine, to be reclassified as reserves. This would in turn raise the R/P ratio. It is therefore important to include the future conversion of resources to reserves when considering how long coal will last as a source of energy. However, the current lack of coal resource data for many countries, together with uncertainties associated with predicting future changes in global coal prices make such calculations rather speculative. Of course, such speculations presuppose that coal continues to be favoured despite its contribution to global warming (Chapter 6).

In Table 2.3, the R/P ratio for the USA is given as 258 years. If half of the coal currently considered a resource, but not a reserve, was to become a reserve in future, how long would this coal last (assuming no change in production figures)?

For the USA, reserves $= 250.0 \times 10^9$ t, and resources $= 457.5 \times 10^9$ t. Converting half of the coal currently considered a resource only to a reserve:

$0.5 \times (457.5 \times 10^9 - 250.0 \times 10^9)$ t $= 103.8 \times 10^9$ t

Therefore total future reserves $= 250.0 \times 10^9$ t $+ 103.8 \times 10^9$ t
$= 353.8 \times 10^9$ t

R/P ratio $= 353.8 \times 10^9$ t$/970.0 \times 10^6$ t yr$^{-1} = 365$ years

Table 2.3 shows that global coal reserves could last into the 23rd century without any further reserves being identified, *if production were to be stabilized at present-day levels*. However, coal is mined in over 100 countries, and in the absence of any cartels that might control production (as OPEC does with oil), constraining future global coal production is unlikely. Therefore, this R/P ratio will most likely change with time. As you will see in the next section, R/P ratios for individual countries can vary dramatically too.

2.5 Coal production in the UK early in the 21st century

This section examines the UK's coal industry in a little more detail, to see how the complex interplay of location, economics and politics has led to the rapid demise of an industry that was once at the heart of the UK's economy.

Figure 2.37 shows production and consumption figures for coal mined in the UK since 1945 as a number of categories. The decline in total consumption shows that the demand for coal in the UK fell steadily since 1955.

● Which markets, shown on this figure, have contributed to this decline?

○ Between the early 1950s and 2003, both the coke and domestic coal markets declined markedly. In fact, the quantity of coal used for both fell by over 90%. Since 1990 the quantity used for electricity generation also fell sharply.

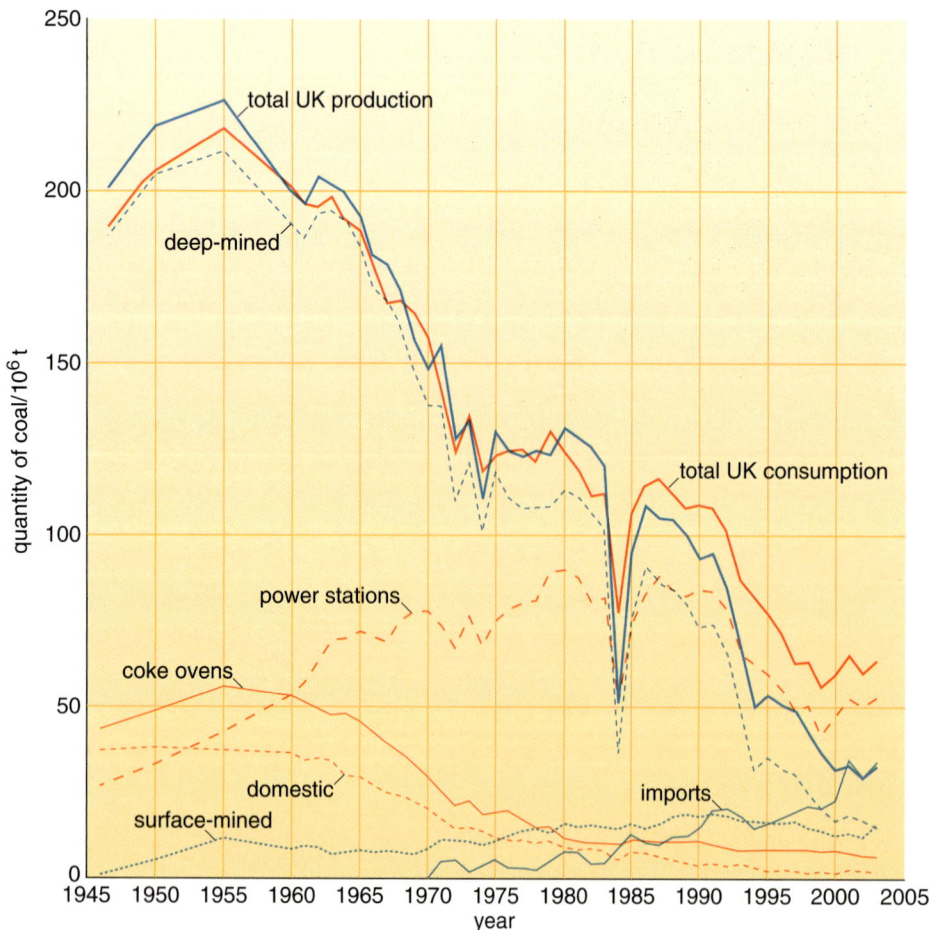

Figure 2.37 The UK's coal production and imports (blue lines) and consumption (red lines) figures for the period 1945 to 2003. (The miners' strike accounted for the drop in production in 1984.)

Coking coal was used extensively for steel-making: the decline in demand between 1955 and 1980 parallels the decline in the UK's steel industry. (Indeed as the globalization of coal and steel markets grew during this time, EU policies aided the decline of both industries in the UK.) The demand for domestic coal reduced over the same period as gas from the North Sea was offered as a cleaner and more convenient fuel from the early 1970s onwards.

Conversely, the demand for coal in electricity generation (Figure 2.37) grew, albeit erratically, and peaked in the 1980s. At first glance it appears that coal sales to the electricity generating market more than compensated for the decline in the coking coal and domestic markets. However, the fact that total UK consumption has exceeded UK production since the 1984–5 miners' strike bears witness to a global change in coal economics. The growth in size of bulk-transport ships and of surface-mining equipment resulted in coal being produced where surface mines work thick seams close to the surface, thousands of miles from potential markets: coal ceased to be a high place-value resource. Such coal can be transported to the UK for less cost than producing the UK's own coal from underground mines. The UK has been an importer of coal since 1970, but statistics suggest that the miners' strike triggered an increasing reliance on imports (Figure 2.37).

In the 1980s and 1990s, the UK's coal producers became progressively dependent on power stations for their survival. However, the lucrative contracts offered to the power companies when the coal industry was privatized in 1994 were insufficient to prevent the UK's electricity generation industry (itself privatized in 1990) from moving away from coal to natural gas. This initiative, at the time dubbed the 'Dash for Gas', was triggered by a change in EU regulations, which for the first time allowed gas to be used to generate electricity. Consequently, the amount of coal used for UK electricity generation dropped from over 84 million tonnes in 1990 to less than 42 million tonnes in 1999. That reduced amount of coal was also increasingly supplied by surface mines in the UK and by imports, rather than by the more expensive underground-mined coal.

Figure 2.38 illustrates how the demand for coal (and indeed oil) for electricity generation in the UK dropped sharply in favour of gas from 1990 to 2003.

Figure 2.38 The balance of fuels used to generate electricity in the UK in 1990 and 2003. Imports refer to electricity imported from Europe.

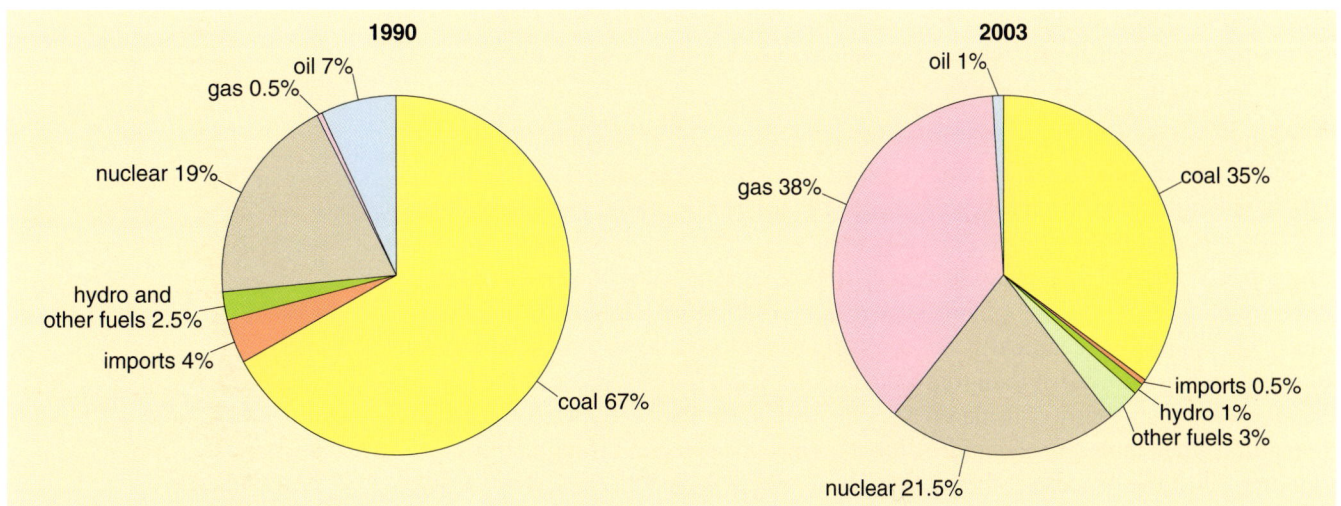

1990: oil 7%, gas 0.5%, nuclear 19%, hydro and other fuels 2.5%, imports 4%, coal 67%

2003: oil 1%, gas 38%, coal 35%, imports 0.5%, hydro 1%, other fuels 3%, nuclear 21.5%

As a direct consequence of the switch to gas, the UK's underground-mined coal industry virtually closed down in the early 1990s. In 1984, there were 170 underground mines; by 1994 the rush to close down less profitable mines in the run up to privatization of the coal industry in that year helped to reduce this figure to just 17. By the end of 2005, there were only seven major underground mines still operating. (Small, independent mines still produce coal for domestic use in South Wales, The Forest of Dean, and even north Cumbria.) Ironically, rationalization of the industry now means that the 9000 workers still employed in it are twice as productive as those of the 1980s and 1990s. We return briefly to this topic in Chapter 6.

Despite coal production being in private hands, government subsidies during periods of low coal price have prolonged the lives of some mines, though political and economic pressures continue to ensure an uncertain future for the seven mines that remained in 2004 (Figure 2.19). At the outset of the 21st century, the UK's major coal producers continue to depend almost entirely on the power companies. Power generators consume coal preferentially when the global gas price is high, with a knock-on effect for coal production (this is the cause of the fluctuations seen in Figure 2.37). With this variable in mind, it has been estimated that by 2010, coal-fired electricity generation could be anywhere between 35 and 80% of current levels, the huge range illustrating the uncertainties of prediction in the energy industry.

The quantity of imported coal continues to grow and in 2001, for the first time, more coal was imported than produced in the UK (Figure 2.37). In 2004 the UK imported about 50% of coal used here, mainly from Columbia and South Africa, together with some from Australia (a sea voyage of some twelve thousand miles), Poland and the USA. The low cost of bulk shipping, the quality of the coal and a competitive price clearly matter much more than the distance from the UK.

The short-term future of the UK's coal industry may ultimately be decided by its own government, which is committed to renewable sources of energy. New EU Directives aimed at reducing NO_x, SO_2 and CO_2 emissions will force power companies to switch to low-sulphur imports and close older coal-fired power stations, further damaging the coal industry. The technology needed to capture NO_x and SO_2 during coal burning does exist, but has yet to be deployed widely. However, the UK's coal industry may ultimately be saved by the need for a secure domestic fuel source, especially with nuclear power facing an uncertain future and North Sea gas production in decline.

The plight of UK coal illustrates dramatically how vulnerable the concept of reserves is to economic and political change. In 1950, it was estimated that the UK's 'remaining coal reserves' were 172×10^9 t. The UK's total coal production during the period from 1950 to 2003 was about 7×10^9 t, which should have left more than 165×10^9 t in place. Why then does Table 2.3 show the UK's reserves to be just 1.5×10^9 t? Although most of the coal from the 1950 estimate is still in place, the country's 'technically and economically accessible' reserves have fallen to less than a hundredth of their mid-20th century value. Whilst most of the 165×10^9 t not currently considered reserves will be available in future, the coal remaining when underground mines close down is lost forever, through both flooding and subsidence. So, the proportion of the UK's coal resources that will constitute reserves in future is a technological, economic and political issue.

From today's perspective, the global electricity generation sector cannot manage without coal. Despite the demise of the UK's underground coal mines, the global future for coal looks more optimistic than for any of the other fossil fuels, simply because there is so much of it.

2.6   Summary of Chapter 2

1   Waterlogged organic matter accumulates in deltaic, coastal barrier or raised mires to form peat. Coal forms by the compaction and decomposition of peat. Chemical changes imposed by increasing temperature and pressure over time determine the coal rank.

2   Coalfields can be classified as either exposed or concealed, depending on whether or not the coal-bearing rocks are hidden by younger strata. In most coalfields, mining commenced in the shallower exposed regions and has gradually extended into the deeper parts of the concealed coalfields.

3   Surface outcrops of rock can indicate the likelihood of finding coal at depth. In remote areas of the world, initial surveys often use data acquired from satellites or aeroplanes, before detailed ground exploration takes place.

4   Drilling boreholes is the only way to determine reserves and coal quality in concealed rock sequences. Geophysical logging records the nature of the strata located in the boreholes without the need to take expensive cores. Calculation of the stripping ratio indicates whether surface or underground mining is most economic.

5   Surface mining is a flexible, cheap system that is currently producing most of the coal exported globally. However, it results in short-term local environmental disturbances; underground mining provides longer lasting problems, owing to spoil heaps, subsidence and acid mine drainage.

6   Most underground coal is extracted nowadays using the longwall method and mechanized systems. Such coal-cutting systems are very inflexible and are incapable of negotiating geological variations in the coal seam. Continued exploration during coal production is therefore necessary to predict these variations and to maintain the continuity of operations.

7   The oldest coalfields are of Carboniferous age, which formed in tropical latitudes. The Permian–Triassic coalfields of the Southern Hemisphere originated in temperate latitudes. Brown coals make up half of global coal reserves; they are mainly Jurassic to Tertiary in age. Extensive Tertiary deposits of low-rank coal are worked in Eastern Europe.

8   Reserves of coal depend on whether or not deposits are economic to mine under current economic circumstances (costs of mining set against the price of coal). Technology, economics and political circumstances often change unexpectedly. So it is possible for reserves to become resources if they fail to be profitable, and vice versa.

9   In 2003, global reserves of 984.5×10^9 t of coal were predicted to last into the 23rd century, at 2003 rates of production. China is currently the largest producer though it has a relatively low R/P ratio.

10 Demand for coal has fallen in the UK over the last 20 years, because of technological, economic and political factors. The UK now imports more coal than it produces.

PETROLEUM

3

Oil and gas seeps have been known since earliest recorded history. Sticky black asphalt was used by the Babylonians as a roofing material, the ancient Egyptians used it to preserve their dead, and Noah supposedly caulked his Ark with it. In Azerbaijan gas seeps have burned for centuries, and therefore it is perhaps surprising that the world's first major underground oilfield was discovered in Pennsylvania, USA only as recently as 1859. That discovery launched an era in which the world became increasingly reliant on cheap energy provided by oil and gas, a reliance assured by the invention of the internal combustion engine in the late 19th century. Only now, as the issues of long-term sustainability and climate change become more apparent, are we beginning to think about unshackling ourselves from that dependency.

This chapter begins by examining the geological characteristics of petroleum and the key ingredients necessary to form oil and gas accumulations. Then there is a brief description of industrial operations during the life-cycle of an oilfield, starting with subsurface analysis and exploration drilling. The chapter also highlights the role of safety and environmental management as an integral part of the petroleum business and concludes with a short review of global resources and non-conventional petroleum.

3.1 The chemistry of petroleum

Petroleum is the term for a complex mixture of hydrocarbons and lesser quantities of other organic molecules containing sulphur (S), oxygen (O), nitrogen (N) and some metals. Hydrocarbons are compounds that contain only hydrogen (H) and carbon (C) atoms and the number of carbon atoms in a compound determines its physical properties. For example, simple compounds such as methane (CH_4), ethane (C_2H_6), propane (C_3H_8) and butane (C_4H_{10}) all have boiling temperatures below 0 °C and are therefore gases under ambient (surface temperature and pressure) conditions. Larger, more complex hydrocarbon compounds ranging from pentane (C_5H_{12}) to hexadecane ($C_{16}H_{34}$) are liquids under ambient conditions, whilst even larger compounds with a high molecular weight form waxy solids.

The molecular arrangement of hydrocarbon compounds is highly variable. The most commonly occurring forms are chemically stable, so-called saturated compounds known collectively as paraffins and cycloparaffins. A second group, in which the bonding arrangement is more complex and potentially less chemically stable, comprises unsaturated compounds called aromatics and alkenes. Aromatic compounds such as toluene ($C_6H_5CH_3$) rarely amount to more than 15% of petroleum but they may impart a pleasant odour, hence their name.

Petroleum occurs naturally in several forms: **natural gas** — mainly gaseous hydrocarbons but also containing variable amounts of carbon dioxide; a liquid, called **crude oil**, that typically contains a very wide range of hydrocarbon compounds; and solid **bitumen**. Bitumen contains the heaviest (in the sense of high molecular weight) and most complex hydrocarbon compounds found in petroleum, and they are relatively enriched in oxygen, sulphur and nitrogen. The composition of typical petroleum samples are shown in Table 3.1. Note that oxygen is a significant impurity in bitumen, but is commonly found only in trace amounts in crude oil and natural gas. In contrast, nitrogen is negligible in bitumen and crude oil, but may constitute up to 15% in natural gas.

Table 3.1 Composition of typical petroleum samples.

Element	Natural gas /weight%	Crude oil /weight%	Bitumen /weight%
carbon	65.0–80.0	85.0	80.2
hydrogen	20.0–25.0	12.0	7.5
oxygen	trace	<2.0	7.6
nitrogen	1.0–15.0	<1.5	1.7
sulphur	<0.2	<3.0	3.0

3.2 Key ingredients for petroleum accumulation

There are several 'ingredients' or geological conditions that are prerequisites for every subsurface accumulation of petroleum. They are petroleum charge, reservoir, seal and trap. We will look at each of these in turn in this section.

3.2.1 Petroleum charge

Petroleum charge is an abstract concept concerning the likelihood that petroleum can form, migrate and accumulate in a body of sedimentary rocks. It depends on interactions that involve a number of factors, i.e. it concerns a dynamic *system* in a sedimentary basin.

An effective petroleum charge system requires:

- A *source rock* rich in organic debris that could potentially generate liquid and/ or gaseous hydrocarbons — petroleum;
- Changes in temperature and pressure through time that induce the organic debris to undergo chemical reactions that produce petroleum fluids: the source rock must *mature*;
- A pathway along which petroleum fluids can *migrate*. As hydrocarbons are less dense than water they migrate upwards, and sometimes sideways toward the Earth's surface, through water-saturated permeable rocks;
- An impermeable rock or *seal*, somewhere along the migration pathway, beneath which the hydrocarbons can become *trapped*.

If all these factors combine, then hydrocarbons start to fill up or *charge* the pore spaces in a *reservoir rock*. To understand petroleum charge geoscientists need to consider the nature of source rocks, maturation and its timing, and migration.

Source rocks are sediments that contain sufficient organic matter to generate petroleum when they are buried and heated. Under normal conditions very little dead plant and animal tissue is preserved in sediments. Higher concentrations tend to occur only in environments where there is unusually high productivity of organic matter, such as in coastal upwellings, shallow seas, mires and lakes. Even then, the organic matter reaching the sediment–water interface must be protected from scavengers or aerobic bacteria. If not, these microorganisms use enzymes to digest and oxidize most of the organic matter completely to produce carbon dioxide and water. Under these circumstances there is clearly no potential for the sediments to preserve sufficient hydrocarbons to constitute a source rock.

You can see, therefore, that the preservation of organic matter under reducing conditions is a common factor underlying the formation of both petroleum source rocks and coal (Section 2.1). However, most petroleum source rocks form under water, unlike coals that are almost entirely products of organic accumulations at the land surface, albeit in very wet conditions. Note that coal does generate methane (Section 2.3.2) and coal deposits have given rise to natural gas resources, for instance those beneath the southern North Sea. So coal can be a petroleum source rock, but only for gas fields. Petroleum source rocks can form on the beds of freshwater lakes and brackish lagoons, but the most important ones are marine.

There have been prolonged but isolated periods in the geological past when ocean water did not circulate as it does now. Under these conditions oxygen did not reach the sea floor, leading to widespread anoxia there. Decomposition by anaerobic bacteria involves chemical processes of reduction that produce methane and hydrogen, along with hydrogen sulphide, carbon dioxide and water, but leave a residue of organic compounds that are enriched in carbon and have high molecular weight. If enough organic matter has been buried, this leads to a concentration of hydrocarbons within the source rock. Since anoxia is characterized by stagnant water — currents would bring in oxygen — source rocks are products of very low energy deposition.

● What kind of sediments are likely to form under such quiet conditions?

○ They will contain very fine grains, mainly clay minerals, and will form mudstones (an example is shown in Figure 3.3a).

Anoxic conditions can also occur on a smaller scale where water circulation is restricted. For example, the present-day Oslo Fjord, Norway has an anoxic bottom layer because a shallow lip of rock prevents water from the Skagerrak from circulating around the fjord. The bottom waters of the Black Sea are also strongly reducing because it is essentially a stagnant saline lake.

Another setting that encourages preservation of organic matter is provided by shallow, often land-locked seas in tropical or subtropical latitudes. Evaporation produces a highly saline surface water layer that is denser than the underlying water column and so it sinks to form a salty layer immediately above the sea floor. Organic material derived from plants and animals that thrive in the normally saline water column sinks onto the salty sea bed. Here it remains undisturbed as only rather specialized bacteria survive in this environment. The Dead Sea is a modern example of such a system.

Organic material in buried sediments is called **kerogen**, a word derived from the Greek for 'wax producer'. The concentration of kerogen in a potential source rock is usually expressed in terms of the percentage, by weight, of organic carbon in the rock. Rocks with more than 0.5% organic carbon may be effective source rocks, but prolific source rocks have more than 5% and occasionally much higher concentrations of kerogen. The world's first commercial petroleum products to be created on a large scale — in 1850 — were from black 'oil shales' that outcrop in the Scottish Midland Valley. These shales, or mudstones, contain more than 15% of kerogen, and when heated in sealed vessels by William 'Paraffin' Young they yielded the light liquid hydrocarbons from which he got his nickname.

Conventionally, kerogen is subdivided into four main types on the basis of its chemical composition, which reflects its original source material. Each type has characteristic ratios of carbon, hydrogen, and oxygen and they each generate contrasting petroleum products when they mature. Table 3.2 highlights the major differences between the four kerogen types in terms of their chemical properties and biological origins. Type I kerogen is comparatively rare as it is derived mainly from algal sources in lake and/or lagoonal environments: the Scottish Midland Valley 'oil shales' used by 'Paraffin' Young contain kerogen of this kind. Type II kerogen, the most abundant, is typically derived from plant debris, phytoplankton and bacteria in marine sediments; it is the common source of crude oil but also yields some natural gas. Type III kerogen comes mainly from remains of land plants found in coals and it principally generates natural gas. Type IV kerogen includes oxidized plant remains and fragmentary charcoal derived from forest fires; it has virtually no petroleum potential being devoid of hydrogen.

Table 3.2 Characteristics of the main types of kerogen.

Kerogen type	H:C ratio	O:C ratio	Origin of organic material	Petroleum products
Type I	1.7–0.3	0.1–0.02	Algae in lacustrine and/or lagoonal environments	Light, high-quality oil and some natural gas
Type II	1.4–0.3	0.2–0.02	Mixture of plant debris and marine microorganisms	Main source of crude oil and some natural gas
Type III	1.0–0.3	0.4–0.02	Land plants in coaly sediments	Mainly natural gas with very little oil
Type IV	0.45–0.3	0.3–0.02	Oxidized and charred wood	No petroleum potential

Box 3.1 summarizes some of the characteristics of the Kimmeridge Clay (150 Ma), which is a mudstone sequence of Upper Jurassic age that is widespread in northern Europe. This world-class source rock is the primary reason why there is a North Sea oil industry.

The process of biological, physical and chemical alteration of kerogen into petroleum is known as **maturation**. Source rocks that experience the right conditions for these processes and can generate petroleum are termed *mature*. Maturation begins within an organic-rich sedimentary layer while it is being deposited. Here a series of low-temperature reactions that involve anaerobic bacteria reduce the oxygen, nitrogen and sulphur in the kerogen, leading to an increased concentration of hydrocarbon compounds. This stage continues until the source rock reaches about 50 °C. Thereafter the effect of elevated temperatures becomes much more pronounced as the reaction rates and solubility of some of the organic compounds increase.

Box 3.1 Kerogen in the Kimmeridge Clay formation

The Kimmeridge Clay Formation is the most important source rock for North Sea oil deposits. It has an average organic carbon content of 5%, rising to 20–30% in the richest 'oil shales' that outcrop along the coasts of Yorkshire and Dorset in England (Figure 3.3a). It has an H:C ratio varying from 0.9 to 1.2.

Bacterially degraded marine algae and degraded humic matter and woody debris of land origin make up about 75% of the total carbon content. Other marine algae, land-plant spores and oxidized land-plant fragments form the remainder. The relative proportion of these constituents varies widely according to the depositional setting of the mudstones. The most organic-rich intervals developed in deeper basins where the highly anoxic bottom waters and high sedimentation rates favoured organic preservation.

● Using the information in Table 3.2, can you suggest what kerogen type characterizes the Kimmeridge Clay?

○ The abundance of marine algae and land-plant debris, coupled with the mid-range H:C ratio, suggests that most of the organic carbon in the Kimmeridge Clay is Type II kerogen.

Since temperature increases with depth in the Earth, heating is naturally achieved by burial of the source rock. The actual temperature reached at a given depth depends on the rate of increase of temperature with depth, the **geothermal gradient**. Figure 3.1 shows the relative proportions of crude oil and gas formed from Type II kerogen buried in an area with a geothermal gradient of about 35 °C km^{-1}. Significant amounts of petroleum only begin to form at temperatures over 50 °C and the largest quantity of petroleum is formed as the kerogen is heated to temperatures between 60 and 150 °C. At still higher temperatures oil becomes thermally unstable and breaks down or 'cracks' to natural gas. Even after maturation, some of the kerogen still remains unaltered as a carbon-rich residue.

● Look at Figure 3.1 and estimate the subsurface temperature and depth at which peak oil generation is achieved.

○ Peak oil generation in a typical Type II kerogen occurs at about 100 °C. In this example, where the geothermal gradient is 35 °C km^{-1}, this corresponds to a depth of about 2850 m.

The most important factors in maturation studies are the amount and type of kerogen, the temperature and time. Maturation rates generally increase exponentially with

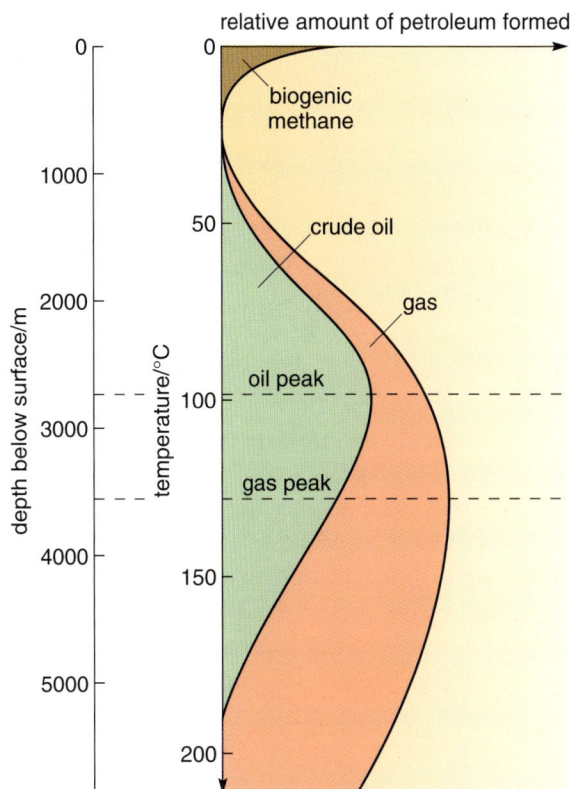

Figure 3.1 The relationship between depth of burial, temperature and the relative amount of crude oil and natural gas formed from Type II kerogen in an area with a geothermal gradient of about 35 °C km^{-1}.

respect to temperature (up to a point) and linearly with respect to time. Thus crude oil can form in old basins with low geothermal gradients ('cold') as well as in young basins where the geothermal gradient is high ('hot'). However, it cannot form in young, cold basins except in trace amounts. It is usually destroyed in old, hot basins, assuming that subsidence has been continuous, because temperature eventually rises to a point where all kerogen and any crude oil formed earlier has been converted into gas.

To illustrate this point it is useful to examine the burial histories of source rocks in three different sedimentary basins (Figure 3.2). The source rocks in the Paris Basin, the North Sea Viking Graben and the Los Angeles Basin are different in terms of age and composition, and each has been subjected to differing burial histories. The point at which petroleum generation starts is known as the **threshold**, and this was reached after 40 million years in the Paris Basin (i.e. about 140 million years ago) when Early Jurassic (175 Ma) source rocks were buried to a depth of 1400 m. In contrast, it took some 80 million years before the Kimmeridgian (150 Ma) source rocks in the Viking Graben started to generate petroleum during early Tertiary times.

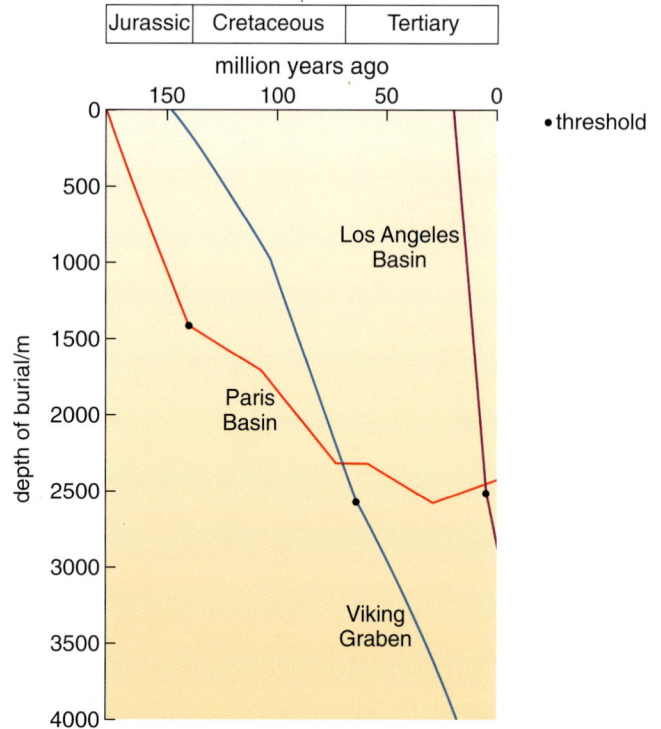

Figure 3.2 Reconstruction of burial histories of rocks from three basins; the Paris Basin in northern France, the Viking Graben in the northern North Sea and the Los Angeles Basin in the USA.

Question 3.1

Examine Figure 3.2 and determine the threshold depth for petroleum generation in the Los Angeles Basin. Then, assuming that this depth equates to a temperature of 120 °C, and ignoring surface temperature effects, calculate the geothermal gradient.

Migration refers to the movement of fluid petroleum through rocks. This process begins with **primary migration**, i.e. the expulsion of petroleum from the source rock. The driving force for this process is the pressure difference caused by the loading effect of overlying rocks. Overburden loading preferentially compacts mudstones, making it difficult for fluids within them to escape. As a result, pressure builds up in them until it is sufficient to drive the water and petroleum into adjacent rocks that are at a lower pressure because they are more permeable (Figure 3.3). In the context of water resources, these rocks would be termed *aquifers* (Smith, 2005, Chapter 3), and at depth they would inevitably be saturated with water, but in petroleum parlance they are **reservoir rocks** (Section 3.2.2). Figure 3.3b shows a potential reservoir rock exposed on land.

Once expelled from the source rock, buoyancy takes petroleum (both liquid and gas) from depth up towards the surface of the Earth because it is less dense

(a)

(b)

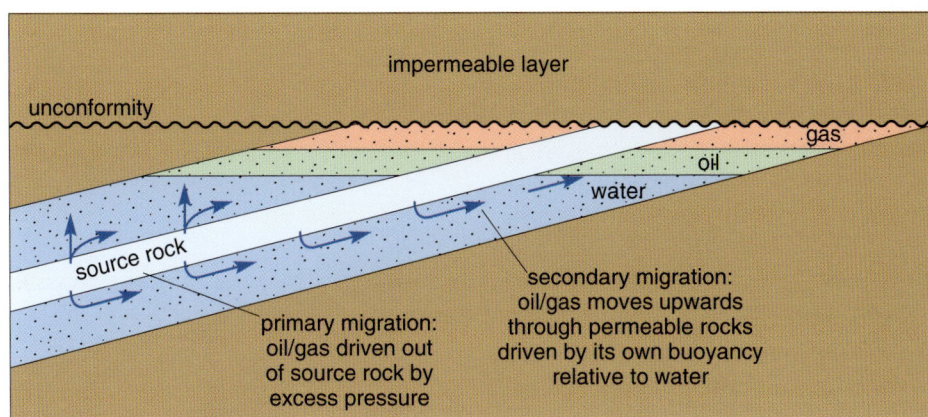
(c)

Figure 3.3 (a) A potential Jurassic source rock exposed in Dorset, the Kimmeridge Clay, whose black colour is due to high kerogen content. (b) A potential Jurassic reservoir rock exposed in Dorset, the Bridport Sand. (c) Migration of petroleum out of a source rock and upwards through a reservoir to a trap.

than pore water and 'floats' on top of it in the reservoir rock. This is known as **secondary migration**, and its effectiveness depends on the permeability of the reservoir rocks and the density and viscosity of the petroleum fluids flowing through them. As Figure 3.3c shows, oil and gas continue to migrate upwards until they are trapped beneath an impermeable rock layer. At that point they segregate according to their density; gas is lighter so it will pool immediately beneath the permeability barrier, whereas oil is heavier and will accumulate beneath the gas. Rocks beneath will be saturated with pore water. Secondary migration serves to concentrate petroleum and by the time it reaches the trap it can occupy more than 90% of the pore volume in the reservoir.

Timing of the petroleum charge relative to the formation of a trap is critical, simply because a trap has to pre-date petroleum migration in order for an accumulation to develop: migration before suitable traps have formed would ultimately result in all petroleum escaping at the Earth's surface. Figure 3.3c reinforces this point by showing that the horizontal impermeable layer both truncated and sealed the dipping reservoir rocks, creating a trapping configuration, *before* migration occurred; otherwise the petroleum could not have been trapped.

As discussed above, petroleum generation may occur only a few million years after the source rock was deposited or tens of millions of years later, depending on the rate of burial and the geothermal gradient. An understanding of basin evolution is vital in this context, not only to determine when potential traps were formed, but to assess the degree to which they were subsequently filled, and the chances of petroleum having escaped (Section 3.2.4).

3.2.2 Reservoir rocks

The properties of a petroleum reservoir rock are very similar to those of an aquifer since both petroleum and water can be contained within and move between its pore spaces and fractures. Sedimentary rocks that are well cemented have only small voids between grains and hence low porosity.

● Which sedimentary rock type is most likely to be a potential reservoir rock?

○ The most porous reservoir rocks are generally well-sorted, poorly cemented sandstones (Figure 3.3b), and these make up some of the most important petroleum reservoirs around the world.

Migrating waters can increase porosity and permeability by dissolving the cement that holds the grains together and widening small fractures that run through the rock. This effect is often enhanced if the waters are slightly acidic. Many limestones are well-cemented and therefore have low porosity, but the calcium carbonate ($CaCO_3$) that makes up the grains and cement is soluble in weakly acidic water. Consequently limestones can form good reservoirs, and in fact limestones hold 40% of the world's resources of petroleum.

The essential properties that describe a reservoir rock are **porosity** (*the void space expressed as a percentage*) and **permeability** (*a measure of the degree to which fluid passes through it, measured in millidarcies, mD*). Another property that is commonly used is the ratio of porous and permeable (net) intervals to the overall reservoir (gross) thickness. This is referred to as **net to gross** and it is important because it recognizes that most sandstone and limestone reservoirs are not entirely homogeneous, but contain intervals or strata that less readily allow fluid flow.

To put these properties in context, Table 3.3 provides reservoir data for 20 oil and gas fields in the North Sea. Note the very wide range of net to gross and permeability values, despite the fact that most of the reservoirs are of the same (sandstone) type. Porosities are typically in the range 15–30%, but the more telling parameter is permeability because that largely determines petroleum flow rates. Permeability cut-offs of 1 mD for gas and 10 mD for light oil are often used as a rule-of-thumb for productive reservoirs; less permeable rocks are not usually capable of sustaining commercial flow rates. Notice also that one of the two Cretaceous Chalk reservoirs in the Ekofisk field exhibits chalk's characteristic properties of high porosity and low permeability — the latter results from very small channels that connect the pores between the tiny calcareous plankton shells that form chalky sediments. The other chalk reservoir in Ekofisk has higher permeability because it has been fractured tectonically.

Table 3.3 Properties of reservoirs within North Sea oil and gas fields. *Note:* A formation is a distinctive sequence of sedimentary rocks in a particular field.

Field	Age/Formation	Reservoir	Net to gross /%	Porosity /%	Permeability /mD	Fluid
Alwyn North	Jurassic/Brent	sandstone	87	17	500–800	oil
Alwyn North	Jurassic/Statfjord	sandstone	65	14	330	oil
Auk	Permian/Zechstein	fractured dolomite	100	13	53	oil
Auk	Permian/Rotliegend	sandstone	85	19	5	oil
Brae South	Jurassic/Brae	sandstone	75	12	130	oil
Britannia	Cretaceous/Britannia	sandstone	30	15	60	gas in liquid form under high pressures
Buchan	Devonian/Old Red	fractured sandstone	82	9	38	oil
Cleeton	Permian/Rotliegend	sandstone	95	18	95	gas
Cyrus	Palaeocene/Andrew	sandstone	90	20	200	oil
Ekofisk	Cretaceous/Chalk	limestone (fractured Chalk)	64	32	<150	oil
Ekofisk	Cretaceous/Chalk	limestone (Chalk)	62	30	2	oil and gas
Forties	Palaeocene/Forties	sandstone	65	27	30–4000	oil
Fulmar	Jurassic/Fulmar	sandstone	94	23	500	oil
Frigg	Eocene/Frigg	sandstone	95	29	1500	Gas
Heather	Jurassic/Brent	sandstone	54	10	20	Oil
Leman	Permian/Rotliegend	sandstone	100	13	0.5–15	gas
Piper	Jurassic/Piper	sandstone	80	24	4000	oil
Ravenspurn South	Permian/Rotliegend	sandstone	39–77	13	55	gas
South Morecambe	Triassic/Ormskirk	sandstone	79	14	150	gas
Scapa	Cretaceous/Valhall	sandstone	—	18	111	oil
Staffa	Jurassic/Brent	sandstone	76	10	10–100	oil
West Sole	Permian/Rotliegend	sandstone	75	12	—	gas

Question 3.2

Using the information in Table 3.3, calculate the average porosity of the five Permian sandstone reservoirs and the three Palaeocene–Eocene sandstone reservoirs. Compare the results and suggest reasons for the marked difference.

3.2.3 Seals

Above permeable reservoir rocks there must be an impermeable layer (known as a **seal** or **cap rock**) to stop migrating petroleum from rising further towards the surface of the Earth. Seals are fine-grained or crystalline, low-permeability rocks such as mudstone, anhydrite and salt. Rock salt is by far the most effective seal, because it is crystalline and therefore impermeable. Seals are also enhanced if they are ductile (ductile deformation prevents the formation of open fractures and joints), substantially thick and laterally continuous; little surprise then that the largest oil fields in the Middle East are sealed by evaporites (Argles, 2005) with these characteristics.

However, seals are rarely, if ever, perfect. Hydrocarbons can migrate through almost all rock types, but at different rates that depend upon any fracturing and microscale fluid-flow, and whether liquids adhere to or are repelled by the surfaces of mineral grains. Many oil and gas fields have active surface seeps of petroleum overlying them that provide a direct indication as to their location. In marine settings seeps may be detected as bubbles of gas rising from the sea bed, or as an oily sheen on the water. On land, plant communities are stunted, surface layers of rock and soil may be altered, tarry residues may encrust the surface, and sometimes there may be active oil seeps. The first oil fields to be developed in the 19th century were located beneath such obvious features. It is thought that ignition (by lightning strikes) of petroleum escaping above the huge oilfields of Iran gave rise to the fire-worshipping Zoroastrian religion. Even odder, the Ancient Greek Oracle at Delphi is thought to have made her prognostications while hallucinating under the influence of escaping natural petroleum gas.

3.2.4 Traps

Petroleum that accumulated as a thin layer at the top of an extensive horizontal reservoir would be uneconomic to extract. That is because many wells, each with only a small rate of production and lifetime, would be needed to extract the petroleum. To be worth working, a sealed petroleum-bearing 'container' or **trap** must be shaped naturally to retain and focus petroleum, rather as the curved upper surface of a balloon traps buoyant hot air. The lower surface of a trap is defined either by a petroleum–water contact or sometimes by another seal. There are many different styles of trap (Figure 3.4) but the most common are **structural traps** in the form of anticlines produced by tectonic processes, by differential compaction of soft rocks above hard, irregular surfaces and by evaporitic salt masses that rise gravitationally. The low density of salt, combined with its ductility, enables it to rise to form domes and intrusive masses. Because they produce distinctive geological and geophysical features, structural traps are the easiest to find.

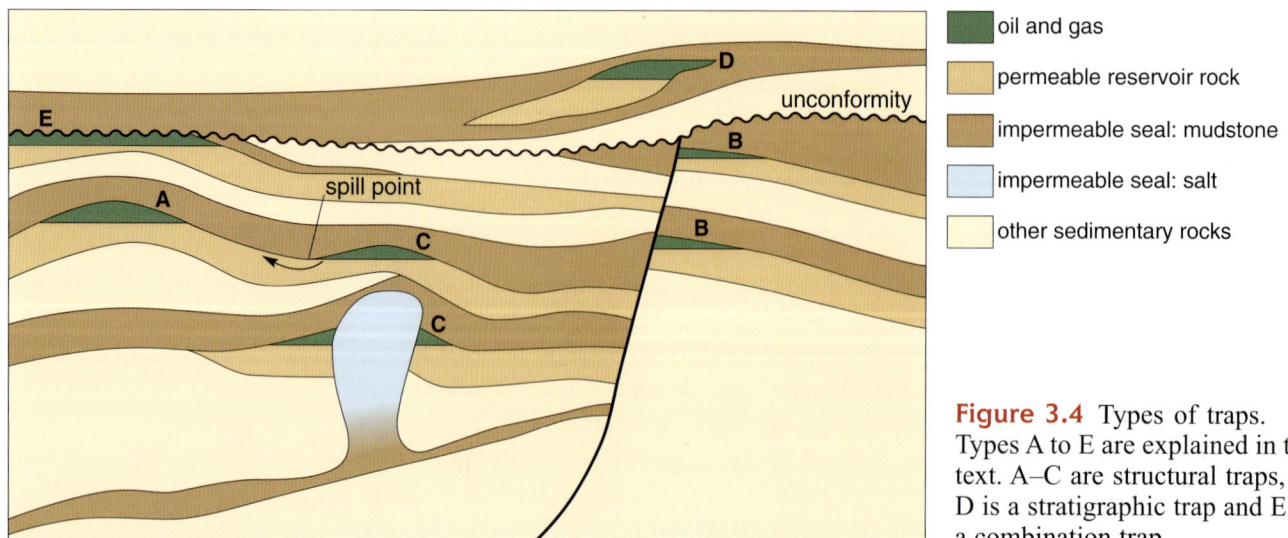

oil and gas
permeable reservoir rock
impermeable seal: mudstone
impermeable seal: salt
other sedimentary rocks

Figure 3.4 Types of traps. Types A to E are explained in the text. A–C are structural traps, D is a stratigraphic trap and E is a combination trap.

About 80% of the world's petroleum reserves are held in structural traps like those shown in Figure 3.4. They include simple anticlines (A), faulted structures that juxtapose reservoirs against seals (B), and traps created at the flank of a salt dome or in the compaction anticline above it (C). Most fields in the North Sea occur in structural traps.

Stratigraphic traps result from lateral changes in rock type and typically consist of discontinuous sandstone bodies encased in mudstone (D). Sometimes referred to as 'subtle traps', they currently contain about 13% of the world's petroleum reserves, but much of the remaining undiscovered petroleum will probably be found in these settings because the more obvious structural traps have long since been exploited.

In practice, traps often form through a sequence of different processes over the course of tens of millions of years. For example, in E the reservoir was first deposited, then folded, uplifted and eroded, before being overlain by a much younger impermeable mudstone. The resulting configuration is appropriately called a **combination trap**. Provided it was intact before the reservoir received a petroleum charge, it forms a valid trap regardless of how long it took to form.

As petroleum accumulation continues it is possible for traps to fill beyond their natural *spill-point*, when petroleum can escape sideways to re-migrate to other traps (Figure 3.4, upper C) or to the Earth's surface where it emerges as oil or gas seeps.

- Suggest another type of trap on Figure 3.4 that might experience such leakage.

- The lowest trap associated with the fault might leak along fractures produced by the faulting, to help charge higher traps.

3.2.5 Combining the ingredients

Having examined the essential ingredients for a petroleum accumulation, this section discusses how knowledge about them is combined to create a **petroleum play**. That is a particularly useful concept, since it consolidates what is known (or not known) about the petroleum potential of a particular level within a basin and forms the basic strategy for oil and gas exploration. A play is defined as a perception or model of how a petroleum charge system, reservoir, seal and trap may combine to produce petroleum accumulations at a specific stratigraphic level. By examining whether each of the play ingredients is both present and effective, it is possible to define parts of a basin where petroleum accumulations can reasonably be expected to exist. This process can be conducted systematically in a given area to generate a **play fairway map** that depicts where the ingredients coexist, even though the precise details of trap location and size may not be known. As more data become available the play becomes better defined, but even when the play is proven by a discovery it does not imply that every trap within the same fairway will contain a petroleum accumulation. It is in the nature of exploration that more often than not geoscientists are wrong with their predictions, but this approach at least helps to reduce their uncertainty.

To illustrate the petroleum play approach an example is provided from the Upper Jurassic of the North Sea (Table 3.4). There are two play types: the Kimmeridgian–Volgian deep marine play and the slightly older Oxfordian–Volgian shallow marine play. Whilst the depositional settings for the two reservoir types are quite different, both plays share the same petroleum charge, seal and trap ingredients. Importantly, notice that the onset of petroleum generation comfortably post-dates trap formation. For simplicity the two plays can be combined and their distribution plotted to create an Upper Jurassic play fairway map (Figure 3.5). This illustrates the close correspondence between the limit of mature Upper Jurassic Kimmeridge Clay source rocks and the fairway, such that the migration pathways between the two are short (less than 15 km) and highly permeable. More than 60 Upper Jurassic fields have been discovered to date beneath the North Sea. They have combined oil reserves of about 2.5×10^9 toe (tonnes of oil equivalent, see Box 1.1) and account for 23% of total North Sea production (Evans et al., 2003).

Table 3.4 Upper Jurassic petroleum plays of the North Sea.

Play	Reservoir	Petroleum charge	Regional seal	Trap	Oilfield examples
Upper Jurassic deep marine play	Kimmeridgian–Volgian slope-apron and submarine-fan sandstones	Upper Jurassic Kimmeridge Clay mudstones generated oil and some gas from Late Cretaceous (65 Ma) times onwards (155 Ma)	Upper Jurassic and Lower Cretaceous mudstones	Structural and combination traps, with a minor stratigraphic component, formed between Late Jurassic and Early Cretaceous (125 Ma) times	Magnus, Brae, Miller, Ettrick, Borg
Upper Jurassic shallow marine play	Oxfordian–Volgian coastal and shoreface sandstones				Piper, Clyde, Beatrice, Hugin, Scott, Fulmar, Troll, Puffin

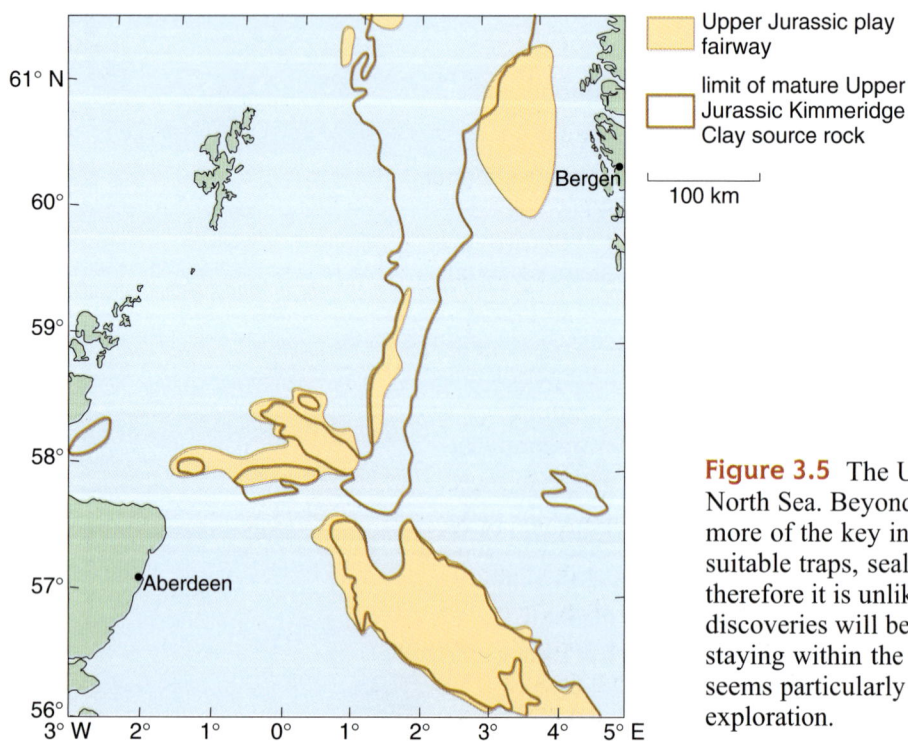

Figure 3.5 The Upper Jurassic play fairway in the North Sea. Beyond the mapped fairway limits one or more of the key ingredients of the play, for instance suitable traps, seals or reservoir rocks, are missing and therefore it is unlikely that Upper Jurassic petroleum discoveries will be made there. The golfing analogy of staying within the fairway in order to be successful seems particularly appropriate in the context of exploration.

3.3 Exploring for oil and gas

It would be prohibitively expensive to explore for oil and gas on a random basis, and most of the effort would be wasted. When geological knowledge was far more limited than it is today, most of the discoveries were beneath quite obvious signs of petroleum seepage at the surface. Eventually, such easy targets ran out, although some are still being discovered. The components in Section 3.2 were gleaned from the knowledge gained by drilling such targets and examining the geology around them. As more has been learned, increasingly sophisticated methods have been developed that increase the odds of making a discovery in less obvious situations. This section covers some of those methods and describes how an exploration well is drilled and evaluated.

3.3.1 Surface detection methods

Field mapping is a well established technique that has contributed to the discovery of billions of barrels of petroleum. In the pioneering era of onshore exploration the search for anticlinal structures at shallow, drillable depths usually began with the recognition of an overlying part of such a structure at the surface. These days, with ready access to various forms of subsurface data, field mapping is more commonly used to assess the structural style of a basin and to provide analogues for concealed reservoirs or source rocks. Far from being an outdated technique, modern fieldwork is becoming increasingly sophisticated as digital data collection is underpinned by Global Positioning Systems (GPS), satellite imagery and digital terrain models. In Norway's Lofoten Islands a detailed analysis of onshore fault and other structural patterns is being extrapolated offshore in order to calibrate three-dimensional (3-D) seismic data in unexplored portions of the Norwegian continental shelf.

3.3.2 Remote sensing methods

Remote sensing involves gathering information of many kinds at a distance from the object of investigation: it gives a regional picture and helps sort the likely areas to follow up from those much larger areas that are less favourable. Satellite, gravity and magnetic methods (see below) are commonly used during the early phase of exploration when a sedimentary basin, or at least a substantial part of it, is not known in sufficient detail to deploy more expensive methods. Their interpretation is simple and can be done relatively cheaply in the office. Satellite images take the form of spectral data over a wide range of wavelengths, from the visible through infrared to microwave (radar). They can sometimes detect unknown petroleum seepages. On land, the presence of a seep is often associated with a change in vegetation or soil colour, especially if the seep is of crude oil, whilst in the offshore setting rising gas bubbles may draw deep water to the surface, giving a cool thermal image. Alternatively, satellites can provide photographic imagery with an extraordinarily good resolution, sufficient to map rock exposures, analyse topography, and to locate roads, habitations and so on.

Gravity surveys are often used to analyse sedimentary basins at the regional scale. Because sedimentary rocks usually have a lower density than crystalline rocks, thick sequences of relatively low-density sediments effectively reduce the Earth's gravitational force and they are characterized by regional gravity lows. Gravity data may be collected on land, at sea or by air and they are particularly useful in areas of difficult terrain, such as jungles and deserts, where access is difficult. Regional airborne **magnetic surveys** can also be used to define the shape and gross structure of a basin and they are often acquired in tandem with gravity surveys. Magnetic rocks cause perturbations in the Earth's magnetic field, whereas non-magnetic rocks have little effect. Sediments are typically poorly magnetic because they do not contain large amounts of iron-rich minerals, whereas igneous rocks such as volcanic lavas often do. So sedimentary basins characteristically have a low, uniform magnetic signature that contrasts markedly with the highly variable magnetic anomalies associated with metamorphic basement rocks and near-surface volcanic intrusions. Where faults juxtapose rocks with different magnetic properties at depth, the faults show up as distinctive linear features.

3.3.3 Seismic data and interpretation

Seismic surveying is by far the most widely used and important method of gaining an impression of the subsurface (Section 2.2.2). Seismic surveys can be acquired at sea as well as on land. The marine method is the most common in petroleum exploration and is shown schematically in Figure 3.6, although the same principles apply to any seismic reflection survey (see Figure 2.9).

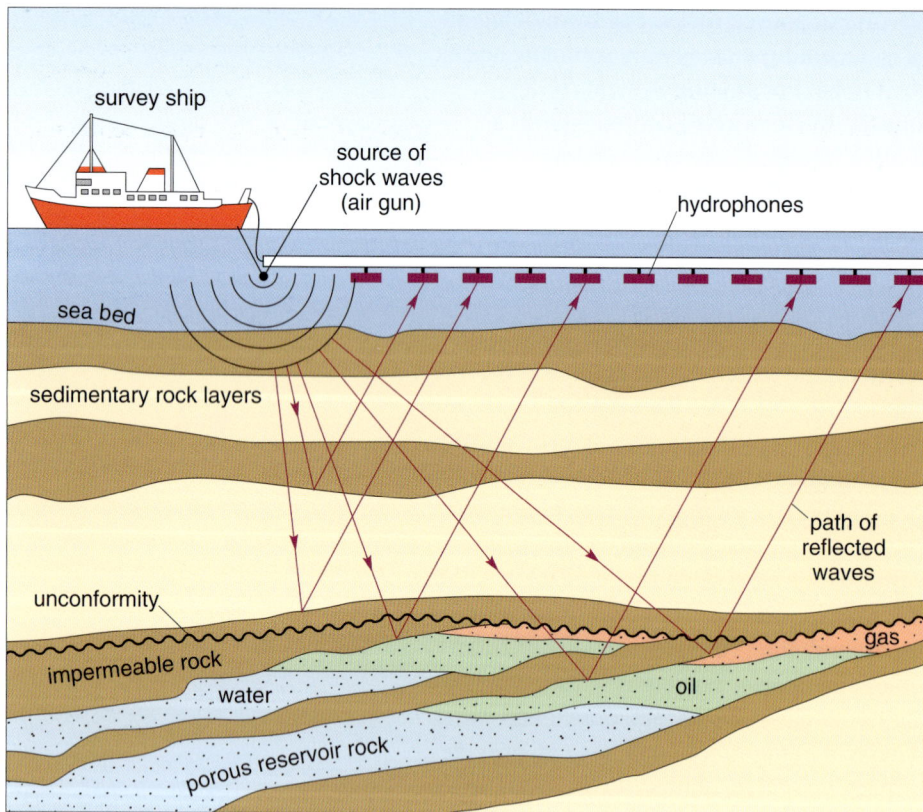

Figure 3.6 Marine seismic acquisition — pulses of sound energy penetrate the subsurface and are reflected back towards the hydrophones from rock interfaces.

Compressed air guns towed behind a boat discharge a high-pressure pulse of air just beneath the water surface. The place of detonation is called the shot point and each shot point is given a unique number so that it can be located on the processed seismic survey. The sound waves (effectively the same as seismic P-waves produced by earthquakes) pass through the water column and into the underlying rock layers. Some waves travel down until they reach a layer with distinctively different seismic properties, from which they may be reflected in roughly the same way that light reflects off a mirror. For this reason such layers are called seismic reflectors.

The reflected waves rebound and travel back to the surface receivers (or hydrophones), reaching them at a different time from any waves that have travelled there directly. Their exact time of travel will depend on the speed that sound travels through the rock: its **seismic velocity**. Other waves may pass through the first layer and travel deeper to a second or third prominent reflector. If these are eventually reflected back to the hydrophones they will arrive later than waves reflected from upper horizons.

The hydrophones therefore detect 'bundles' of seismic waves arriving at different times because they have travelled by different routes through the rock sequence. Computer processing allows the amalgamation of recordings from all the shot points, filtering out unwanted signals of various sorts. The final result is a two-dimensional (2-D) seismic section (such as shown in Figure 2.10). By using closely spaced survey lines or hydrophones arranged in a grid it is possible to produce 3-D seismic datasets. These are usually interpreted on a PC workstation and colours are normally used to enhance the image and aid interpretation. The data can be viewed in any orientation in order to create a 3-D visualization of selected horizons (Figure 3.7).

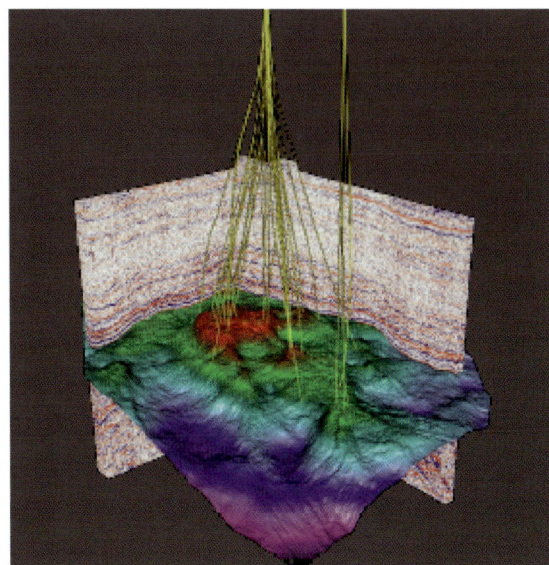

Figure 3.7 A 3-D view of the Palaeocene reservoir in the Nelson Field, North Sea. The image is derived from a 'cube' of closely spaced 3-D seismic data, onto which the paths of the production wells are superimposed. Bright colours in this perspective view relate to depths to a particular reflecting boundary. Reds and greens are structurally highest, where petroleum may be trapped.

Seismic data of all forms (2-D or 3-D) are displayed with the horizontal axis indicating geographic orientation and distance, whereas the vertical axis is calibrated in time (see Figure 2.10). The time, measured in seconds, records how long it took the seismic wave to travel from shot to reflector and then back to the hydrophone, so it is described as **two-way travel time (TWT)**. Further processing and the incorporation of seismic velocity data allows TWT to be converted into depth. Depth-converted seismic data is the mainstay of exploration since it provides a meaningful basis for all subsequent interpretation.

- What would happen to seismic waves if there was a strongly reflective layer, such as an igneous sill or salt body, in the shallow subsurface?

- It would tend to reflect most of the seismic waves back towards the surface and reduce the quality of seismic imaging beneath it.

Interpreting seismic sections is something of a 'black art', requiring both experience and a certain amount of interpretative flair. At the outset, interpretation involves tracing continuous reflectors on 2-D sections in order to build up a plausible structural representation of the subsurface. In the context of an initial exploration programme to find possible traps this is often sufficient.

● Look at the 2-D seismic section in Figure 2.10. Even though it was produced to explore for coal seams it contains lots of information that might help the petroleum explorationist. What kinds of trap shown in Figure 3.4 might be present in that section?

○ The unconformity might have associated combination traps (E on Figure 3.4). There is a large anticline at lower left (see C on Figure 3.4), and reservoir rocks might be in contact with seals along the fault (C on Figure 3.4) that extends vertically downwards in the centre of the seismic section.

Good quality 3-D seismic data provides sufficiently fine resolution for exhaustive processing and analysis to help in managing and developing known oilfields (Box 3.2).

Box 3.2 Applications of 3-D seismic data

Modern 3-D seismic data can be used for many purposes other than simply defining trap geometry. Sometimes it is possible to identify the presence of petroleum directly, particularly dispersed gas which tends to dissipate seismic waves and produce an ill-defined 'shadow zone' above a leaking trap. The changes in acoustic properties across a gas–water or gas–oil contact may also be detected as a horizontal reflector that conforms to the geometry of the trap.

More commonly, however, seismic data are used to map rock characteristics at a variety of scales. Starting with the recognition of distinctive reflector geometries and seismic sequences, and then by applying a range of seismic techniques, depositional environments can be mapped over a very wide area. As drilling progresses and data on rock properties (such as seismic velocity and density) become available, increasingly sophisticated reservoir descriptions can be developed. These commonly include an assessment of lithology, the amount of petroleum that is present, fluid type and porosity.

Interpretation of 3-D seismic data is an enormously varied and rapidly developing area of petroleum exploration that is beyond the scope of this chapter. However, if you are interested to know more, there are many useful sources (see Further Reading).

Seismic technology has been transformed since the 1980s. Today, 3-D seismic, rather than single 2-D sections, is routinely used for exploration purposes in offshore environments because the data can now be acquired quickly and cheaply. New processing techniques and improved computerized visualization tools add clarity to the data, helping to provide an unparalleled impression of the subsurface. The emphasis in exploration is to reduce the risk of drilling a dry hole and wasting a great deal of investment. This can only be achieved by integrating all the appropriate types of data, and with thoughtful analysis.

3.3.4 Exploration drilling

When seismic data highlight a suitable prospect, the next step is to drill into the reservoir in order to establish whether or not petroleum is trapped, and, if it is, to establish how large the accumulation might be. There are several types of drilling rig, ranging from relatively small ones as deployed on land (Figure 3.8a) that can be dismantled and transported by truck or helicopter, to large offshore units (Figures 3.8b–d) that are capable of working in a range of water depths and sea conditions. An offshore *jack-up rig* is a barge with lattice steel legs that can be raised and lowered (Figure 3.8b). It is towed into position by tugs and its legs are lowered to the seabed before the barge is raised 10–30 m out of the water to create a stable drilling platform. They usually operate in water depths up to 200 m.

Drilling in greater water depths requires a floating unit and the *semi-submersible rig* is the most common and versatile type (Figure 3.8c). The working platform is supported on vertical columns that are attached to submerged pontoons. Once in position, the rig is anchored to the seabed and the pontoons are flooded with water to submerge them beneath wave-level. The lower the pontoons are beneath the water, the less likely they are to be affected by wave action. This makes them

(a)

(b)

(c)

(d)

Figure 3.8 Mobile drilling units can operate on land (a) or in a variety of water depths. Jack-up rigs (b) are used in water up to 200 m deep, whilst semi-submersible rigs (c) and drill ships (d) can operate in much deeper water.

stable in rough seas. Some semi-submersible rigs have computer-controlled positioning propellers, rather than anchors, to keep them in position and they can be used in water depths down to 1000 m or more.

Drill ships resemble conventional ships and they can move easily around the world (Figure 3.8d). They too have dynamic positioning that allows them to stay on location with remarkable accuracy in all but the most severe storms. Since they are not ballasted they can be unstable in high seas but their advantage is that they can drill in water depths in excess of 2000 m.

● What type of rig would be used to drill in the Amazonian rainforest and what preparation would be required before drilling commenced?

○ Components of a land rig could be taken into the rainforest by river boats or helicopters and assembled on-site. Before that process began it would be essential to survey the drilling site, determine the best access route, clear the site sensitively and safeguard local water supplies from any risk of contamination. In ecologically sensitive areas the cost of site preparation and restoration may exceed the drilling costs.

Drilling for oil and gas is a sophisticated and very expensive process. Wells often penetrate over 3000 m into sedimentary rock; the deepest exceed 6500 m. At such depths the fluid pressures in the rock formations are so high that a dense drilling mud is continuously pumped into the borehole to counter-balance the pressure. This significantly reduces the possibility of an uncontrolled surge of petroleum to the surface, a situation that is graphically described as a 'blowout'. The enduring image of rig workers celebrating beneath a gushing fountain of crude oil in the pioneer days of exploration distorts reality, since blowouts and the release of associated toxic gases such as hydrogen sulphide (H_2S) are very dangerous. Every modern well is fitted with hydraulic rams that instantly isolate the borehole if excess pressures cause the well to flow. The other useful functions of drilling mud are to lubricate and cool the drill bit, to circulate rock fragments (cuttings) back to the surface and, in some cases, to power a turbine that rotates the drill bit.

3.3.5 Well evaluation

To some extent, well evaluation is similar to evaluation of coalfields (Section 2.2.2). Traditionally an exploration well is evaluated at discrete stages by withdrawing the drill bit, lowering instruments (colloquially known as 'tools') down the hole on a steel cable and then hauling them slowly back to surface. This process is known in the petroleum industry as **wireline logging** (see also *borehole logging* in Section 2.2.2) As the tools are withdrawn they record the properties of the rocks that surround the well and the fluids in them. Nowadays this approach is supplemented by measurements that are made while drilling is in progress, which has the advantages of providing near instantaneous data and incurs none of the expense of halting the drilling process.

The rock properties that are of interest include those used for identifying lithologies and small-scale structural or sedimentological features. Other tools help estimate porosity, permeability, pressure and fluid content. None provide a completely definitive description of the borehole wall, but in combination the data acquired by wireline logging provide sufficient information to determine whether further evaluation is justified.

The most useful geological data are derived from pieces of rock recovered from specific depth intervals. They range in size from small fragments of rock (**drill cuttings**) produced as the drill bit cuts into rock, to thumb-size and larger (5–15 cm diameter) cores of solid rock that are retrieved with special tools. These provide the basis for a detailed description of the reservoir, although cores may also be taken in mudstones to gain biostratigraphic and/or geochemical information.

Some exploration wells, particularly those that encounter significant volumes of petroleum, justify an extensive evaluation programme that is designed to recover fluid samples from selected intervals down the well. The fluids (oil, gas and water) are captured *in situ* at reservoir temperature and pressure, and then brought to the surface in a small sealed chamber for analysis. Less commonly, the fluids may be sampled by allowing them to flow to the surface. Such *well testing* may continue for several days. During that time it is possible to draw some preliminary conclusions about the nature of the reservoir, flow rates and the commercial potential of the petroleum accumulation.

Question 3.3

Exploration is an expensive activity that can quickly lead to 'gambler's ruin' — betting more money than you win — unless there is a proper understanding of risk and potential reward. At the outset it is vital to decide where *not* to explore. List some of the fundamental geological, technical and commercial factors that you might use to reject certain parts of the world from exploration.

3.4 Petroleum production

Once quantities of oil or gas have been discovered by exploration drilling, the next step is to carry out an appraisal programme to determine whether the accumulation is worth developing into a producing field. To justify the move into production, the field must contain enough petroleum to repay the huge cost of development, finance day-to-day operations and still make a profit. This section outlines the steps required to translate the excitement of an offshore discovery into a commercial product at the refinery, a process that may take several years because it involves a huge amount of data collection and additional drilling.

3.4.1 Appraising the discovery

During the appraisal stage, the size of an oilfield discovery must be established as accurately as possible and the most cost-effective way to produce petroleum from the reservoir(s) sought. Geologists and engineers focus on the reservoir in particular, by attempting to provide an improved definition of the trap geometry and considering whether or not the reservoir is segmented by barriers to lateral flow, such as faults or impermeable layers that will require wells to be drilled into each segment. This work is normally underpinned by a more detailed 3-D seismic survey (Figure 3.7) acquired on a closely spaced grid (with less than 50 m between shot points) to provide maximum resolution. Such surveys give much more detail about the trap configuration, the depth to reservoir units and, to a certain extent, the nature of the reservoir rocks themselves.

Further drilling will improve the knowledge of how reservoir fluids (oil, gas and water) and boundaries between them are distributed. Rock properties are used to calibrate the seismic data, thereby allowing specific reservoir rock types to be mapped in areas beyond existing wells. The resulting reservoir model provides the basis for calculating the volume of petroleum trapped in the reservoir, and differing production scenarios can be assessed to determine the likely reserves that are recoverable during the life of the field.

Running in tandem with the technical work there is also a complex web of commercial, regulatory and sometimes political issues to be resolved. For example:

- How much will it cost to build, install, operate, maintain and, eventually abandon, the production facilities?
- Is the production facility to be located in an environmentally sensitive area or a shipping lane?
- How will the petroleum be transported from the well to its point of use?
- Are new tax laws envisaged that will impact profitability?
- Might the field be confiscated if the political regime changes?

The decision to commit to developing a new field will only be made when the bulk of these issues are satisfactorily resolved. The go-ahead then relies on approvals being granted by the owner(s) of the asset, governing bodies and financial underwriters. It quickly becomes evident that large-scale appraisal projects are hugely complex undertakings that require careful data integration and a multidisciplinary approach.

3.4.2 Development options

The approach to field development is as varied as petroleum accumulations themselves, so what follows is a brief summary. Most major petroleum field development projects in the 1990s and early 21st century were located offshore, and often presented the challenge of very deep water (over 1500 m) and elevated reservoir temperatures and pressures. One reason for long-term optimism about the future of the petroleum industry lies in its growing ability to access remote and difficult resources safely. The pace of technological innovation is rapid, and that emphasizes why the perception of global petroleum resources and reserves will always change; what is inaccessible today may not be tomorrow.

Various deep-water development systems are illustrated by Figure 3.9. Conventional *fixed platforms*, fabricated from welded pipe and pinned to the sea floor with steel piles, are commonly used in water depths up to 500 m. They may act as a focus for *sub-sea satellite developments* that draw petroleum from a cluster of remote wellheads — drilled from floating rigs — that are often tens of kilometres away and in water depths up to 2500 m. Sub-sea systems do not have the facility to drill, only to extract and transport petroleum back to the host platform to which they are connected by pipelines and remote control systems.

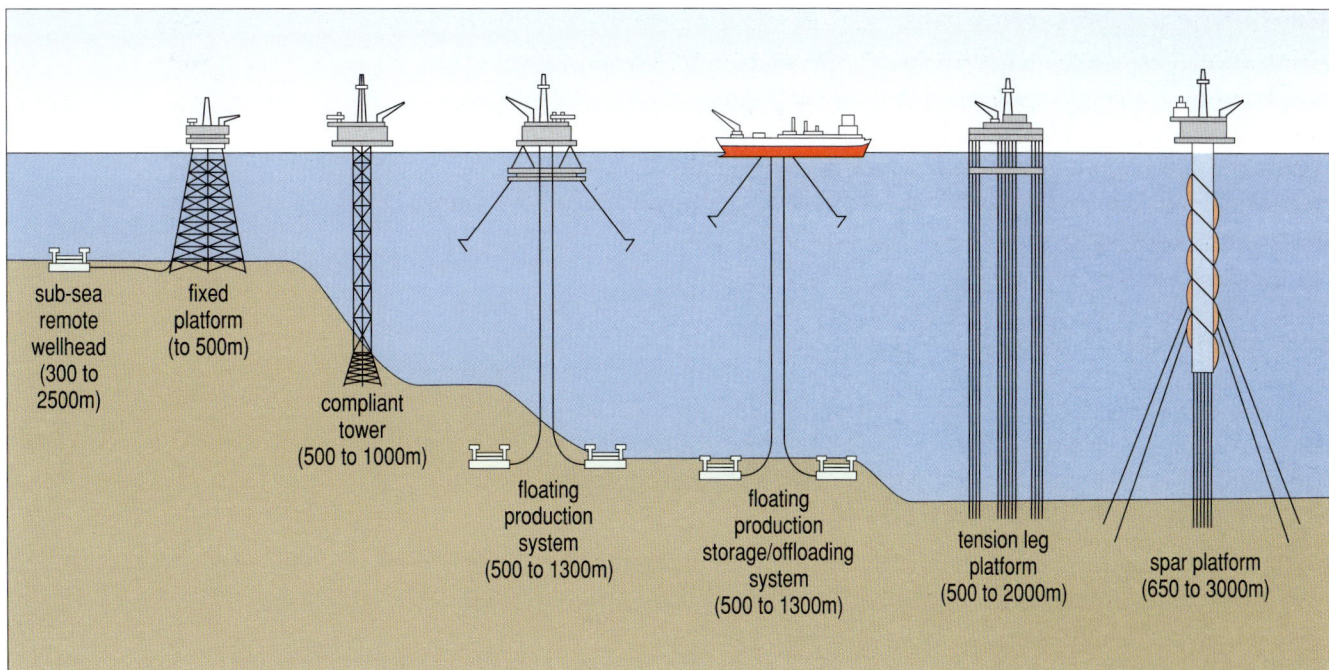

Figure 3.9 Offshore development schemes.

Compliant towers are much like fixed platforms, but bigger. They consist of a narrow tower, attached to a foundation on the sea floor and extending up to the platform. The tower is flexible, as opposed to the relatively rigid legs of a fixed platform. This flexibility allows it to operate in much deeper water, as it can 'absorb' much of the stress exerted on it by the wind and waves at the sea surface.

Floating production systems employing ships or semi-submersible rigs have the advantage of mobility. They can be moved from field to field as reserves are depleted and are generally used to develop smaller fields that cannot support the cost of a permanent platform.

In very deep water, *tension leg platforms* work by tethering the floating platform to the sea bed by jointed legs kept in tension by computer-controlled ballast adjustments. *Spar platforms* are among the largest offshore facilities in use. These huge platforms consist of a large cylinder supporting a fixed rig platform. However, the cylinder does not extend all the way to the sea floor, but instead is tethered to the bottom by a series of cables. The cylinder serves to stabilize the platform in the water and allows for movement to absorb the force of potential hurricanes.

● Can you suggest which development options might be best suited to the following petroleum provinces: (a) the long established, shallow water North Sea (<500 m); (b) the ultra-deep Gulf of Mexico (up to 2 km deep)?

● Sub-sea satellite developments from fixed platforms and floating production systems are cost-effective in the North Sea where the latest discoveries are relatively small and there is plenty of existing infrastructure in the form of fixed platforms and pipelines. (b) Modern generation tension leg and spar platforms are increasingly being deployed in the Gulf of Mexico as developments move into ever increasing water depths.

3.4.3 Production techniques

In order to develop offshore fields economically, numerous directional wells radiate out from a single platform or from several sub-sea wellheads to drain a large area of the reservoir. This allows each well to produce as much petroleum as possible at economic rates. Wells which deviate at more than 65° from the vertical and reach out horizontally more than twice their vertical depth are known as *extended reach wells* (Figure 3.10). Where reservoirs are thin or suffer from low permeability it may be appropriate to drill production wells at more than 80° from the vertical and these are called *horizontal wells*. The flow rate from a horizontal well may be more than five times that from a vertical well, thereby justifying the higher cost of drilling a well with a complex geometry. In order that wells that deviate from the 'standard' vertical drilling can be guided precisely through layered reservoirs, real-time information about the location and inclination of the drill bit is transmitted back to surface. This allows the driller to 'steer' the bit assembly to intersect particularly productive zones.

All fluid petroleum is confined underground at high pressure, which provides a natural 'drive' for production, rather like artesian water supplies (Smith, 2005).

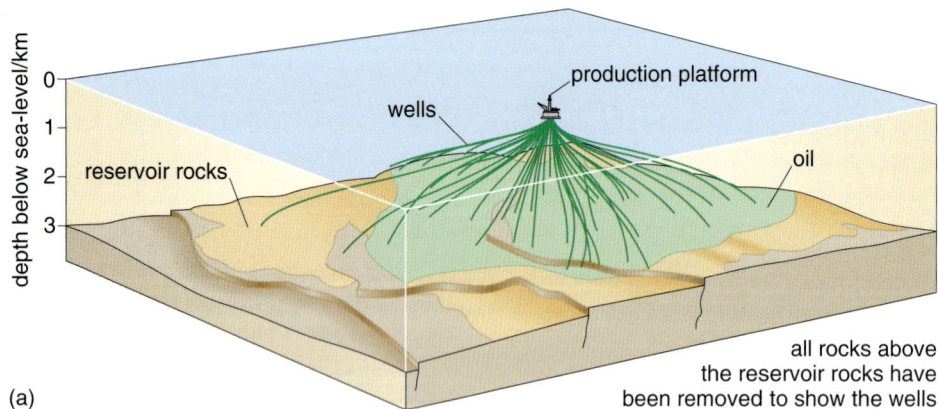

(a)

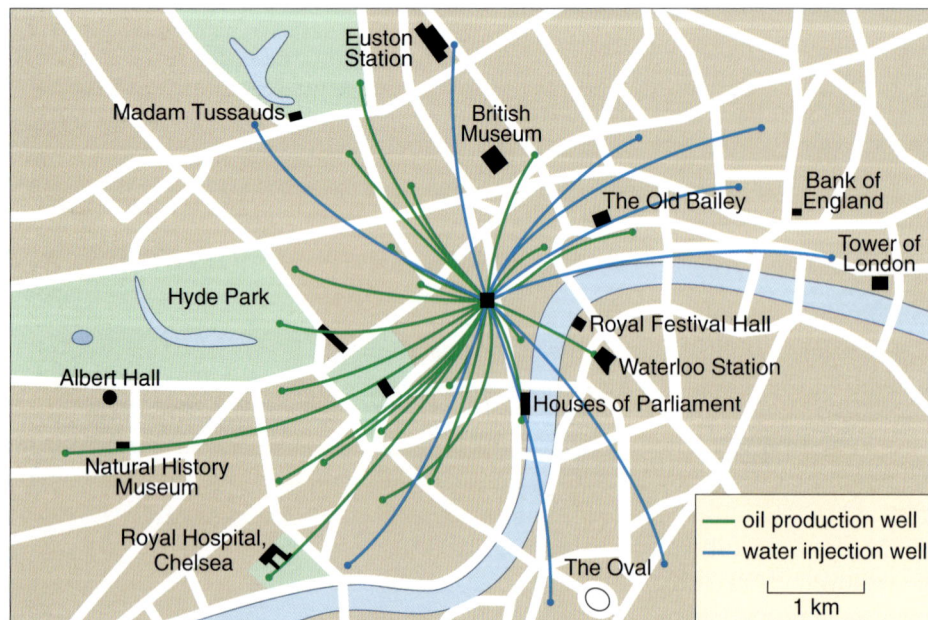

(b)

Figure 3.10 (a) Wells at a variety of angles extract petroleum from all parts of a large, saturated reservoir. (b) Superimposition of the plan of the wells shown in (a) over central London gives a graphic expression of the area that can be exploited from a single production platform by deviated drilling into a reservoir.

During the early stages of production, getting these fluids to the surface safely means allowing a controlled release of fluids under pressure. To prolong extraction later in the life of an oil or gas field, it usually becomes necessary to maintain the pressure underground by injecting pressurized water or gas, or both, into the reservoir.

When production begins, during **primary recovery**, pressurized fluids within the reservoir rise up the borehole and reach the surface. As the pressure is released, any gas dissolved in the oil comes out of solution, to rise and escape along with the oil. As production continues, the pressure of the petroleum remaining in the reservoir begins to fall. This fall in pressure and the loss of dissolved gas increases the viscosity of the oil, so that it will not flow so readily. Typically only 5–30% of the petroleum in the reservoir is brought to the surface during the primary recovery stage.

As the natural drive of the petroleum dwindles, **secondary recovery** techniques are needed for continued production. These techniques maintain reservoir pressure by injecting gas into the gas-cap that often lies above the oil, thus forcing the oil downwards (Figure 3.11a), or by flooding water into the aquifer below the oil to force it upwards (Figure 3.11b). In some reservoirs both gas and water may be injected at the same time or alternately, increasing the recovery of petroleum to 25–65% of the volume contained in the reservoir. The gas required for injection may be derived from the production stream (the fluids that emerge from the well, which comprises gas, oil and water) itself, or from an adjacent field. Similarly, the water for injection may be water from a producing well or sea water.

In order to improve recovery still further, chemical or biological additives may be added to the injected water, or steam may be pumped into the reservoir, in order to reduce the viscosity of the crude oil. Secondary recovery methods can result in over 70% of the initial oil being recovered, but the processes are expensive and for many smaller fields the amount of extra oil recovered may not be worth the investment.

The percentage of petroleum that can actually be recovered from a reservoir is a function of both fluid and reservoir properties, as well as the method of

Figure 3.11 Oil and gas production techniques. When the natural pressure within the reservoir has dissipated, the drive can be maintained by injecting (a) gas into the gas cap at the top of the reservoir, or (b) water into the aquifer beneath the oil. In some reservoirs both techniques are used at the same time.

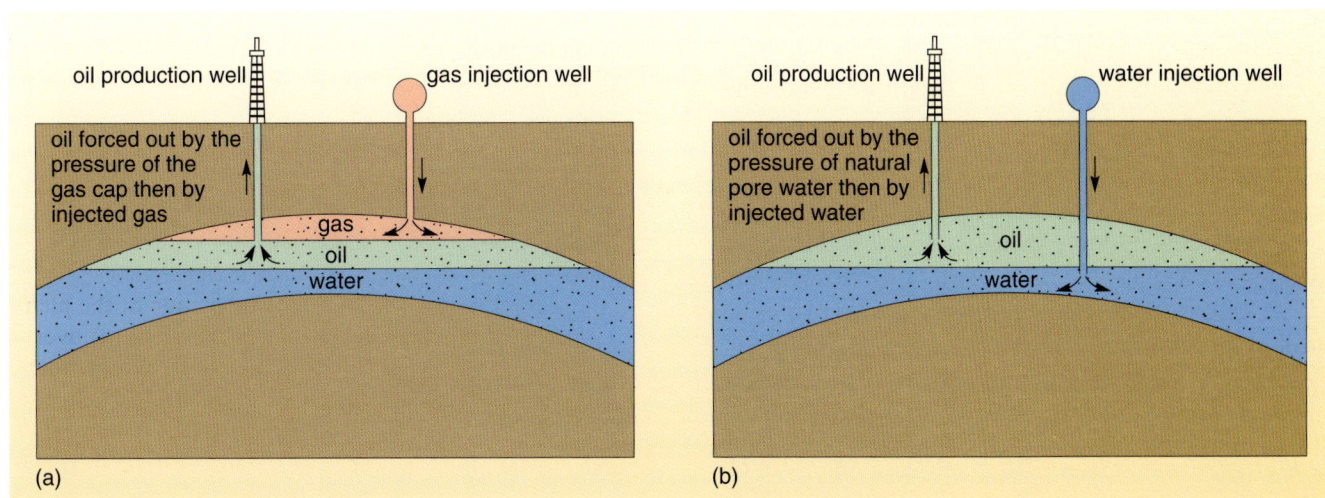

(a)

oil production well   gas injection well

oil forced out by the pressure of the gas cap then by injected gas

gas
oil
water

(b)

oil production well   water injection well

oil forced out by the pressure of natural pore water then by injected water

oil
water

extraction. Viscous, waxy oils are more difficult to extract than light, mobile oils, and low-permeability, segmented reservoirs yield less petroleum than good quality, homogeneous ones, even using secondary recovery. Much oil can be left behind if the displacing fluids follow a few discrete pathways rather than flushing out the oil uniformly from the bulk rock. Even with modern techniques the percentage of recovered petroleum varies enormously: in North Sea oil fields it varies from around 10% up to 70% for the best reservoirs, with the average typically in the range of 30–40%. For gas fields, percentage recovery is generally much higher, with figures in the 70–80% range, because gas is many orders of magnitude less viscous than oil.

- Imagine that you are the Managing Director of Spoof Oil, a small, entrepreneurial company that owns a 100 million barrel oil field with a primary recovery of 25%. Studies indicate that an alternating water and gas injection scheme would cost $80 million to install, but would increase recovery to 45%. Would you make the investment if forecasts of future oil prices are likely to remain above $30 per barrel?

- Yes. Spoof Oil can access a further $(0.45 - 0.25) \times 100 = 20$ million barrels by installing the secondary recovery scheme for $80 million, a cost of $4 per barrel. Allowing for operating costs and taxes there are still very significant profits to be made while oil price remains high.

During the course of field production the amount of new dynamic data that becomes available rises exponentially and it allows an improved description and visualization of the reservoir. Constant interaction between reservoir engineers and geoscientists is required to ensure that modelled outcomes are matched by production performance. Specific initiatives such as novel drilling strategies, time-lapse (4-D) seismic surveys and well-stimulation programmes may be used to maximize recovery and manage the long-term decline. It should be clear that the incentive to produce an additional 5% of reserves from a large field is very significant, particularly in an environment of rising petroleum prices.

3.4.4 Getting petroleum ashore

Most offshore oil and all offshore gas are transported to shore by pipelines; the safest, most cost-effective and environmentally friendly solution to transporting large volumes of petroleum without interruption. Pipelines may be buried if the seabed conditions are suitable or they may rest on the seabed and be covered with rock and gravel to provide protection.

Where seabed topography makes pipelines vulnerable or where they cannot be justified on economic grounds, tankers transport oil from production platforms or storage systems. Storage systems may be within massive fixed platforms or in floating 'spar' type offshore terminals. More commonly, storage is provided within tankers themselves and these are known as *floating, production, storage and offloading (FPSO)* vessels (Figure 3.9). FPSOs may remain on location at a single field for many years, offloading stabilized crude to a shuttle tanker at regular intervals. Alternatively the FPSO can move between one or more fields and the shore terminal, and be redeployed as each field is abandoned.

3.5 Safety and the environment

Safety and the environment have increasingly become matters of prime concern to the petroleum industry. Losses of life, particularly offshore, and large oil spillages increasingly raise outcries and make headline news. As with the mining industry, governments in many countries have legislated to ensure that companies conform to acceptable norms of conduct.

3.5.1 Safety issues

Following the fire on the North Sea Piper Alpha platform in 1988, which killed 167 people, the industry implemented safety improvements, most notably the *Offshore Installations (Safety Case) Regulations 1992*, which changed the approach to management of safety worldwide. The regulations require the operator or owner of every fixed and mobile installation operating in UK waters to submit a safety case to the Health and Safety Executive (HSE).

Safety cases are required at the design stage for fixed installations and cover all subsequent operations and decommissioning. The safety case gives full details of the arrangements for managing health and safety and controlling major accident hazards on the installation. It must demonstrate, for example, that safety management systems are in place, that risks are identified and reduced as reasonably practicable, and that there are provisions for safe evacuation and rescue.

In the UK, current safety legislation sets out the objectives that must be achieved, but allows flexibility in the choice of methods or equipment that may be used by companies to meet their statutory obligations. The HSE employs a team of inspectors who are responsible for enforcing the regulations; their work includes regular inspection visits to offshore installations and investigation of incidents. They have the authority to shut down an installation and prosecute if necessary.

Figure 3.12 shows the safety performance of the UK offshore petroleum industry between 1996/97 and 2003/04. The trend in the number of reported injuries resulting in more than 3-days sick leave (called 'over-3-day injuries') shows an encouraging decrease, but the trend in 'combined fatal and major injuries' is considered far from acceptable. In 2003/04 there were 3 fatalities and 48 major injuries among 18 793 workers, the main causes of which were handling, lifting

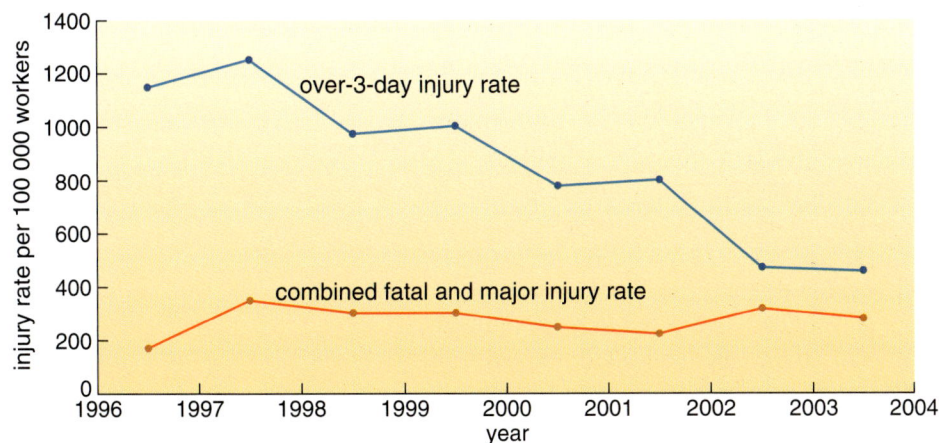

Figure 3.12 Frequency of all significant injuries (including fatalities) among workers in the British offshore petroleum industry. *Note*: The statistics are expressed by convention per 100 000, whereas no more than 20 000 people work in the industry.

and carrying. Whilst safety legislation and management commitment is clearly vital, the challenge of improving safety performance is largely met by workers feeling responsible for their own safety as well as that of their colleagues. In this sense employee commitment to safety is an attitude of mind rather than a taught discipline, although it can be enhanced by training and incentive schemes.

3.5.2 Environmental management

Management of the environmental impact of projects is not only a legislative requirement, but also good business. It is cost-effective, provides a competitive advantage, responds to the public demand for scrutiny, and allows operations to continue in an area. Environmental management, like safety, is now an integral part of all phases of the petroleum business, from early exploration activity through to decommissioning a petroleum field.

Base-line studies and environment impact assessments are commonly used schemes for describing the potential impact of a project on the local environment. A **base-line study** describes the initial natural state of the flora, fauna and land/ seabed conditions prior to any activity. Such a benchmark allows future changes to be identified and provides a reference point if restoration or improvement is required. Such studies are often conducted by independent scientific specialists in order to provide rigour and objectivity.

Environmental impact assessments (EIAs) are detailed ecological studies that are linked to the planned activities of a particular phase of work. They build on the findings of the base-line study and aim to develop or use specified techniques and procedures to minimize the impacts on the environment. These measures may range from using 'soft-start' airguns at the start of a seismic survey in order to alert nearby marine mammals, which can be disoriented by repeated loud noises, to avoiding drilling in fish spawning grounds and reducing the use of oil-based drilling muds. Effective EIAs involve widespread cooperation and consultation amongst the industry and stakeholders in order to achieve the best possible outcomes. As an example, the Atlantic Frontier Environmental Network (AFEN) focuses on the Atlantic waters to the west of Orkney and Shetland, and involves a consortium of oil companies, government bodies and conservation organizations.

The most significant environmental risks from petroleum production come from *oil spills*. Despite precautions, accidents do occur. Perhaps the best documented case history is provided by the *Exxon Valdez*, which ran aground in 1989 in Prince William Sound off Alaska. Some 37 000 tonnes of oil were spilled (roughly equivalent to 125 Olympic-sized swimming pools), to affect more than 2000 km of coastline. Whilst this spill was not the highest ever in terms of volume, it is widely considered the worst in terms of damage to the environment, because of the rugged shoreline and abundance of wildlife. The clean-up (*remediation*) process cost more than 2 billion dollars and took several years. Today there is only very localized evidence of the spill.

Two much larger spills, from the *Braer* (85 000 tonnes) and *Sea Empress* (72 000 tonnes), caused far less environmental damage and had only short-lived effects. The *Braer*, which foundered in 1993 on the Shetland Isles, was carrying a relatively light and easily biodegradable crude oil which was quickly dispersed into the water column by storm-force winds and high seas. Similarly, the *Sea Empress* was transporting a light crude oil when she grounded approaching the oil refinery at Milford Haven, South Wales in 1996. Although more than 100 km of outstanding coastline were initially polluted, the combination of efficient clean-up operations and rapid natural dispersion restored their visually aesthetic appeal within six months. The point is that whilst oil spills are never good news, their long-term effect is rarely enduring.

Despite the increasing number of large tankers carrying crude oil around the world, the number of large spills shows a significant decrease over the last several decades (Figure 3.13). This is encouraging but it does not provide grounds for complacency and maritime safety systems continue to be improved. Nevertheless, many analysts contend that more oil is 'spilt' each year by the deliberate flushing of tanks at sea than is lost by accident.

Aside from oil spills, there are several other significant environmental concerns for the petroleum industry. **Gas venting and flaring** has traditionally been used to dispose of excess gas in fields where no containment or transport facilities existed. This practice releases large amounts of methane and carbon dioxide into the atmosphere, both of which are greenhouse gases. The World Bank estimates that the annual volume of natural gas being vented and flared is about 100 billion cubic metres, enough fuel to provide the combined annual gas consumption of Germany and France. There is now an increasing effort to make commercial use of excess gas, as in the Clair field west of Shetland. That field was inaugurated in 2005, some 27 years after it was first discovered. With an estimated 5 billion barrels of oil in place it was considered one of the largest undeveloped resources on the UK continental shelf. The oil is being exported to the Sullom Voe Terminal in Shetland via a 105 km pipeline and the gas will be transferred to the Magnus field and re-injected there to enhance oil recovery.

One of the most promising schemes for dealing with greenhouse gases is known as **gas sequestration**. This involves injecting carbon dioxide into a depleted underground reservoir and monitoring the integrity of the trapped gas with time-lapse seismic data. Successful trials in Norway indicate that this technique has the capacity to make significant reductions to the carbon dioxide emissions

Figure 3.13 Annual number of large oil spills (over 700 tonnes) worldwide. (The horizontal bold red lines represent the 10-year averages.)

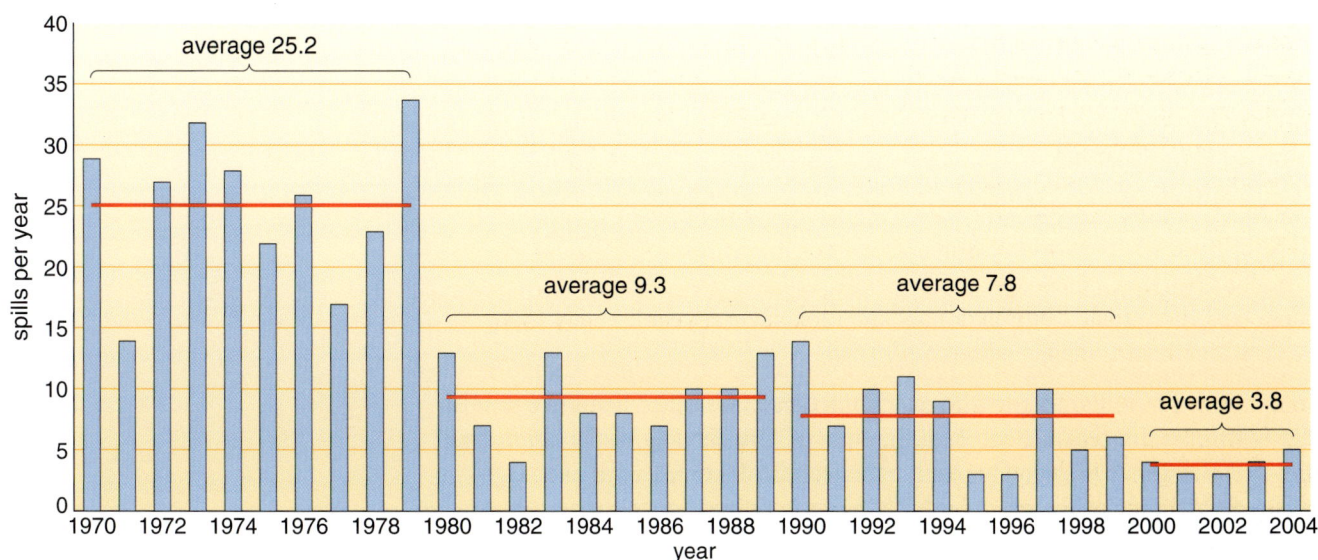

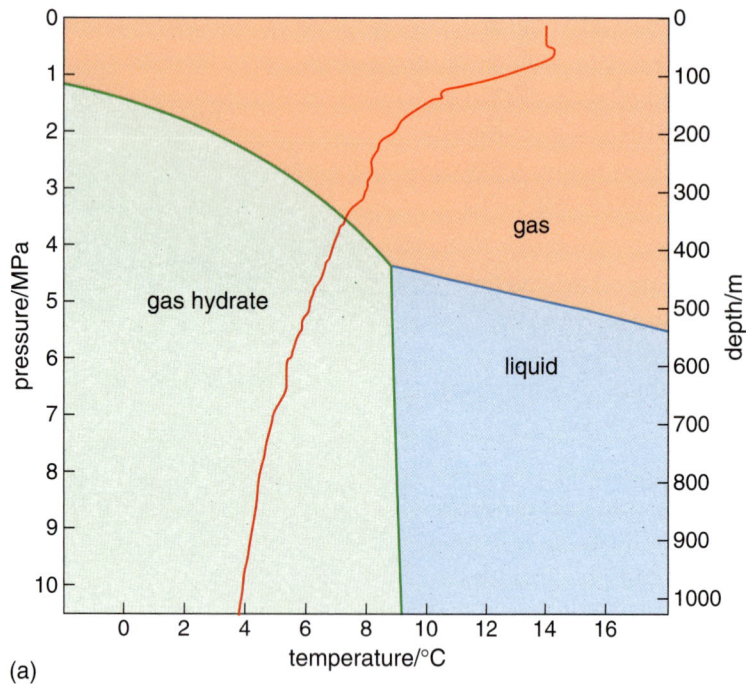

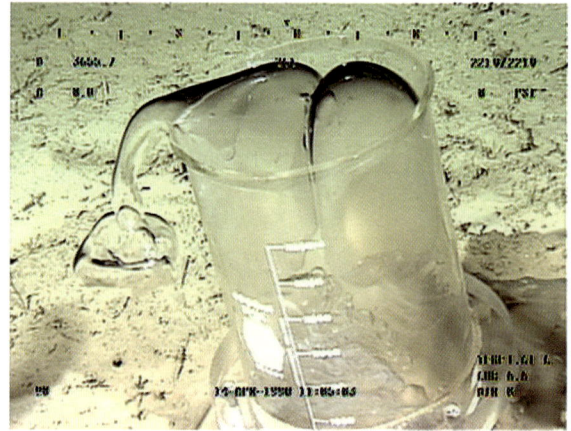

Figure 3.14 Results from experiments conducted at the Monterey Bay Aquarium Research Institute, California to test the feasibility of sequestering carbon dioxide in deep ocean basins. (a) A typical ocean water column temperature profile (solid red line) for Monterey Bay, California overlain on a diagram showing physical states in which carbon dioxide occurs at different pressures and temperatures. (b) Liquid carbon dioxide being poured onto the sea bed at a depth of around 900 m.

throughout northern Europe. Interestingly, at pressures of a few atmospheres and temperatures below 30–40 °C, carbon dioxide forms a stable liquid that is denser than water (Figure 3.14). At temperatures lower than 10 °C it can combine with water to form an ice-like substance known as a gas hydrate (see also Section 3.7.2). Provided the temperature in a depleted petroleum reservoir is below that where pressured carbon dioxide is only stable as a gas, storage can be indefinite. Methane can also be stored in similar settings, or in underground salt caverns, and recovered according to demand. The key issues now appear to be cost and legislation, rather than feasibility.

The safe disposal of waste products such as drill cuttings and oily water is subject to increasingly tight regulations and innovative solutions have emerged. For offshore installations this may involve transporting all waste to the shore for proper treatment and recycling; for example, drill cuttings are commonly recycled as cat litter, fertilizer or used in the construction of footpaths. Alternatively they are cleaned on-site and re-injected underground. On a far larger scale there remains the challenge of **decommissioning** the 6500 offshore installations worldwide as they come to the end of their productive life. Most will be completely removed from their current location and brought to shore for reuse or recycling. The remainder will be examined on an individual basis to establish what is technically feasible and safe to remove. The debacle of the *Brent Spar* (Box 3.3) illustrates the need to resolve the technical, commercial and environmental debate before embarking on a prescribed course of action.

Box 3.3 The *Brent Spar* incident

In 1995 the giant oil company, Royal Dutch/Shell, needed to decide how to decommission a disused North Sea oil storage platform called the *Brent Spar*. Nearly 150 m long, this enormous structure had been in service for 20 years and had the capacity to store 50 000 t of crude oil. Shell considered various disposal options and two made it through to final consideration: deep-sea disposal and on-shore dismantling. These are summarized in Table 3.5.

Table 3.5 Two possible disposal options.

Deep-sea disposal	On-shore dismantling
Tow the *Brent Spar* to the North Atlantic.	Tow the *Brent Spar* into a deep harbour
Use explosives to sink the platform in deep water	Decontaminate the structure
Allow the structure to settle on the sea bed	Dismantle and reuse the materials
Recognize that there will be local pollution for 12–14 months	Dispose contaminants safely onshore
Technically the easiest option	Technically complex and with a greater hazard to the workforce
Cost estimate about £10 million	Cost estimate about £40 million

Shell decided on the deep-sea disposal option and gained the necessary permissions from the Government and regional authorities. Greenpeace, the environmental group, argued that Shell were underestimating the amount of contaminants that remained in *Brent Spar* and saw no reason to pollute the ocean when an alternative existed. They further argued that *Brent Spar* would set a precedent for deep-sea disposal in the future. Greenpeace orchestrated a successful campaign that influenced public opinion against Shell's preferred option and it extended to a boycott of Shell petrol stations in parts of Europe.

Faced with growing opposition, Shell decided to review its options and towed the *Brent Spar* to Norway. Some 2 years later they dismantled it and recycled parts as a foundation for a new ferry terminal. The *Brent Spar* episode will be remembered as a bruising conflict in which both protagonists were accused of manipulating technical and emotional opinions to their own advantage. For example, Greenpeace have since admitted that their counter-arguments over the amount of contaminants remaining in *Brent Spar* were flawed. It is clear that the issue of how North Sea structures will be abandoned in the future demands full cooperation between all the stakeholders and a firm hand from the legislators.

The combination of legislation and good practice has led to a significant reduction in the environmental impact of the petroleum industry over the last decade. Quite properly, this shows no sign of slackening. However, the lasting damage done to the environment by petroleum is not primarily caused by ongoing operations or oil spills, but through society's deliberate use of petroleum products as fuels, as explained in Chapter 6.

3.6 Oil and gas reserves

Exploration companies need to understand how much petroleum remains to be found in a given area or play before they commit significant expenditure to new ventures. More generally, reserves and their depletion in different parts of the world have profound political implications for ensuring future energy supplies: petroleum resources lie at the centre of global political affairs.

3.6.1 Estimating reserves

Estimating the amount of petroleum in a field can be achieved in a variety of ways and with differing levels of accuracy, according to the amount of data that is available. The reserves estimate for a virgin basin — one that has yet to be drilled — may be based simply on the supposed richness of the source rock or comparisons with analogous basins that contain petroleum. Conversely, in an established play that is peppered with wells and seismic data, it should be possible to define the undrilled prospects and determine how likely they are to contain given reserves.

- In recent years a virgin basin to the north of the Falkland Islands in the southern Atlantic has begun to be evaluated, and several exploration wells have been drilled. Media reports (and some oil companies) claim that more than 5 billion barrels of oil will be discovered here in due course. How would you assess the validity of their claim?

- Any estimation of reserves that is based on sparse data must be treated with caution. It would be more correct to quote the range of likely outcomes (e.g. 0 to over 5 billion barrels) at this early stage, rather than one possible outcome. Billion-barrel oil provinces normally have prolific source rocks, large structural traps and excellent reservoir rocks, so the diligent geoscientist should seek such evidence.

It is also possible that the reports are designed to generate enthusiasm and encourage potential investors, particularly because the initial drilling campaign off the Falkland Islands was allegedly disappointing.

One particularly useful approach to reserves estimation is based on the observation that the largest discoveries are normally made early in the life of a play, because they are in the biggest structures that get identified and drilled first: they are the least risky and income from them allows costs to be recovered quickly. Then, as the number of exploration wells increases, so the average volume of reserves proven by each discovery diminishes. This typical pattern is illustrated in Figure 3.15 where, at point A, the first major discovery defines the play and reserves estimates increase rapidly. Over time, when point B is reached, only smaller fields within the play remain undiscovered and the rate of reserves additions declines. If only one play exists in this basin, then the total reserves discovered over time will approach the estimate at point B*. This trend, described by the curve A–B–B*, is referred to as a **creaming curve** by analogy with skimming the cream off the top of the milk.

However, if a second play is discovered (Figure 3.15, point C), reserves estimates increase rapidly again, and the cumulative reserves for the basin as a whole will approach point D* in time. Discovery of a third play, at point E, will increase the

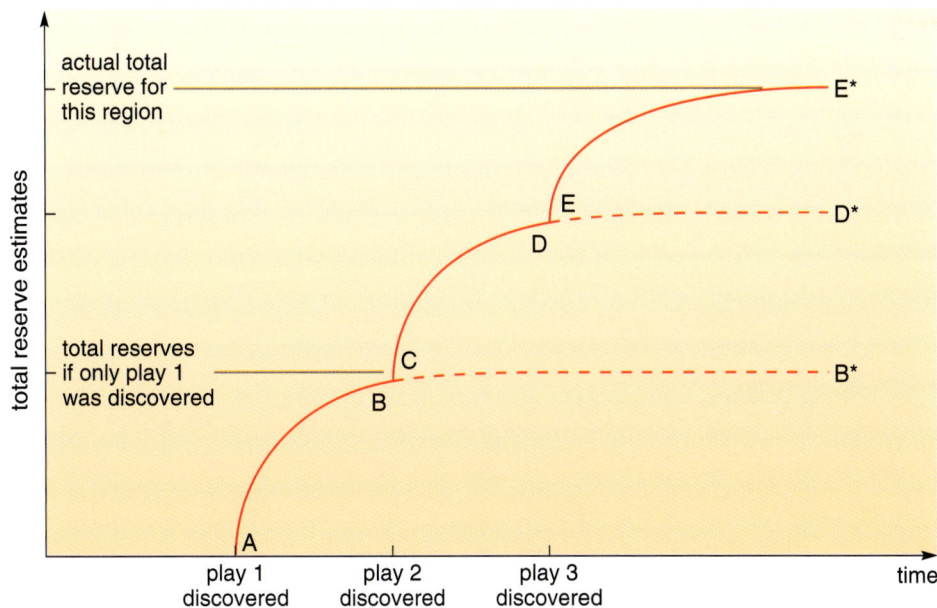

Figure 3.15 Hypothetical creaming curves for new plays discovered within a basin. See text for discussion of the figure.

reserve estimate still further. The discrete creaming curves describe the evolution of plays over time and give a measure of their contribution to the basin as a whole.

Question 3.4

Examine Figure 3.16, showing the creaming curves for the Norwegian sector of the North Sea.

(a) Which is the most productive play and how much resource has it yielded?

(b) How many targets were drilled in the Upper Jurassic play before the largest discovery was made?

(c) Do you think it is likely that the Palaeocene play will yield a single discovery in excess of its current reserves?

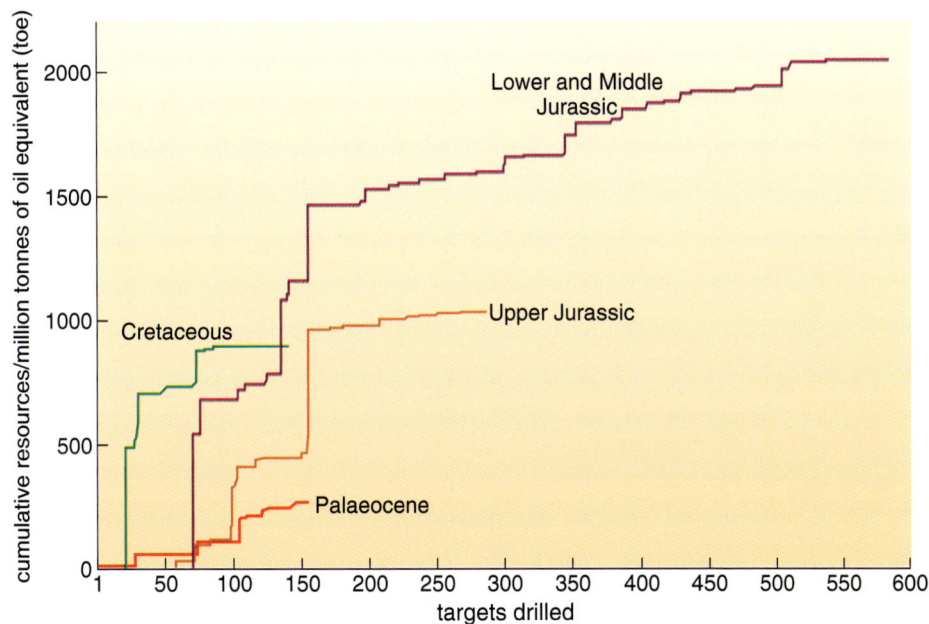

Figure 3.16 Creaming curves for the Norwegian sector of the North Sea. For use with Question 3.4.

It is clear therefore that assessments of reserves rely upon the current state of knowledge within a basin or play. Many basins that appear to be thoroughly explored continue to provide surprises as new play tests are conducted, and thus it is useful to remember the explorer's adage:

> 'We usually find oil in new places with old ideas. Sometimes, also, we find oil in an old place with a new idea, but we seldom find oil in an old place with an old idea. Several times in the past we thought we were running out of oil whereas we were only running out of ideas.'

(Parke A. Dickey)

3.6.2 Reserves categories and reporting

There are inherent difficulties in estimating petroleum reserves accurately, not only within a given area or play (as described above), but also within a single field. Reserves never equate solely with physical measurements such as the petroleum-saturated pore volume in a trap, but instead are influenced by a combination of technological, commercial, and sometimes political, factors, as are all other physical resources.

Petroleum reserves are an estimate of future cumulative production from known fields, and they are typically defined in terms of a probability distribution into 'proved', 'probable' and 'possible' categories. A probability cut-off of 90% is often used to define **proved reserves**, meaning that there is *a better than 90% chance that they will be produced over the lifetime of the field*. Although there is no single technical definition of proved reserves, a commonly used description is as follows:

> The estimated quantities of petroleum which geological and engineering data demonstrate with reasonable certainty to be recoverable in future years from known reservoirs under current economic and operating conditions.

Probable reserves are often considered to have *a better than 50% chance of being technically and economically producible*, whilst **possible reserves** are those which are estimated to have *a significant, but less than 50% chance of being technically and economically producible*.

In general, a proportion of a field's probable and possible reserves tends to get converted into proved reserves over time, as experience from its operating history reduces the uncertainty around what remains in the reservoir. This is an aspect of the phenomenon referred to as 'reserves growth'. It follows that petroleum companies have to be very careful in assigning their reserves to the proper category since most value is attached to those that are proven (see Box 3.4).

3.6.3 The global picture

The global occurrence of petroleum is very patchy and there are sound geological reasons for this. The most significant is the distribution of continental and oceanic crust, because source rocks, the prerequisite for any petroleum system, are confined to continental crust, including continental shelves. Elsewhere, and mainly concealed beneath the world's great oceans, vast areas of oceanic crust have no source rocks and therefore no petroleum potential. Similarly, igneous and most metamorphic rocks cannot source and rarely host petroleum, so areas where they predominate, such as Scandinavia and the Canadian Shield, are poor in petroleum resources.

Box 3.4 Reserves reporting

During the course of a turbulent 18 months of financial reporting between 2003 and 2005, Royal Dutch/Shell Group announced its fifth reduction of proven reserves. The 'write-downs' amounted to 8×10^8 toe, or 30% of the company's total reserves originally reported for 31 December 2002. Regulatory authorities fined Shell about $150 million for committing stock market abuse and breaching the reporting rules. Shell has also agreed to commit a further $5 million to developing and implementing an internal compliance programme.

The importance of proper reserves reporting is to allow investors and the public at large to put a value on the assets of oil companies and to make comparisons between them. However, this only works if all companies report reserves to the same standards. Financial accounting rules, and thereby reserves reporting, vary from country to country.

The United States Securities and Exchange Commission (SEC) define the accounting standards that must be adhered to by companies listed on the New York Stock Exchange. Their definition of 'proved reserves' was first published in 1978. This has created numerous problems for companies because technology has not stood still in the meantime. As an example, advances in 3-D seismic, direct hydrocarbon detection and refined reservoir models, all allow far greater precision in establishing in-place and recoverable reserves than was possible a quarter of a century ago.

The SEC engineers themselves state that 'it is difficult, if not impossible, to write reserve definitions that easily cover all possible situations'. Nevertheless, the thrust of the SEC definition remains that companies may disclose only proved reserves that have been demonstrated by actual production or conclusive formation tests to be economically and legally producible under existing economic and operating conditions.

In contrast, petroleum-rich countries generally have one of the following two features:

1   Particularly prolific petroleum basins within their borders. The top five countries in terms of their share of proved world oil reserves (as at end 2004) are: Saudi Arabia 22.1%, Iran 11.1%, Iraq 9.7%, Kuwait 8.3% and the United Arab Emirates 8.2%.

2   Large continental or continental shelf areas, which are statistically more likely to contain sedimentary basins with the key ingredients for petroleum. For example, the five largest countries in the world (by area) contain the following share of total proven world oil reserves (as at end 2004): Russian Federation 6.1%, Canada 1.4%, China 1.4%, United States 2.5% and Brazil 0.9%.

There are specific features of the geology of the Middle East that make it so richly endowed with petroleum. The region contains several world-class source rocks ranging in age from Palaeozoic to Tertiary, with very thick reservoirs and seals above them, in enormous, low-relief anticlines. In addition, most of its reserves were easily discovered because of the simplicity and sheer size of the traps.

ENERGY: FOSSIL FUELS, NUCLEAR AND RENEWABLES

World oil statistics

According to BP's *Statistical Review of World Energy*, which is generally taken as a reliable source, world proved crude oil reserves at end 2004 were estimated at 1188.6 billion barrels or 161.6 billion toe (1 barrel is equivalent to 0.136 toe); Figure 3.17 and Table 3.5 show the breakdown of this total by region. Note that Europe and Eurasia, Africa, and South and Central America each have about 10% of world reserves, but they are overwhelmed by the Middle East which has 61.7%.

One useful measure of assessing reserves is the reserves-to-production (*R*/*P*) ratio (Section 2.4.5). If the reserves remaining at the end of any year are divided by the production in that year, the *R*/*P* ratio is the length of time that those remaining reserves would last if production were to continue at that level. The world oil *R*/*P* ratio rose sharply during the 1980s because significant new discoveries in the Middle East outpaced the steady growth in world production. Since reaching a peak of 43.7 years in 1989 it has hovered around

Table 3.5 Proved oil reserves (at the end of 2004, in billions of toe).

Region	Reserves /10^9 toe	Share /%	*R*/*P* ratio /years
Middle East	99.8	61.7	82
Europe and Eurasia	18.9	11.7	22
Africa	15.3	9.4	33
South and Central America	13.8	8.5	41
North America	8.3	5.1	12
South Asia and Pacific	5.6	3.5	14
global total	161.6	100.0	40

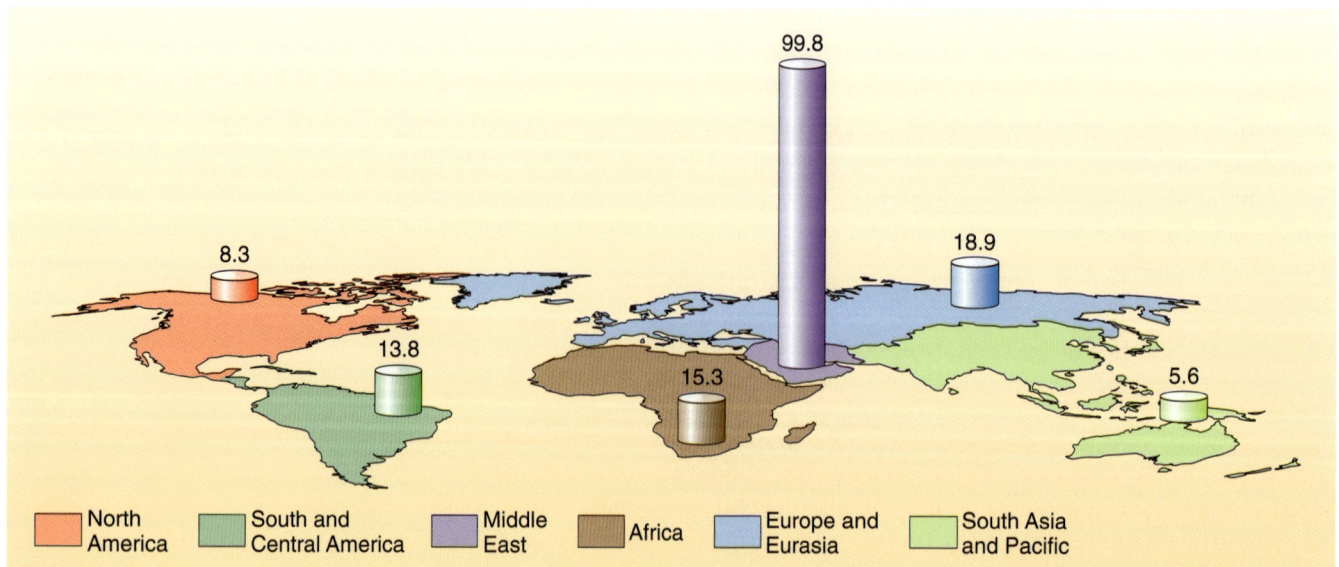

Figure 3.17 Proved world oil reserves by region at the end of 2004 in billions of tonnes of oil equivalent (10^9 toe).

Box 3.5 Petroleum and units of measurement

The petroleum industry has a somewhat lax attitude towards standardization of units. Whereas scientists have adopted the SI system universally, relics of the past prevail amongst oil-industry production engineers and statisticians. As you will know, crude oil is still sold by the *barrel*; a unit of volume that was defined by US coopers in the 19th century as 42 US gallons. Unfortunately, the US gallon differs from the Imperial gallon formerly used in the UK (1 barrel = 35 Imperial gallons). In SI units a barrel has a volume of about 0.16 m^3. Since the density of oil varies from 0.79 to 0.97 $t\,m^{-3}$, with an average around 0.84 $t\,m^{-3}$, expressing oil in terms of mass is rather vague. We use the average density in converting reserves quoted in barrels into tonnes of oil equivalent (toe).

Natural gas might seem an easier material as regards units, and the unit used most commonly is the cubic metre (usually in multiples of a trillion or $10^{12}\,m^3$), although in the US cubic feet are still commonly used. The problem arises when the energy content of natural gas, and those of other energy resources, are needed for comparison with that of oil. It is common practice to convert volumes into toe, and many global statistics use the toe. Again there is imprecision, as different oils have different energy contents (in joules, J) and so too do different 'varieties' of other fossil fuels, including natural gas. It would be convenient to compare every kind of energy source in terms of the fundamental unit of energy, the joule. You will appreciate that is not possible in the case of fossil fuels, so we stick with toe for oil, and m^3 for natural gas, but retain barrels of crude oil in places, because we hear of the changing price of oil in terms of barrels on such a regular basis.

the 40-year mark. Whilst this figure conceals some strong regional differences (Table 3.5) it supports the notion that reserves are sufficient to bridge the gap between current demand and a transfer to alternative energy sources in the future.

World oil consumption continued to rise inexorably, and reached 80 million barrels per day during 2004. Growth was a global phenomenon, with consumption in all regions rising above the 10-year average on the back of a strong world economy. In particular, Chinese oil consumption rose by just under 16%. The Middle East accounted for 41% of world crude oil *exports* in 2004 (about 21% of Middle East production is consumed there). The USA accounted for about 26% of global *imports*, and 19% of US imports were from the Middle East (Canada, South and Central America accounted for 27%). European imports accounted for 22% of global oil trading (26% of Europe's oil imports came from the Middle East, and 50% from the former Soviet Union and North Africa). The other major industrialized part of the world, SE Asia, received 78% of its imports from the Middle East. It is quite clear why the Middle East is an area of such great political concern, and that it will remain so for a long time.

World gas statistics

At the end of 2004, world proved gas reserves were estimated at 179.53 trillion cubic metres (179.53×10^{12} m³). Figure 3.18 and Table 3.6 show the distribution of this total by region and it is notable that the Middle East, and Europe and Eurasia are far more equitable in terms of gas reserves than for oil. Together they account for 76.3% of the world reserves.

The world gas R/P ratio has risen over the last 20 years despite a 75% increase in gas production. This is because the 1990s was a particularly successful decade for gas discoveries in Russia and the Middle East and these cumulative reserves have outpaced production. With the exception of North America, all regions are well endowed with natural gas and have R/P ratios sufficient for several decades.

World gas consumption rose by 3.3% in 2004 to reach 2.69×10^{12} m³. Growth was robust outside North America where consumption stagnated in the face of high prices and industrial restructuring. That said, North America still accounted for 29% of world gas consumption, outstripped only by Europe and Eurasia (41%).

Table 3.6 Proved gas reserves by region (at the end of 2004).

Region	Reserves /10^{12} m³	Share /%	R/P ratio /years
Middle East	72.83	40.6	>100
Europe and Eurasia	64.02	35.7	61
South Asia and Pacific	14.21	7.9	44
Africa	14.06	7.8	97
North America	7.32	4.1	10
South and Central America	7.10	4.0	55
global total	179.53	100.0	67

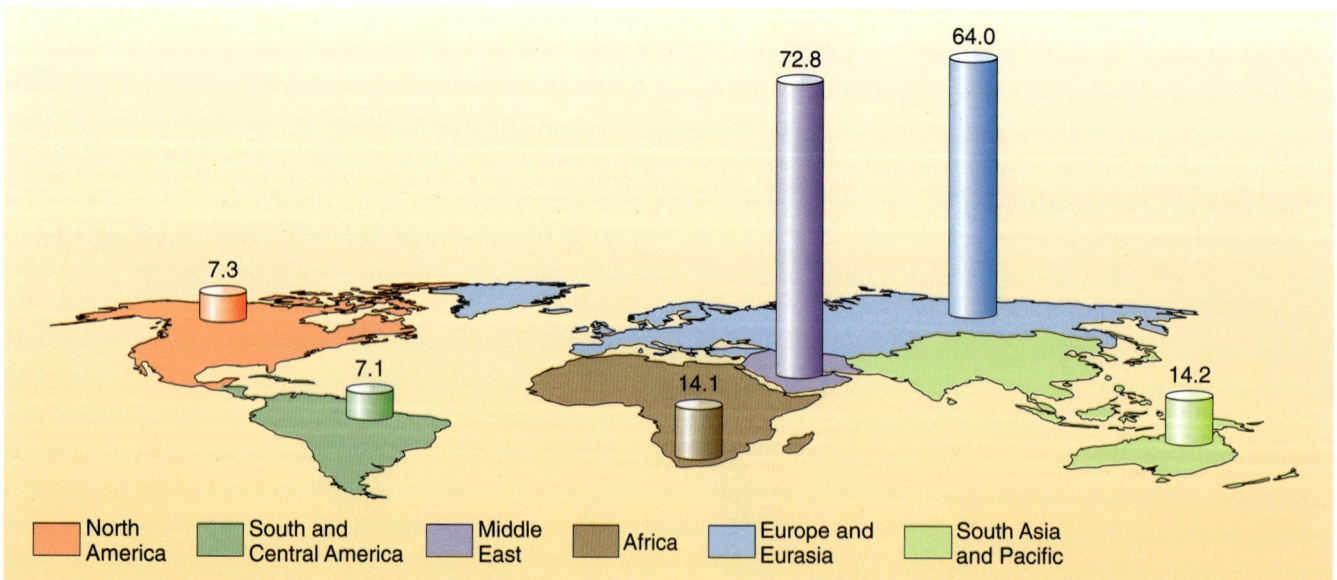

Figure 3.18 Proved gas reserves by region at end 2004. Volumes are in trillion cubic metres (10^{12} m³).

The largest exporters of gas by pipeline were the Russian Federation and Norway, mainly to countries such as Germany, Italy, France and Turkey that have a particularly strong dependency on imported gas. The North American market was sustained by the net import of 93 billion m³ of gas from Canada to the USA. The principal movement of liquified natural gas (LNG) was in the South Asia and Pacific region where Japan and South Korea imported 60% of total world LNG, principally from Indonesia, Malaysia and the Middle East.

3.6.4 The UK context

For the sake of comparison, it is interesting to note that at the end of 2004 the UK had *proved* reserves of 4.5 billion barrels of oil (611 million toe) and 590 billion m³ of gas. This implied a *R/P* ratio of only about 6 years and a UK contribution of less than 0.5% to the world share of proven oil reserves, and about the same percentage for UK gas. These rather stark statistics conceal the fact there are probably still very significant reserves to be exploited. Remaining oil reserves at the end of 2004 were estimated to be between 23–31 billion barrels (3.2 to 4.2 billion toe), which is about the same volume that has been produced to date (Figure 3.19). Some of these reserves are proved because they are associated with producing fields, whilst the remainder fall within the probable and possible reserve categories as they are ascribed to undeveloped discoveries and exploration potential.

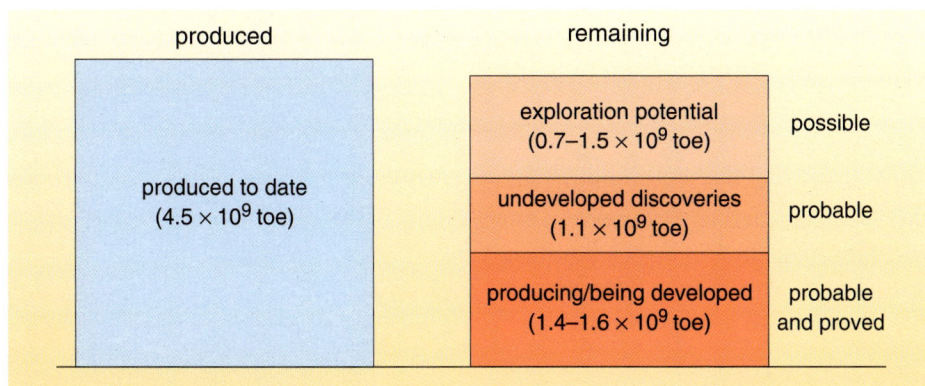

Figure 3.19 Oil reserves estimated to be beneath the UK continental shelf in 2004, compared with production up to 2004. Note that the *proven* reserves are included within the producing/being developed category, some of which are *probable* reserves because of remaining uncertainties. Undeveloped discoveries are mainly *probable* reserves, while the exploration potential covers *possible* reserves.

The exploitation of remaining reserves presents a major challenge to all stakeholders (operators, government, contractors, trade unions). Fields are smaller and more complex, unit costs are high and infrastructure may not be accessible — all these factors will determine the economics of development and dictate the lifespan of the British North Sea petroleum fields.

Whereas the UK passed its peak production in 2000 it is expected to remain self-sufficient in oil until 2009–10. UK gas production currently meets over 90% of demand and is forecast to fulfil 60% of demand in 2010. Thereafter, the North Sea will still sustain meaningful production for several more decades.

● What difference would a decade of consistently high oil prices make to the North Sea fields?

○ It would allow the relatively small discoveries that have been made in recent years to be developed and would sustain the production from existing fields through increased drilling and improved recovery factors. These measures should prolong the economic life of the North Sea fields and reduce the UK's need for importing oil and gas.

You should bear in mind, however, that the price of oil is mainly determined by the rate at which oil is produced from the vast oilfields of the Middle East, where costs are much lower than for North Sea fields. Much of the variability in oil price depends on political decisions in Middle Eastern oil-producing countries, and the extent to which political (and other) pressures from major oil importers affect those decisions.

3.7 Non-conventional sources of petroleum

Non-conventional sources of petroleum, such as oil sands, heavy oil and gas hydrates, greatly exceed the world's entire endowment of conventional petroleum. Yet, because of technological, commercial and environmental constraints to production, non-conventional petroleum currently accounts for only about 5% of global consumption. This huge imbalance will slowly change as the inevitability of declining conventional petroleum reserves and increasing prices hits home.

3.7.1 Oil sands

If a near-surface rock has good reservoir properties, large volumes of oil can flow into it from mature source rocks buried deep below. Exposure of crude oil to air and bacteria close to the surface degrades it to thick, viscous bitumen. Over time, tens of metres of rock from the surface downward can become completely impregnated with bitumen, forming a deposit known as **oil sand**.

Oil sand is composed of bitumen, sand, clays and water. Bitumen, in its raw state, is black and thicker than treacle. It requires treatment to make it fluid enough to transport by pipeline and to be usable by conventional refineries. The process involves large-scale surface strip-mining of enormous volumes of oil sand (Figure 3.20a). The sands are then heated to between 35–80 °C to separate and chemically change the bitumen to lighter hydrocarbons using water-based extraction methods. The upgraded product consists of light and heavy oils that are blended to produce a light crude oil with a low sulphur and nitrogen content.

The world's largest producer of crude oil from oil sand, Syncrude, is based in the Athabaska area of northern Alberta in Canada. Their product is called Syncrude Sweet Blend, and in 2004 it accounted for about 10% of Canada's total crude oil production. With billions of barrels recoverable using current technology, the Athabasca deposit constitutes a resource for decades to come. Importantly, the drive to reduce operating costs to their current level of around US$10 per barrel has also been accompanied by significant reductions in sulphur emissions, water abstraction and power usage during the upgrading process.

(b)

(a)

Figure 3.20 Large-scale extraction of oil sand in northern Canada.
(a) Satellite image of the Syncrude operation at Fort McMurray, Alberta. Active and near-future operations are at lower left and top centre. The image is about 15 km across. (b) To operate continuously, oil-sand mines need the world's largest excavators. This is the Krupps Bagger 288, a bucket wheel reclaimer, which is the largest land vehicle ever built. It is on its way to a lignite mine in Germany: similar machines operate in the Canadian oil-sand mines.

When oil sands occur at depths that are too great for surface mining, *in situ* extraction involves injecting steam and/or hot carbon dioxide to lower the viscosity of the oil and enable it to be pumped to the surface.

Petroleum accumulations that will not flow to the surface under natural reservoir pressure are referred to as **heavy oils**. They are characterized by high viscosity that increases with their density, low hydrogen/carbon ratios, low gas/oil ratios and significant sulphur, asphalt and heavy-metal contents. Heavy oils can form for a variety of reasons, such as kerogen composition, level of maturity, depth of burial and exposure to water, air or bacteria. The economics of heavy oil production typically suffer from high extraction costs and a discounted value because of their inferior quality. As a result, the vast remaining global resource of heavy oils is still very much underexploited.

3.7.2 Gas hydrates

The temperature and pressure of the deep oceans are controlled, respectively, by deep, cold currents that move from polar latitudes along the sea floor and by the mass of the overlying water column. Consequently, at depths exceeding

300–500 m the sea floor is at a temperature of around 1–2 °C and a pressure that is several hundred times greater than atmospheric pressure. Under these physical conditions, gases such as methane (CH_4) and carbon dioxide (CO_2) can combine with water to form solid, ice-like crystalline compounds known as **gas hydrates** (Section 3.5.2, Figure 3.14). Clearly, economically interesting gas hydrates are those which contain proportionally far more hydrocarbon gases in their structure than CO_2. Depending on the geothermal gradient, the base of the hydrate stability zone may extend to depths of more than 1000 m beneath the ocean floor.

Hydrocarbon gas hydrates can form in deep ocean water but they are not found as a carpet on the sea floor. This is because such hydrates have a lower density than that of seawater (850 kg m^{-3} compared with 1025 kg m^{-3}). As soon as they form, they float upwards and turn back into methane and water in the lower pressures and warmer temperatures of the upper layers of the ocean.

However, within the sediments just beneath the sea floor, crystals of hydrocarbon gas hydrate form and move upwards buoyantly. Frequently the rising crystals form a 'log-jam' within the pore space of the sediment and once this occurs, more gas hydrate crystals become trapped in the pore spaces beneath. Eventually, all available pore spaces in the sediment become completely filled by gas hydrate crystals that readily absorb methane into their lattice. Fully saturated gas hydrates can hold up to 200 times their own volume of methane, creating a zone that is denser than seawater and thus gravitationally stable.

The gas required for formation of gas hydrates comes from two principal sources: biogenic and thermogenic. Biogenic gases are those produced *in situ* by bacterial breakdown of organic matter contained within the sea-floor sediment. The dominant biogenic gas is CH_4 (>99%) with traces of CO_2 and H_2S. Such gases typically form in oceanic areas that have relatively high rates of sedimentation and plenty of organic matter, such as the coastal margins of North America and the North Pacific Ocean. In contrast, thermogenic gases are those produced by the maturation of kerogen at much greater depths and elevated temperatures, as described in Section 3.2.1. Thermogenic gas hydrates contain significant amounts of ethane, propane and butane, and they occur in petroleum-rich provinces, such as the Gulf of Mexico and Caspian Sea, where leakage to the surface is common.

Gas hydrates do not form only in the oceans, but also in deep lake sediments and onshore permafrost zones across Arctic Canada and Russia. Current estimates suggest that gas hydrates globally may contain $1-5 \times 10^{15}$ m^3 of methane, a figure that dwarfs the remaining proven reserves of conventional gas. But there are some issues of both concern and potential, summarized by Figure 3.21. On the positive side, the potential of methane hydrates as a major strategic energy reserve is obvious and much research is being conducted to develop appropriate extraction techniques. This extends to considering whether methane production could be combined with CO_2 disposal, thus addressing the twin challenges of this century — reducing the emissions of greenhouse gases and providing a low-carbon fuel to replace oil and coal.

On the negative side, it is now recognized that gas hydrates are a potential geohazard. Dissociation of hydrates at the base of the gas hydrate stability zone

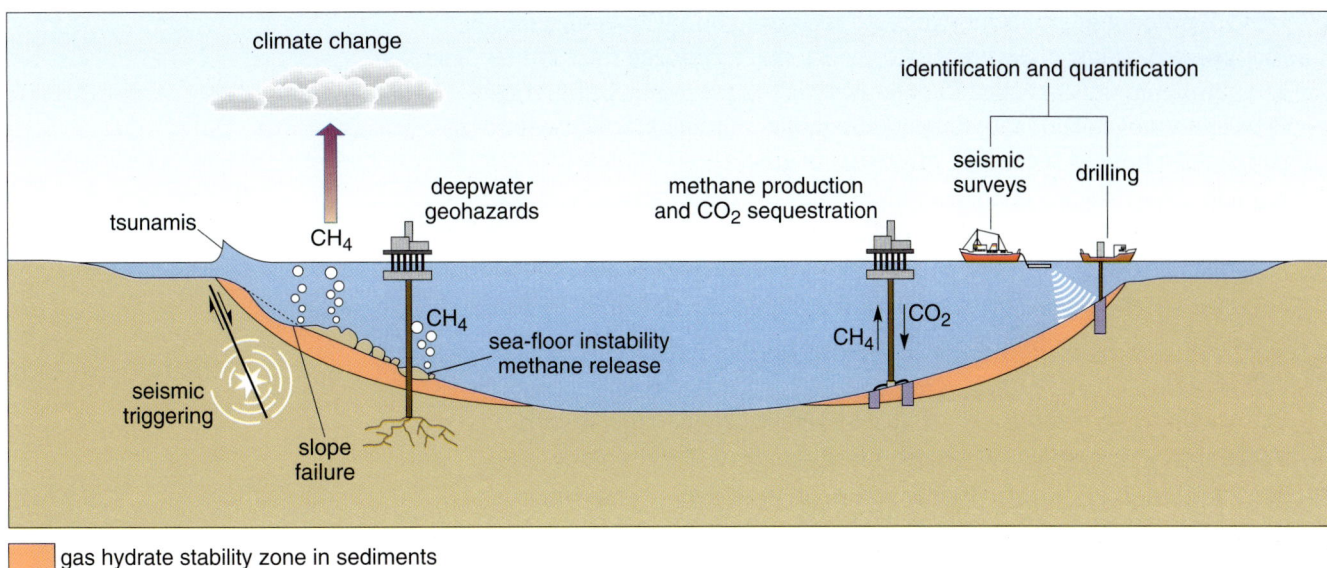

gas hydrate stability zone in sediments

Figure 3.21 Illustration of the major issues concerning gas hydrates in sea-floor sediments. (Left) Earthquakes trigger gas-hydrate instability that in turn triggers massive slumping of sea-floor sediments and tsunamis. Installing large structures on the sea-bed might result in rapid release of gas and instability of their foundations. Any release of methane promotes global warming. (Right) The huge potential for developing gas hydrates as resources — they are readily discovered by their distinct 'signatures' on seismic sections.

(Figure 3.21) can cause increased pore-fluid pressures in under-consolidated sediments, forming a zone of weakness and a site of potential sea-floor failure. Slope failure can threaten underwater installations and, in extreme cases, generate tsunamis. It has even been suggested that during periods of climatic warming such as we are experiencing at present, onshore hydrates become destabilized, liberate methane to the atmosphere and thus accelerate global warming.

Being such a dominant factor in global affairs and also a source of greenhouse gases, petroleum is discussed further in Chapter 6, in the context of the future of energy resources as a whole.

3.8 Summary of Chapter 3

1   Hydrocarbons are compounds composed mainly of carbon and hydrogen. In addition to hydrocarbons, petroleum contains significant quantities of oxygen, nitrogen, sulphur, nickel and vanadium, plus minor quantities of a host of other elements. Petroleum may be liquid (crude oil), gaseous (natural gas) or solid (bitumen).

2   The necessary ingredients for any accumulation of petroleum — petroleum charge — are source, maturation, migration pathway and trap. As knowledge about these factors develops during the exploration and development of a petroleum-rich basin they help define petroleum plays, which are realistic strategies for directing further development.

3   An effective source rock contains sufficient organic matter, in the form of kerogen, to produce fluid hydrocarbons when heated (matured) to temperatures above the generation threshold of about 50 °C.

4   Fluid petroleum migrates for two reasons. During burial and compaction, fine-grained, clay-rich sediments lose pore space between their grains. Water and petroleum thus expelled from source rocks start primary migration. Petroleum moves buoyantly in response to the pressure gradient within permeable strata as secondary migration begins.

5   For petroleum to become concentrated in an economic oil or gas field, suitable rocks must be charged with petroleum fluids. These reservoir rocks must not only be porous enough to contain substantial amounts of petroleum but must also be permeable enough so that fluids can flow in to accumulate, and out during extraction. Almost all reservoir rocks are sandstones or limestones.

6   Permeable reservoirs must be capped by impermeable seals, usually mudstones or evaporates in one of several kinds of trap.

7   Structural traps are formed by the deformation of sedimentary rocks and include anticlinal domes and fault traps. Other examples of structural traps include those associated with salt masses that have moved upwards because of their low density, sometimes to 'intrude' other sediments. Stratigraphic traps occur where there is a barrier caused by lateral permeability variation in sedimentary rocks.

8   The main method of determining whether an area has potential traps for petroleum is seismic exploration. Seismic sections provide images of the subsurface. Once detected, a potential trap can be mapped in detail using 3-D seismic data to define its shape and the thickness of petroleum-bearing parts of the reservoir. Porosity and permeability of the reservoir rock determined by direct measurements of exploration-well samples then allow the volume of oil and gas that can be recovered to be estimated.

9   The most significant local environmental risks from petroleum production come from accidental spillages during transportation.

10   Primary recovery methods produce at best only 30% of the oil present. Secondary recovery techniques that pump pressurized water and gas into the reservoir can boost the amount to ~65%, and more still can be recovered by injecting steam or chemical and biological additives into the reservoir to reduce viscosity. The feasible recovery has an important bearing on the estimates of a field's reserves.

11   In 2004, global petroleum reserves were 162 billion toe of oil (1189 billion barrels) and 180 trillion cubic metres of gas.

12   Petroleum reserves in the British North Sea are minute (0.5%) compared with those of the Middle East, which controlled 62% of the world proved reserves of crude oil in 2004.

13   Large amounts of oil and natural gas are locked into oil sands and gas hydrates, but, apart from the Canadian oil sands, they do not yet constitute a globally significant economic resource.

NUCLEAR ENERGY

The transformation of radioactive uranium and, in some instances, thorium isotopes provides vastly more energy per unit mass of fuel than any other energy source, except nuclear fusion (see Chapter 6), and therein lies its greatest attraction. As you read in Section 1.5, the key to that remarkable fact is the conversion of matter (with mass, m) into energy (E), according to Einstein's famous equation $E = mc^2$, where c is the speed of light (3×10^8 m s^{-1}).

The potential of nuclear fuels for energy production became a reality when the first experimental atomic pile, built by Enrico Fermi and Léo Szilárd at the University of Chicago, began functioning in December 1942. That led to the manufacture of fissionable material for the first atomic weapons. The use of nuclear power for electricity production expanded rapidly in the 1960s, a period when the costs of building nuclear power stations and of purchasing the uranium fuel were thought to be less than for fossil fuel plants. The nuclear industry received a boost in the early 1970s, when fossil fuel prices rose abruptly during the oil crisis of 1974: following the Yom Kippur war of late 1973, oil producers in the Middle East quadrupled the price of their crude oil almost overnight.

During the 1980s, however, the costs of building nuclear power stations rose inexorably as stringent safety requirements grew, especially following the accident at Three Mile Island in Pennsylvania (1979) and the much larger one at Chernobyl (1986) in the Ukraine. In addition, the cost of decommissioning nuclear reactors and of developing secure repositories for radioactive waste had not been taken into account in the initial cost–benefit analyses. These considerations have loomed progressively larger as older nuclear power stations approach the end of their useful lives and as the volume of waste grows year by year. The global rate of expansion of the nuclear industry had slowed almost to a standstill by the early 1990s, and as demand for uranium fuel fell, so did its price. In other words, the fuel got cheaper as the power stations became more expensive.

In the early 21st century, however, with growing concern about global warming the environmental advantage of nuclear power over fossil fuels is becoming increasingly recognized: it produces no greenhouse gases. It also produces no acid rain, unlike coal and to a lesser extent oil.

By 2003 over 400 nuclear reactors were generating electricity globally, producing over 350 GW (Table 4.1). This amounts to about 16% of global electricity generating capacity. An additional 31 nuclear reactors with an additional generating capacity of 25 GW are currently in construction globally. In the UK many reactors are of an early design and only a few of the younger nuclear power stations match the 1 GW capacity of most fossil-fuel stations. Nevertheless, almost a quarter of UK electricity generating capacity is nuclear (Figure 4.1); we consider the advantages and limitations of this situation shortly.

Figure 4.1 The 1.2 GW Sizewell 'B' pressurized water reactor on the Suffolk coast. *Note*: The dome is the top of the containment vessel (Figure 4.4c).

Table 4.1 Nuclear reactors and generating capacity in 2003 arranged in order of number of operating reactors.

	Reactors in operation	Power capacity/ GW	% national electricity generation		Reactors in operation	Power capacity/ GW	% national electricity generation
USA	104	98.3	19.9	Hungary	4	1.8	32.7
France	59	63.4	77.7	Mexico	2	1.3	5.2
Japan	53	44.1	25	Argentina	2	0.9	8.6
Russian Federation	30	20.8	16.5	Brazil	2	1.9	3.6
UK	27	12.1	23.7	Lithuania	2	2.4	79.9
South Korea	19	15.9	40.0	Pakistan	2	0.4	2.4
Germany	18	20.6	28.1	South Africa	2	1.8	6.0
Canada	16	12.1	12.5	Armenia	1	0.4	35.5
India	14	2.6	3.3	Netherlands	1	0.4	4.5
Ukraine	13	11.2	45.9	Romania	1	0.7	9.3
Sweden	11	9.5	49.6	Slovenia	1	0.7	40.4
Spain	9	7.6	23.6	*Grouped by region*			
China	8	6.0	2.2	Europe and Russian			
Belgium	7	5.8	55.5	Federation	209	171.7	
Czech Republic	6	3.5	31.1	North America	120	110.4	
Slovakia	6	2.4	57.4	Asia and Pacific	102	73.8	
Taiwan	6	4.9	21.5	South and Central			
Switzerland	5	3.2	39.7	America	6	4.1	
Bulgaria	4	2.7	37.7	Africa	2	1.8	
Finland	4	2.7	27.3	**Global totals**	**439**	**361.9**	**~16**

4.1 Nuclear reactions, reactors and power generation

The nuclear reactions that produce energy from uranium fuel are complex (Section 4.1.1), but in principle the generation of electricity is no different from that in fossil fuel power stations (Figure 4.2). The fuel (fossil or nuclear) is used to heat water to above its boiling temperature in a boiler, and the resultant high-pressure steam drives turbines that generate electricity. (This was briefly discussed in Box 1.2.) The steam is then condensed and returned as water to the boiler. This is a closed circuit, and the steam never comes in contact with the fuel.

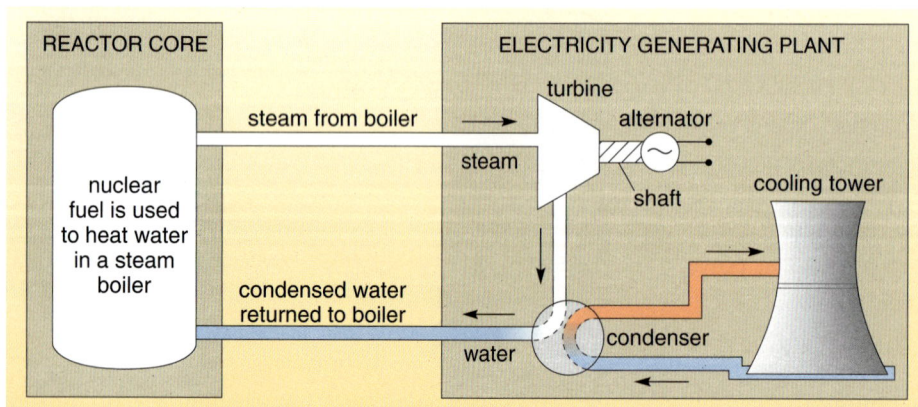

Figure 4.2 The basic design of a nuclear electricity generating plant using a burner reactor (Section 4.1.2).

4.1.1 Nuclear fission

Every atom has a *nucleus* consisting of positively charged *protons* and electrically neutral *neutrons*. Protons and neutrons have virtually identical mass and the total number of protons and neutrons defines the **mass number** of a particular atom. The number of protons in the nucleus is the **atomic number** and this quantity is always the same for each particular chemical element. However, some elements have several **isotopes**, each with different numbers of neutrons, but with the same number of protons. In full notation an isotope is represented by its chemical symbol preceded by a subscript number showing the element's *atomic number* (the number of protons in its nucleus), whereas the superscript is the *mass number* of the isotope (the sum of protons and neutrons in its nucleus). Uranium has two isotopes of interest:

uranium-235, which has 92 protons and 143 neutrons, written as $^{235}_{92}U$;

uranium-238 ($^{238}_{92}U$), which also has 92 protons but 146 neutrons;

i.e. both isotopes have the same atomic number (92) but different mass numbers (235 and 238). *Note*: In equations for nuclear reactions we use this formal notation, e.g. $^{238}_{92}U$, but otherwise the simpler name of an isotope, e.g. uranium-238.

These isotopes have such large nuclei that they are inherently unstable. They spontaneously break down, or decay, by two possible processes:

1 **radioactive decay**, a process that emits **alpha particle**s, which are equivalent to the nuclei of helium atoms, with two protons and two neutrons (^4_2He);

2 **nuclear fission**, a much less frequent process, in which the whole nucleus breaks apart, releasing energy.

● But how does nuclear fission of uranium release energy?

○ The main products of nuclear fission are nuclei of other elements whose combined mass is slightly less than the mass of the parent uranium nucleus. The 'missing' mass is converted into energy according to the relationship $E = mc^2$, where m is the 'missing' mass, and c is the speed of light. Because the speed of light is so enormous (3×10^8 m s^{-1}), when the number is squared, even a small loss of mass converts into a huge amount of energy.

Of the naturally occurring radioactive elements, only uranium and thorium undergo *spontaneous fission* and release energy on this potentially vast scale. The other elements *decay* by emitting either alpha particles (as defined above), or beta particles (equivalent to electrons), or positrons (electrons with positive charge), or gamma rays (very short wavelength electromagnetic radiation).

The isotopes produced by fission may also be radioactive and will themselves decay radioactively until stable nuclei are formed. Radioactive decay can last a long time: the *half-lives* (Box 4.1) of most natural radioactive elements are measured in thousands of years (e.g. carbon-14) to billions (10^9) of years (e.g. uranium-238).

The energy produced by natural *radioactive decay* is the *kinetic energy* of the emitted particles, which is converted into *thermal energy* (heat) when they collide with other nuclei. The heat produced by natural decay of radioactive elements that

Box 4.1 Radioactive half-lives

The pace of *decay* of natural radioactive isotopes can be visualized from each isotope's **half-life**. This is the time it takes for half the total number of atoms to undergo a nuclear transformation (Figure 4.3). The half-life of uranium-235 is 704 million years (Ma), meaning that half the original atoms will have decayed in this time. Half those remaining will decay in the next 704 Ma, so that three-quarters have decayed after 1408 Ma, seven-eighths in 2112 Ma, and so on. For uranium-238 the half-life is 4468 Ma, so the Earth (which is about 4600 Ma old) has lost a much greater fraction of its uranium-235 than its uranium-238 atoms since it formed. (Natural uranium today contains only 0.7% of uranium-235.)

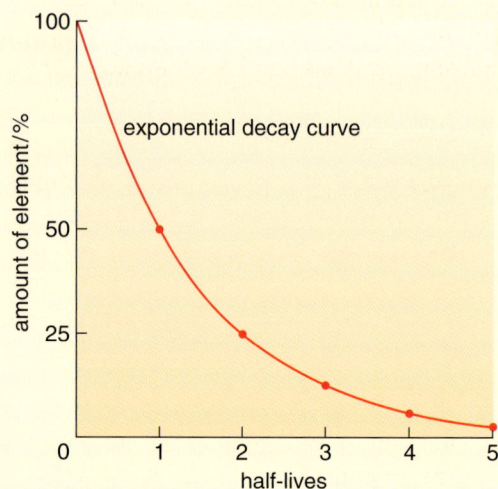

Figure 4.3 The concept of half-life, which leads to exponential decrease (decay) of a radioactive element.

are dispersed in low concentrations in the Earth's outer layers (crust and mantle) is the source of the Earth's internal heat. This heat is indirectly tapped when we exploit geothermal power (see Chapter 5). Radioactive decay of uranium is far too 'dilute' a source of energy to exploit directly for electricity generation, as Question 4.1 illustrates.

Question 4.1

The rate of heat production by natural radioactive *decay* of uranium-238 is $3000 \ J \ kg^{-1} \ yr^{-1}$. Roughly how long would it take for 1 kg of uranium-238 to produce the same amount of energy as is contained in 1 kg of coal, which is $2.8 \times 10^7 \ J$?

So, natural radioactive decay cannot be used to fuel power stations. The ability to undergo nuclear *fission* is what makes uranium a concentrated energy source (Section 1.6.1) — but only after some technological intervention.

Natural fission of uranium nuclei occurs much less frequently than radioactive decay. In its natural state, uranium-238 undergoes one nuclear fission for roughly every million alpha emissions involved in its radioactive decay. So, for nuclear fission to become a viable energy source, the fission rate must be greatly increased. Before describing how this is done, you need to understand a little more about uranium-235 and uranium-238.

Natural uranium today consists of 99.3% uranium-238 and 0.7% uranium-235. A greater proportion of uranium-235 was present early in the Earth's history, but its relative amount has decreased because its half-life is about ten times shorter than that of uranium-238 (i.e. it decays faster — see Box 4.1). Despite its much lower abundance, uranium-235 is the isotope which provides the fissionable fuel in nuclear reactors.

The occasional natural fission of uranium atoms releases neutrons. Nuclei of uranium-235 may capture these low-energy or **slow neutrons**, each capture increasing the mass number of uranium-235 to produce uranium-236:

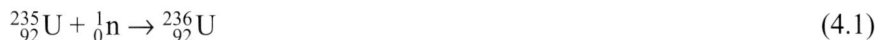

$$^{235}_{92}U + ^1_0n \rightarrow ^{236}_{92}U \tag{4.1}$$

uranium-236 is highly unstable and breaks down by fission:

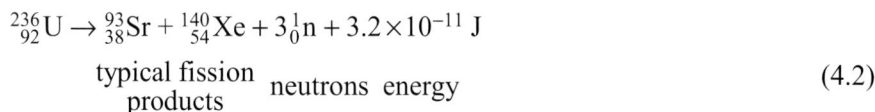

$$^{236}_{92}U \rightarrow ^{93}_{38}Sr + ^{140}_{54}Xe + 3^1_0n + 3.2 \times 10^{-11} \text{ J} \tag{4.2}$$

$$\underset{\substack{\text{typical fission} \\ \text{products}}}{} \quad \underset{\text{neutrons}}{} \quad \underset{\text{energy}}{}$$

There are three important things to note about this reaction.

1   Although in *numerical* terms the mass numbers and atomic numbers total the same on both sides of Equation 4.2 (236 and 92 respectively), in terms of precise *mass*, the sum of the products on the right does not quite equal 236. There has been a small loss of mass in the fission reaction, and this has been converted into energy according to $E = mc^2$.

2   The main fission products (strontium-93 and xenon-140) are radioactive and decay to stable daughter products, releasing thermal energy in the process.

3   More neutrons are produced — three for every atom of uranium-236 that undergoes fission.

If these neutrons are captured by more nuclei of uranium-235 the reaction will continue. That produces still more neutrons and more energy in an *uncontrolled* **chain reaction**, as in atomic weapons, unless it is controlled in some way, as in a power station. Such a chain reaction can only begin, however, if sufficient atoms of fissionable uranium-235 are present in a small volume, as in a reactor core, a nuclear bomb and very rarely in extremely rich uranium deposits: it needs a **critical mass** for the chain reaction to be self-sustaining. In a nuclear reactor, uranium atoms are bombarded with neutrons (^1_0n). This increases the rate of fission, thereby releasing energy much more rapidly.

Equations 4.1 and 4.2 are the basis for calculating the amount of energy produced by uncontrolled fission: the **fission energy** of uranium. A kilogram of uranium-235 contains 25.5×10^{23} atoms. The fission energy available from one atom of uranium-235 (after conversion to uranium-236 and subsequent fission) is 3.2×10^{-11} J (Equation 4.2). So the amount of fission energy available from 1 kg of uranium-235 will be:

$$(25.5 \times 10^{23} \text{ atoms}) \times (3.2 \times 10^{-11} \text{ J per atom}) \approx 8.2 \times 10^{13} \text{ J}$$

The energy released in a year by *fission* of the 0.7% of uranium-235 in 1 kg of natural uranium is 5.7×10^{11} J, i.e. around 2×10^8 times that released by radioactive *decay* of both isotopes. The same amount of energy would be released by burning about 20 t of coal over the same time.

So why not exploit fission energy from the much more abundant isotope, uranium-238? The principal reason is that uranium-238 is more stable than uranium-235; slow neutrons just 'bounce off' its nucleus. Only high-energy **fast neutrons** can penetrate the uranium-238 nucleus to be captured. This creates uranium-239, which in turn is transmuted to neptunium-239 and then to plutonium-239 by loss of electrons:

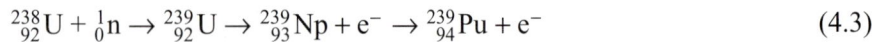

$$^{238}_{92}U + ^{1}_{0}n \rightarrow ^{239}_{92}U \rightarrow ^{239}_{93}Np + e^- \rightarrow ^{239}_{94}Pu + e^- \tag{4.3}$$

Plutonium-239 is unstable and undergoes fission in a similar manner to uranium-235, releasing more neutrons and an amount of energy similar to that in Equation 4.2. This production of plutonium-239 also poses a major hazard in the waste from conventional nuclear power stations, as well as the risks from various radioactive fission products (Section 4.3).

Clearly, by exploiting the more abundant uranium-238 isotope in this way, the potential of naturally occurring uranium as an energy resource can be increased — in fact by almost 150 times — compared with using just uranium-235. However, establishing an efficient chain reaction based on uranium-238 requires a supply of fast neutrons, and this involves much more advanced and complex technology, as you will see in Section 4.1.2.

4.1.2 Nuclear reactors

A critical mass of uranium is necessary for nuclear chain reactions (Equations 4.1 to 4.3) to occur. A smaller concentration of uranium, whether in the form of a single nuclear fuel rod, or in most uranium ore deposits, is not in danger of spontaneous fission. In nuclear reactors the chain reaction is controlled by slowing down neutrons and absorbing any in excess of those needed to keep the reaction going at the required rate.

Most operational nuclear reactors make use of the reaction in Equation 4.2. They are called *burner reactors*.

Burner reactors

The object of a **burner reactor** is to use up or 'burn' as much uranium-235 as possible. For some reactor designs the uranium-235 content of natural uranium is increased chemically to produce **enriched uranium**, which contains up to 3% uranium-235 (compared with the naturally occurring 0.7%). The chain reactions in Equations 4.1 to 4.3 produce neutrons with a wide range of energies. Fast neutrons are slowed down using a **moderator** — graphite, ordinary water or *heavy water* (deuterium oxide — deuterium is an isotope of hydrogen whose nucleus contains a neutron as well as a proton) — so that they can cause fission of uranium-235. The reaction rate is adjusted using **control rods** made of boron that absorb neutrons. These rods can be raised or lowered into the reactor core to increase or decrease the heat output.

Figure 4.4 shows the basic design features of three typical burner reactors in use today. All have **cores** encased in steel vessels surrounded by concrete shields. Transferring heat away from the reactor core involves three stages:

1 The core coolant, either carbon dioxide gas or water depending on the reactor type, is heated by nuclear fission in the core.

2 The heat from the core coolant is transferred to a closed steam–water circuit, which includes the turbine plant.

3 The steam in the turbine circuit is condensed to water on the low-pressure side of the turbine by a cooling water circuit that is open to the environment.

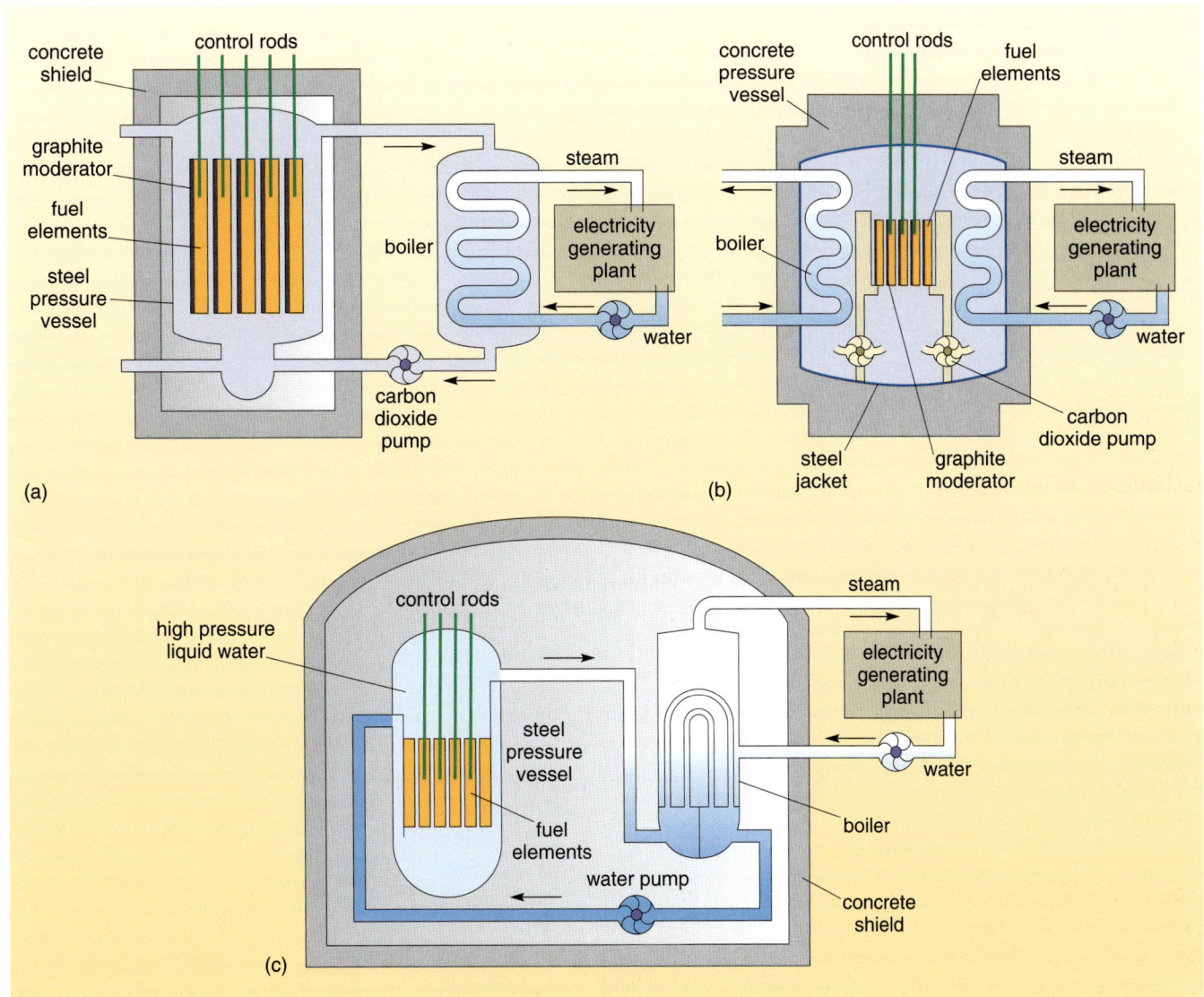

Figure 4.4 Key features of typical burner reactors: (a) Magnox; (b) advanced gas-cooled; (c) pressurized water reactors. *Note*: The water that carries heat to the steam boiler in (c), being pressurized, does not itself turn to steam (see Figure 5.4).

The requirement for large volumes of cooling water means that nuclear reactors must be sited in coastal locations or near large rivers or lakes.

The main contrasts between the three main burner reactor types are:

Magnox reactors are fuelled by metallic uranium containing the natural proportion of uranium-235 (i.e. 0.7%) held in tubes of a magnesium alloy (Magnox). The moderator is graphite, the coolant is carbon dioxide and the operating temperature is about 400 °C.

Advanced gas-cooled reactors (AGRs) are very similar to Magnox reactors (they have the same moderator and coolant), except that uranium oxide, enriched in uranium-235 (2.3% instead of 0.7%) and packed in stainless steel tubes, is used as a fuel. The operating temperature, 800 °C, is higher than that in a Magnox reactor, leading to greater conversion efficiency and output by increasing steam pressure. Although the capital cost of building AGRs is very high, they have many additional safety features over Magnox reactors and, because of their greater efficiency, produce cheaper electricity.

Pressurized water reactors (PWRs) are the most widely used globally (Section 4.1.4). The fuel contained in zirconium alloy tubes is uranium oxide enriched in uranium-235, and pressurized water is both coolant and moderator. They operate at 300–400 °C. The UK Sizewell 'B' plant is of this kind (Figure 4.1).

So, even the most enriched fuel in burner reactors still consists of 97.7% uranium-238. A very small proportion of this is converted into plutonium-239 (Equation 4.3), which is extracted from spent uranium fuel rods. But the bulk of the uranium in burner reactor fuel rods is never used. Indeed, the wastage is even greater than that: not all of the uranium-235 is 'burned' — as fission products build-up, they interfere with the chain reaction.

However, there is another way of extracting fission energy from the more abundant uranium-238.

Fast breeder reactors

If fast neutrons produced in the chain reactions are not moderated or absorbed, the rate of conversion of uranium-238 into plutonium-239 (Equation 4.3) can *exceed* the fission rate of plutonium-239. Reactors that use fast neutrons in this way are called **fast breeder reactors**.

Their main fuel is uranium-238, together with an initial charge of plutonium-239 which is needed to start the chain reaction that 'breeds' plutonium-239 from uranium-238. (This charge of plutonium-239 is obtained from the spent fuel rods of burner reactors.) The attraction of the fast-breeder concept is the excess plutonium-239 that they produce. About 60% of the fuel used in breeder reactors is converted into useful energy, compared with the 0.5–2% in burner reactors. Not only are breeder reactors more efficient, they generate more plutonium than they consume and can use uranium-238 in the waste from burner reactors.

Fast breeder reactors seem very attractive in theory, but only a handful have ever been constructed. In the UK (the Dounreay reactor), France and the United States all fast breeder programmes have been closed. Japan is considering restarting its own programme, which was closed after a fire in 1995, while India is currently building its first prototype fast breeder reactor (this is being designed to use thorium-232 as a nuclear fuel).

The dearth of fast breeder reactors testifies to formidable technological problems. The greatest of these is that the intense heat generated by the fast-neutron chain reaction requires the use of liquid sodium metal as the coolant, rather than water or CO_2 (Figure 4.5). The sodium coolant burns fiercely if any oxygen leaks in at the 600 °C operating temperature. Concerns about fast breeder reactors also centre on their fuel, for three reasons:

- the concentration of fissile material is greater than in a burner reactor — there is a greater chance of fission becoming uncontrollable;
- the necessity for reprocessing nuclear waste (discussed in Box 4.5);
- the central role (and production) of plutonium which is an essential component of nuclear weaponry.

Given this last concern, security has now become a more important issue if such material is not to fall into the hands of terrorists.

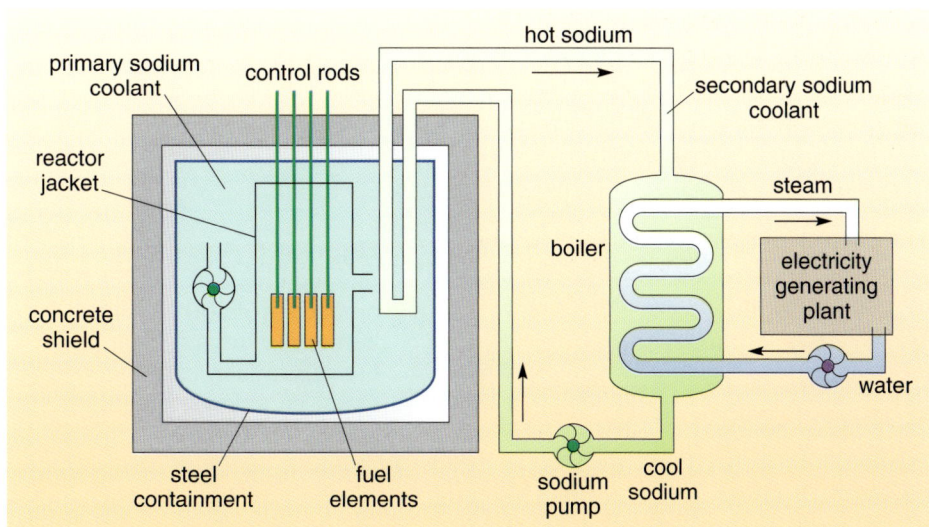

Figure 4.5 Essential features of a fast breeder reactor.

4.1.3 Fuel requirements for nuclear reactors

The fission energy of uranium-235 is 8.2×10^{13} J kg^{-1}. Setting this figure in the context of fuelling a reactor depends on:

1   *The thermal efficiency of the reactor*   For modern burner reactors, this is similar to that of modern fossil fuel power stations, about 35% (i.e. 65% of the available energy is lost within the reactor–coolant–turbine system).

2   *The power output*   A typical value for a burner reactor is 1 GW. which corresponds to an annual energy output of 3.2×10^{16} J yr^{-1}.

3   *The lifetime of the reactor*   This is about 30 years, i.e. about the same as for fossil fuel power stations.

The fuel requirements and products of burner reactors are summarized in Figure 4.6a. Fuel elements (rods) do not last for 30 years, and the total fuel requirement for a typical 1 GW burner reactor is about 4800 t of natural uranium over a 30-year period. Some 600 t of natural uranium are required as an initial fuelling charge (less if the fuel has been enriched in uranium-235), and an average of about 145 t yr^{-1} is used to replace spent fuel rods. Spent fuel rods consist of *depleted* uranium, which has lost most of its uranium-235 component and is mainly composed of uranium-238. About 2–3% comprises fission products from uranium-235, and a further 1% is plutonium-239. These of course have to be disposed of somehow (Section 4.3.2). Depending on a 1 GW burner reactor's characteristics, up to 50 t of plutonium-239 will have been produced after 30 years. Yet both uranium-238 and plutonium-239 are potential fuels for a fast breeder reactor.

The fact that less than 5000 t of natural uranium (containing 0.7% uranium-235) is needed to fuel a 1 GW burner reactor over a 30-year lifetime shows the much higher 'energy density' of uranium relative to fossil fuels. A modern coal-fired power station of comparable output needs to burn something like 10 000 t of coal *per day* (Figure 4.7). The total volume of waste to be disposed of during the life of a power station is thus also much less for nuclear than for coal, though it is far more hazardous.

The comparison between fast breeder reactors and fossil fuels is yet more favourable, in terms of the required masses of fuel (Figure 4.6b). The fission

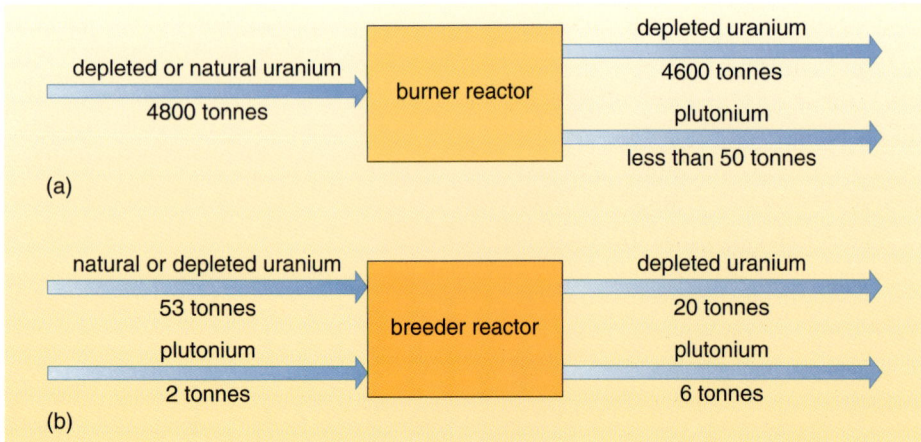

Figure 4.6 A summary of the fuel requirements and spent fuel products from (a) a typical burner reactor and (b) a fast breeder reactor, based, in both cases, on a power output of 1 GW for 30 years.

energy of uranium-238 is about the same as that of uranium-235, but a breeder reactor uses about 60% of the fuel. Operating at 35% efficiency, a 1 GW fast breeder reactor would need 53 t of uranium-238 during its 30-year lifetime (Figure 4.6b), about 100 times less than the fuel requirement of an equivalent burner reactor, and potentially available from burner reactors' spent fuel. In 30 years, a 1 GW burner reactor produces enough uranium-238 in spent fuel rods to fuel around 90 breeder reactors with the same capacity. It also produces enough plutonium-239 (Equation 4.3) to charge 25 of them with the 2 t needed to start their nuclear chain reaction. Each breeder would also produce enough to charge three more with plutonium. So, in theory at least, a global breeder reactor programme could generate much more electricity than do existing burner reactors by using the stockpile of waste from them, without the need to mine uranium ore until some time in the future. Unlike burner reactors, which require continual refuelling during their active life, a successful breeder reactor requires little fuelling after its initial charge. Figure 4.7 illustrates the annual supply of mined energy resources required to service 1 GW power stations using coal, burner and breeder technologies.

Figure 4.7 Visualization of the annual coal and uranium ore requirements for different kinds of 1 GW power station: (top) coal-fired — 3.7×10^7 t of coal; (middle) nuclear fission, burner reactor — 5.6×10^4 t of uranium ore with an average grade of 0.3% uranium; (bottom) nuclear fission, breeder reactor — theoretically no requirements after the initial charge.

4.1.4 The growth, decline and future of nuclear power

The Calder Hall Magnox reactor near Sellafield fired the UK's first commercial nuclear power station in 1956, and launched an early UK lead in global nuclear developments. By 1960 six commercial reactors were operating, and Magnox technology had been exported to Italy and Japan. The UK Magnox building programme was complete in 1971 with eleven stations, each producing between 245 MW and 840 MW. Figure 4.8 shows the distribution of different reactor types in the UK.

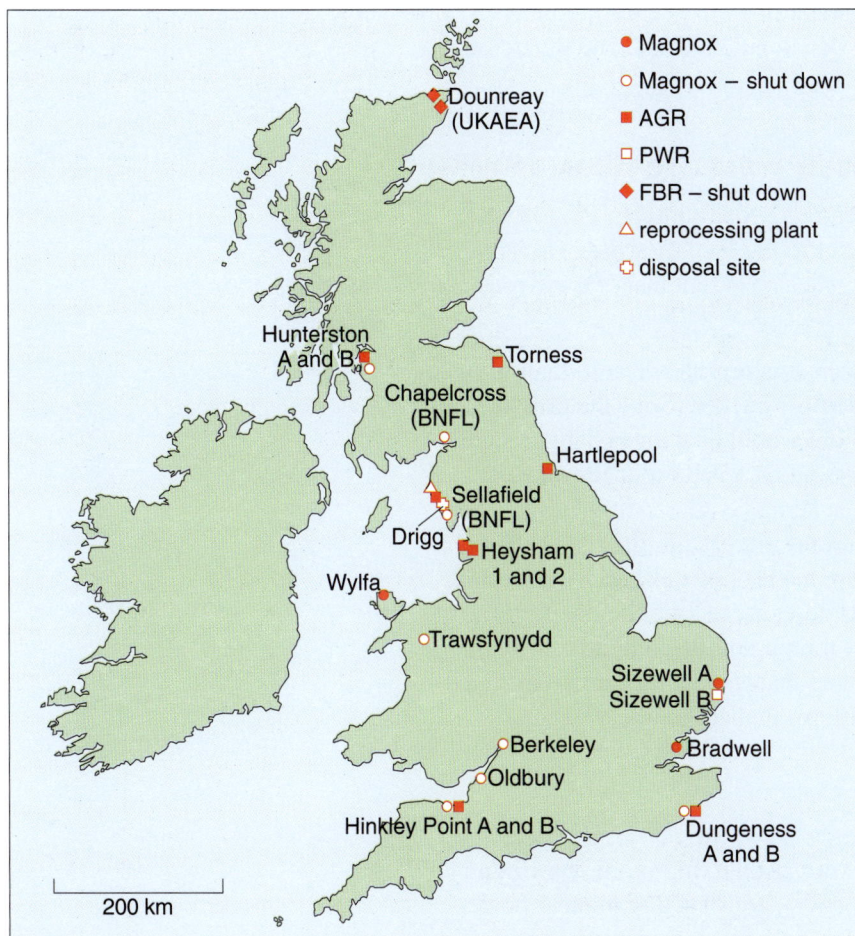

Figure 4.8 Locations of the most important nuclear sites and power stations in the UK as of 2005. (Reprocessing of spent nuclear fuels and waste disposal are considered in Section 4.3.2.)

Advanced gas-cooled reactors (AGR) were designed to use enriched uranium fuel and higher operating temperatures, thus increasing power output efficiency from 30% to a maximum of 40%. After a long history of technical difficulties and delays, eight 1 GW AGR reactors, originally designed and planned in the early 1970s, were still in service in the UK during the early 21st century.

However, US-designed PWRs became by far the most attractive reactor design for nuclear power plants in other countries. Globally some 240 PWRs are operating successfully, including Sizewell B in the UK, leaving only the UK with AGR stations. But PWRs have not been without their problems (see Box 4.2). Despite their proliferation, concerns over the operational safety of PWRs dramatically slowed the global growth rate of nuclear power during the 1980s. Utility companies in the US had stopped ordering PWRs by 1978, even before the Three Mile Island incident, because of concern about the relative long-term economics of nuclear and fossil fuel stations. Similar worries in Europe were exacerbated by the catastrophic reactor accident in 1986 at Chernobyl in Ukraine (Section 4.3.1), and by 1990 global nuclear developments had virtually come to a halt everywhere except in France and Japan.

Box 4.2 Safety and pressurized water reactors (PWRs)

In the early 1970s, about 1% of manufactured zirconium-clad PWR fuel rods leaked, releasing fission product gases to the cooling water. However, by 1992 fuel fabrication had improved to such an extent that leaking rods were extremely rare.

In PWRs, fuel rods are packed into a much smaller reactor core than in AGRs: the fuel is about 40 times more densely packed. So if the pressurized cooling system were to fail, the reactor core would heat up to danger level much more rapidly than in an AGR. Fears about PWRs reached a peak following the Three Mile Island incident in Pennsylvania during 1979, where safety mechanisms failed to prevent a serious increase in the temperature and pressure of the reactor core, resulting in a leak of radioactive materials. The reactor developed a gas bubble in its core, increasing the danger of a further rise in temperature and ultimately melting of the containment vessel (Figure 4.4c). The accident report indicated that the damage arose from a combination of technical and operator failures that are unlikely to be repeated, rather than from fundamental design weaknesses, yet there is still considerable opposition to PWRs.

Concerns about the cost of nuclear power were also an important contributory factor in its decline; £15 billion had been spent on constructing nuclear power stations in the UK alone, and the price for decommissioning reactors and managing waste is still unknown. The sheer scale of the decommissioning problem can be illustrated by the nuclear industry's own argument that 'costs will be considerably reduced' if the closure and decommissioning timetable were spread over 135 years instead of 100 years, and if reactor cores and pressure vessels were not dismantled but buried instead. This is what was done at the Berkeley Magnox station in Gloucestershire (Figure 4.8), which closed in 1989. By 1998 the UK government confirmed these economic doubts with the view that 'at present nuclear power is too expensive to be economic for new capacity and in current circumstances it is unlikely that new proposals for building nuclear plants will come forward from commercial providers'. However, policies change with circumstances. Nuclear reactors provide a means of generating electricity without contributing directly to CO_2 emissions, so their increased use is one possible response to combat climate change.

By the mid-1990s the future of nuclear power in the US and most of Western Europe (France excluded) was economically and politically uncertain, even though the industry itself was strong scientifically and its technology still improving. On the other hand, in France, Japan and other countries of the Pacific Rim (Table 4.1), brighter economic prospects were matched by a favourable political situation. In the former Soviet Republics and Eastern Europe, where scientific and technical developments had lagged somewhat behind those in the West, the economic and political picture was less clear. But nuclear power is potentially a very lucrative market, and by 1993 nuclear industries in several Western countries, faced with declining business at home, had begun competing vigorously to acquire a share of business in more favourable areas of the world.

Because of these uncertainties, forecasts of future requirements for nuclear-generated electricity must be viewed with caution. The International Atomic Energy Agency (IAEA) suggested in 2003 that nuclear generating capacity globally will have grown from 362 GW in 2003 to about 400 GW by 2010, an approximate 11% increase in capacity. The IAEA view takes into account decommissioning of old power stations at the end of their lifetime. Notwithstanding such growth, the proportion of nuclear power against other energy sources is likely to remain static. With anticipated growth in global energy demand, IAEA predictions estimate a rise in total nuclear generating capacity to between 427 and 512 GW by 2020.

Whether these estimates are met, or even exceeded, will depend on factors such as:

- the extent to which global demand for electricity continues to grow;
- public perception of the dangers and costs of nuclear power generation;
- whether governments perceive the threat of global warming as real enough to warrant serious attempts at curbing greenhouse gas emissions from fossil fuels;
- the extent to which alternative energy sources can meet demand.

In any event, nuclear fuels will continue to be needed. After Question 4.2, we turn to how they can be supplied.

Question 4.2

(a) In burner reactors, a moderator is used to slow down the neutrons and the position of the control rods is continually adjusted to absorb excess neutrons. Why it is necessary (i) to slow down neutrons and (ii) to absorb the excess?

(b) Figure 4.5 shows that there is no moderator in a fast breeder reactor. Why is this?

(c) All uranium-based reactors breed plutonium but some do so faster than others: why is this?

(d) Why does the uranium in ore deposits not all disappear by spontaneous fission?

4.2 The geological occurrence and extraction of uranium

Just how readily available are uranium resources, and do their distribution and cost impose restrictions on nuclear power generation? Compared to a coal-fired power station a nuclear power station requires far less fuel in terms of mass. You have seen that a 1 GW burner reactor requires 5000 t of natural uranium over 30 years, whereas a comparable modern coal-fired power station needs 10 000 t of coal every day. However, uranium does not occur naturally in metallic form, nor in the concentrations required for reactor fuel. Its average abundance is only about 3 parts per million (ppm) by mass in the continental crust — equivalent to just 0.4 cm^3 of pure uranium dispersed evenly through a whole cubic metre of granitic rock. Uranium occurs at concentrations up to 100–10 000 ppm in minor minerals in granites, such as zircon ($ZrSiO_4$), apatite ($Ca_5(PO_4)_3(OH)$) and titanite ($CaTiSiO_5$). For mining and extraction of uranium to be economic, it must occur at much higher abundances than this.

⬤ A typical ore grade for uranium is about 0.3% or 3000 ppm in rock. By how much must uranium be concentrated above its average abundance in continental crust to form an ore of this grade?

⬤ $$\frac{\text{typical ore grade}}{\text{average crustal abundance}} = \frac{3000 \text{ ppm}}{3 \text{ ppm}} = 1000 \text{ times}$$

The amount of an ore containing 0.3% uranium that must be mined to produce fuel for the 30-year lifetime of a 1 GW burner reactor is 1.7 million tonnes (i.e. 5000 t/0.003 = 1.7×10^6 t). Nevertheless, even that huge quantity is only half the mass of a year's supply of coal to a modern power station.

4.2.1 Uranium occurrence and ore deposits

In igneous rocks, uranium is more abundant in granites (~3.5 ppm) than in basalts (~1 ppm). The large size of the uranium atom prevents it from easily entering the structures of common rock-forming minerals, so it is an *incompatible* element that tends to remain in magmas until a late stage of crystallization, when it enters minor minerals, or even the uranium oxide, uraninite (UO_2). In suitable circumstances, following fractional crystallization of uranium-rich granitic magma, uraninite may be sufficiently abundant in pegmatite sheets and veins to form a **disseminated magmatic uranium deposit**. Alternatively, when rocks rich in uranium become hot enough to start melting, uranium, being an incompatible element, preferentially enters the melt. If only a small amount of melting occurs, and if such melts crystallize as pegmatite veins before further melting takes place, they may be particularly rich in uranium. One such occurrence is at Rossing in Namibia, where a relatively low grade of ore (~0.03% uranium) has nevertheless been economic to mine because of the huge size of the deposit and the advantages associated with economies of scale. It is said that a year's production of uranium from Rossing could meet the UK's entire power needs for eight years.

Richer occurrences of uranium result from its solubility in water, in particular, hydrothermal fluids — hot, usually saline groundwater that circulates within the

Earth's crust. Hydrothermal fluids may circulate through many cubic kilometres of crust for thousands of years, being driven by heat from deep within the crust. The hot water rises through joints, fractures and permeable rocks, drawing in cooler water as part of a convection system. Hydrothermal fluids may dissolve certain components of rocks in one environment only to deposit them in another, as a result of chemical processes within the fluid or its reaction with surrounding rocks. For example, hydrothermal fluids that flow through a granite may dissolve dispersed uranium, migrate until a suitable change in solution chemistry occurs and then deposit and concentrate uranium over a long period. The key to understanding the behaviour of uranium in aqueous solutions is that uranium in its oxidized U(VI) state forms soluble ions (e.g. uranyl, UO_2^{2+}), but in its reduced U(IV) state it tends to be insoluble, when it precipitates uranium minerals, such as uraninite, in **hydrothermal vein deposits**.

Some of the world's richest uranium occurrences were formed by hydrothermal processes. Many occur at or below unconformities that are cut by faults (Figure 4.9), as in the Athabasca Basin of Saskatchewan, Canada, where numerous **unconformity related uranium deposits** have been discovered. These include the MacArthur River (~24% uranium) and Cigar Lake (~19% uranium) deposits, which are probably the richest in the world. Whether the source of the uranium in such settings is fractured metamorphic basement rocks beneath the unconformity or as a result of 'leaking' from the overlying sediments down faults is uncertain. Whichever, the uranium deposition was clearly associated with a change in oxidation state of the fluid, when large quantities of oxidizing fluids containing soluble uranium ions became reduced and precipitated uraninite.

Similar oxidation–reduction reactions are responsible for uranium deposits in quite different geological settings in the Wyoming, South Dakota and Colorado Plateau areas of the western US. These are **sandstone-hosted uranium deposits**. One form of these is the 'roll-front' deposit, which typically comprises crescent-shaped zones of uranium ore minerals (Figure 4.10) in sandstones. They formed where oxidizing groundwater containing soluble uranium became reduced by carbonaceous fossil plant material in the permeable sandstone. Locally, whole tree trunks have been replaced by the mineralizing solutions. The uranium in the groundwater is derived by weathering of uranium-bearing rocks, such as granites, in upland areas. It is then transported as dissolved ions in oxygenated surface water, which infiltrates and migrates through permeable sandstone strata. Where these oxidizing waters come into contact with carbonaceous material they are reduced, and deposition of insoluble uranium minerals (along with vanadium and copper) starts.

Figure 4.9 Cross-section of an unconformity related deposit. Mineralization has occurred where oxidized uranium-bearing fluids have been reduced near a major unconformity between metamorphic basement below and sedimentary rocks above, which was faulted after deposition of the sediments.

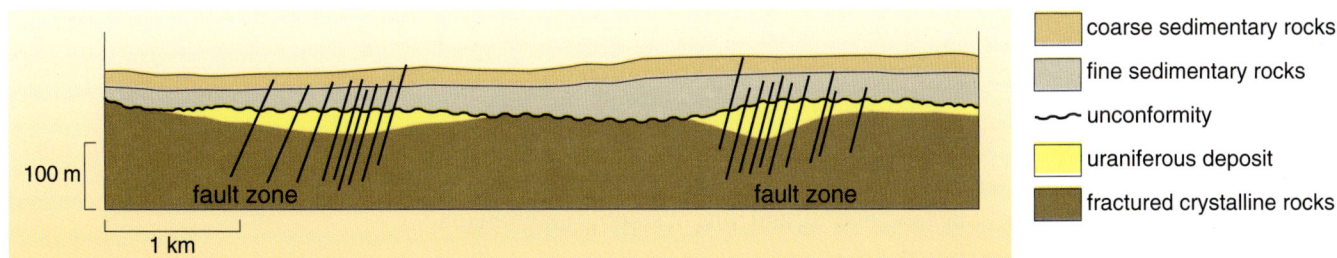

coarse sedimentary rocks
fine sedimentary rocks
unconformity
uraniferous deposit
fractured crystalline rocks

100 m

fault zone

fault zone

1 km

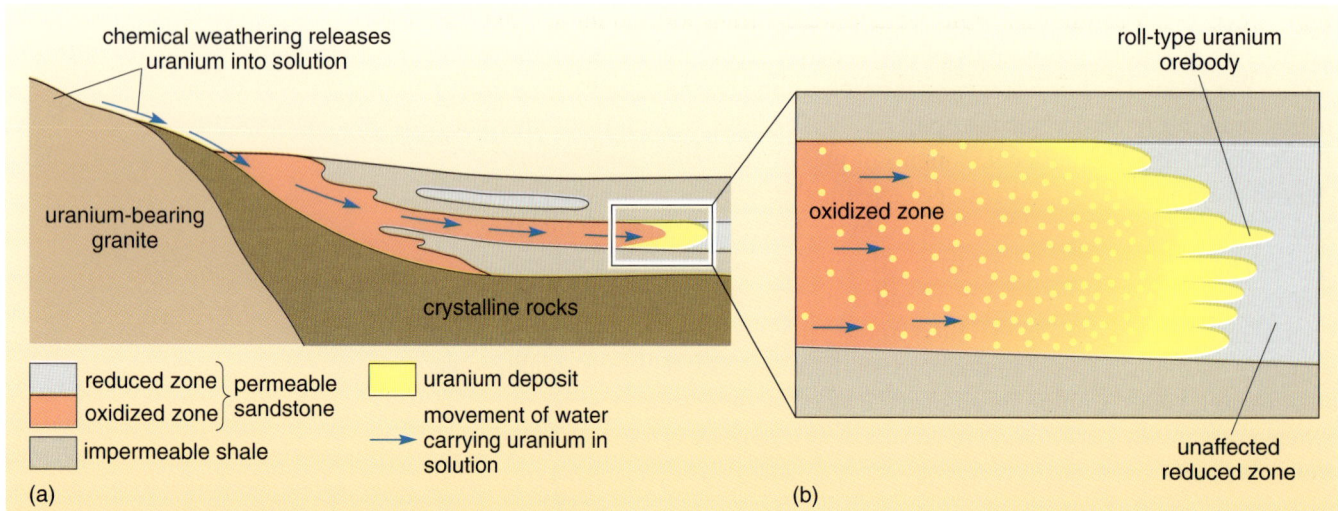

(a)   (b)

(c)

Figure 4.10 (a) A cross-section showing the setting for roll-front type uranium concentration and the relationship between mineralization and the oxidation state of the fluids in the sandstone hosting the ore. (b) Roll-type uranium orebody produced by groundwater flow (from left to right) in sandstone containing organic matter. (c) Roll-front uranium ore deposit.

Continued flow of oxidizing groundwater dissolves uranium minerals from the upflow side of the deposit and redeposits uranium under reducing conditions on the downflow side, thus developing the arcuate form of the deposit (Figure 4.10c).

Some important uranium deposits consist of sedimentary grains of uranium-bearing minerals in **quartz-pebble conglomerates**, such as those of the Witwatersrand Basin in South Africa and Blind River, Elliott Lake in southern Canada. These conglomerates are unusual. Not only do they contain rounded grains of uranium-bearing minerals, principally uraninite, that were deposited in river gravels in exactly the same form as when they were originally eroded, but also sedimentary grains of gold and pyrite.

● What is unusual about the occurrence of sedimentary uraninite and pyrite grains?

○ Oxygenated surface waters tend to dissolve uranium: weathering today rapidly breaks down uraninite (and pyrite), and ions of soluble uranium (and iron) are carried away in solution.

● Under what conditions could surface waters transport uraninite grains?

○ Only if the surface water and the Earth's atmosphere were non-oxidizing.

These conglomerates were deposited more than 2000 million years ago, and there is abundant geological evidence for the Earth's atmosphere containing very low oxygen until about that time. Consequently, uranium minerals were not broken down by weathering as readily as they are today. So, rapidly flowing rivers derived from granitic rocks could have transported and deposited high-density uraninite, to concentrate it in sands and gravels where river energy slackened. Many of these deposits contain very low concentrations of uraninite but are economic to mine because of the gold they contain.

There are many other different types of uranium mineralization in addition to those mentioned here, so it is no surprise to learn that uranium ore deposits are well distributed globally (Figure 4.11).

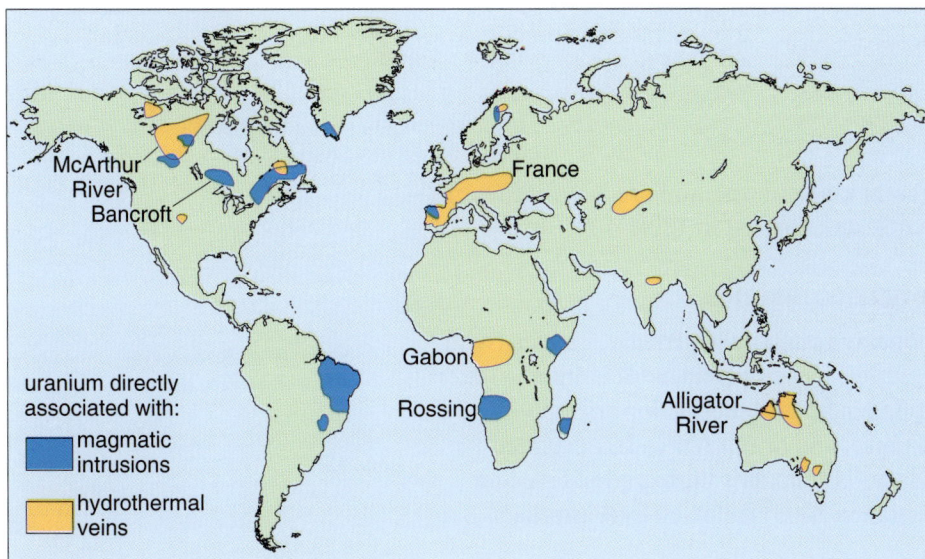

(a)

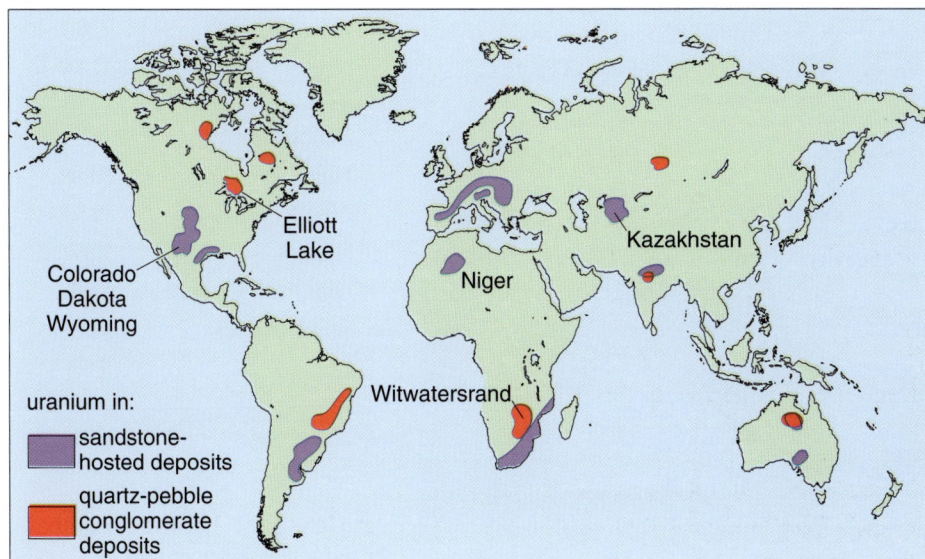

(b)

Figure 4.11 Globally important areas that are enriched in uranium. (a) Uranium directly associated with magmatic intrusions (blue) and with hydrothermal veins (orange). (b) Uranium in sandstone-hosted deposits (purple) and quartz-pebble conglomerate deposits (red).

Table 4.2 The main categories of uranium ore deposits.

Type	Uranium grade/% uranium	Typical deposit	Largest known deposits/tonnes uranium	Percentage of global resources
disseminated magmatic	0.03–0.1	Bancroft, Ontario, Canada; Rossing, Namibia	100 000	unknown
unconformity-related	0.3–25	Athabasca Basin, Saskatchewan, Canada; Alligator River, Northern Territory, Australia	200 000	33
sandstone-hosted	0.05–0.4	Colorado Plateau, Wyoming and South Dakota, USA; Niger; Kazakhstan	40 000	18
quartz-pebble conglomerate (detrital)	0.01–0.12	Witwatersrand, South Africa; Elliott Lake, Ontario, Canada	150 000	13

The relative importance of these different kinds of uranium ore deposit in terms of the typical grade and size of deposit is summarized in Table 4.2.

4.2.2 Uranium production and economics

Table 4.3 lists the major uranium-producing countries. Currently, Canada (with 29% of global supply in 2003) is the world's largest producer of uranium, followed by Australia (21%), both having increased production since about 1980, whereas production from the USA, France, and South Africa has declined (Figure 4.12). The largest single producer is Canada's high grade McArthur River deposit (16.3% of global production in 2003) followed by Australia's Ranger mine (12%).

Table 4.3 Global production of uranium, 2003.

	Production/tonnes
Canada	10 457
Australia	7572
Kazakhstan	3300
Niger	3143
Russian Federation*	3150
Namibia	2036
Uzbekistan	1770
USA	857
Ukraine*	800
South Africa	758
China*	750

* Estimated production.

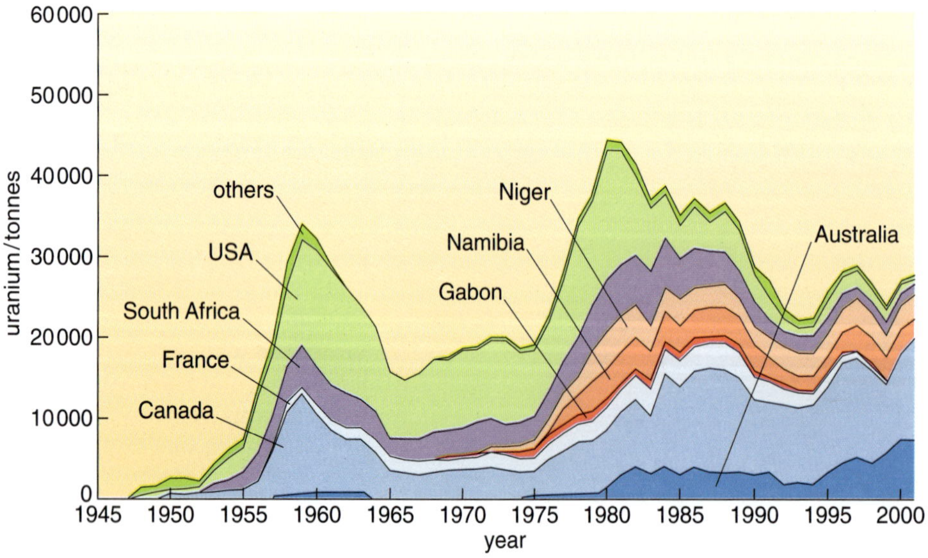

Figure 4.12 Long-term global production of uranium.

Uranium is mined at the surface (mostly lower grade, sandstone-hosted and disseminated deposits), in deep underground mines (mostly higher grade deposits, especially unconformity-related) and by *in situ* **leaching** (**ISL**) or solution mining (see Box 4.3). In 2003 about 50% of global production was mined underground, 30% at the surface and 20% by ISL. These proportions vary depending on the economics of mining and the types of deposit available for exploitation. In recent years some particularly high-grade deposits have been discovered in Canada, e.g. McArthur River and Cigar Lake, where underground mining uses remote methods which help minimize hazards for workers. Elsewhere, it is still economic to mine very low-grade deposits at the surface (e.g. Rossing) or by ISL (e.g. Inkai, Kazakhstan).

Where ore is extracted by mining, it is first crushed and much of the worthless rock separated, then the uranium is leached into solution. This process leaves a massive amount of waste rock which must be disposed of. The solution is then

Box 4.3 *In situ* leaching (ISL)

In recent years there has been a growing trend to extract uranium by ISL. It can be used when uranium deposits are below ground in a layer of permeable rock that is confined by impermeable layers (Figure 4.13). The procedure involves pumping a liquid (such as sulphuric acid) that dissolves (leaches out) uranium down boreholes into the deposit and extracting the resultant uranium-bearing solution (Figure 4.13), which is then processed to extract the uranium. ISL has several advantages over conventional mining:

- it is less hazardous for employees;
- it is cheaper;
- there is no solid waste for disposal;

but it also has disadvantages:

- there is a risk of groundwater contamination by leaching solutions;
- there is a need to store waste sludge and water after uranium recovery;
- it is not possible to restore natural conditions to the leaching zone after operations have finished.

ISL has been used extensively in the USA (93% of US uranium production in 1996) and in eastern Europe, the Russian Federation and Australia. It is a cheap method of mining, appropriate for low-grade ores hosted in permeable rocks, particularly sandstone-hosted deposits.

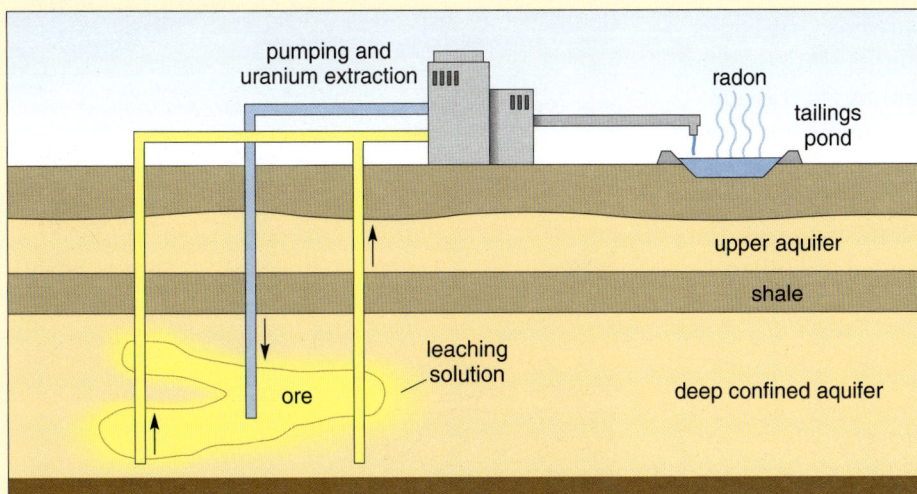

Figure 4.13 Schematic diagram of uranium extraction by *in situ* leaching.

chemically processed by passing it through ion-exchange columns that selectively extract uranium, then re-dissolving the uranium in another solution under different chemical conditions. This purified solution is treated with ammonia, which precipitates uranium as an ammonium uranyl salt. This compound is dried and then heated in air to break it down to an impure form of uranium trioxide known as 'yellowcake' (Figure 4.14).

Solid waste from either underground or surface mines (generally called *tailings*) always contains traces of the minerals being worked, because mineral processing never achieves complete separation. The amount of waste produced by conventional mining of low-grade uranium ores is prodigious. Processing an economically profitable ore containing 0.5% uranium to yield 1 t of uranium creates about 200 t of waste, which contain traces of uranium and possibly other radioactive and toxic elements. Although tailings may contain only 5–10% of the original uranium in the ore, up to 85% of the radioactivity emitted by this waste is from the radioactive gas, radon (see Box 4.4). So, mining and processing uranium ore transfers material from a relatively safe underground location to deposit it at the surface, where hazardous constituents are more readily dispersed in the environment.

As with any other form of mineral extraction (Webb, 2006), processing uranium ore requires massive amounts of water. Eventually, this water carries generally fine wastes from ore crushing and separation operations as slurry to settling ponds (called tailing ponds; Figure 4.15). Contamination from tailings can spread over large areas as wind-blown dust, slurry can be eroded and transported by water, and water can seep into groundwater or surface streams: uranium is highly soluble under surface oxidizing conditions. After the collapse of the Soviet Union and former Eastern Bloc countries, it emerged that uranium mining areas in these countries had higher than normal incidence of cancers, particularly lung cancer and childhood leukaemia, as well as various respiratory diseases. Lax safety standards are thought to have led to similar

Figure 4.14 The end product of uranium mining, uranium trioxide or 'yellowcake' after drying.

(a)

(b)

Figure 4.15 (a) Waste from uranium ore processing is fed into a tailings pond where solids settle out and water evaporates. A heavy-duty plastic liner prevents seepage of potentially hazardous water. (b) The open pit at the top left in (a) after mining had finished, showing regrowth of local vegetation around the lake.

Box 4.4 Radon hazards

Radon is a highly radioactive gas produced from the uranium decay chain. Although radon-222 has a short half-life of only 3.8 days, it disintegrates to form alpha particles and *solid* radioactive daughter products. Once inhaled these radioactive particles can lodge permanently in the lungs, causing damage to lung tissue and long-term disease. Risks to miners are now minimized by the use of powerful ventilation but they are not eliminated.

Radon is not only a hazard to miners. Release of mining waste into the environment creates a long-term source of radon, as its immediate parent, radium-226, has a half-life of 1600 years and its parent in turn, thorium-230, has a half-life of 80 000 years. Although ventilation reduces the hazard locally, as in buildings where radon may collect having emerged from uranium-bearing rocks beneath their foundations, the spread of radon is a hazard to the population at large. Although the risk from small doses of radon may be very small, the effect over time on a large population can be significant.

Radon is emitted by uranium decay from *any* rocks, particularly those that contain higher than normal uranium concentrations, i.e. some granites, limestones and shales, and even coals. Housing with foundations on even moderately uranium-rich rocks can accumulate radon in their foundations or cellars. UK building regulations now require assessment of radon risk from this source, and protective measures for any new constructions in radon-prone areas.

---

consequences in other uranium mining areas (e.g. Australia, India, Namibia), but this is an emotive issue and allegations are not easy to prove. Nowadays, compliance with legal requirements has ensured that more mining companies have safety controls in place. Long-term isolation of wastes is a priority, and sites are engineered in a similar fashion to landfill sites (Argles, 2005), with impermeable liners at the base and as a cover. In the USA uranium mining wastes have even been relocated at great expense to make them safe.

Underground mining is a high-cost operation, as shafts have to be sunk and tunnels drilled in order to access ore and install equipment below ground, and the ore has to be raised to the surface. Surface mining in contrast is a lower cost operation, with the ore directly accessible and often mined by high capacity equipment, which gains enormous advantages from economies of scale. With significant cost differentials you might wonder why underground uranium mines continue to operate.

● What single factor can make some underground mines more profitable than surface mines?

○ Ore grade. A higher concentration of an element in ore requires less ore to be mined and processed to produce the same quantity of the saleable commodity. This offsets the high costs of underground mining.

So underground mining is generally only viable for high-grade ores, but open-pit mining can be used for low-grade ores that are accessible at the surface. Some low-grade ores, in suitable locations and in suitable rocks, can be extracted underground by ISL (Box 4.3) at significantly lower cost.

A minimum economic grade of 0.3% uranium is usual for open-pit mining, but there are exceptions. The Rossing mine in Namibia continues to operate despite a grade of only 0.03%. That mining continues there partly reflects economies of scale associated with the huge size of the deposit. Because the infrastructure is already in place from a period of very high prices for uranium, the true cost of continued operation is much reduced. However, long-term contracts for supply from the Rossing mine were agreed when prices were higher than US$25 lb^{-1}. Note that the uranium market quotes uranium prices quaintly in US$ lb^{-1} for 'yellowcake' (U_3O_8). At prices lower than US$25 lb^{-1}, few low-grade open-pit mines are viable.

Figure 4.16 gives uranium prices, nuclear reactor requirements and production statistics for almost six decades, during which attitudes to nuclear power have swung from one extreme to the other. Uranium provides an excellent example of how prices and markets influence resource industries, in ways that have nothing to do with geology.

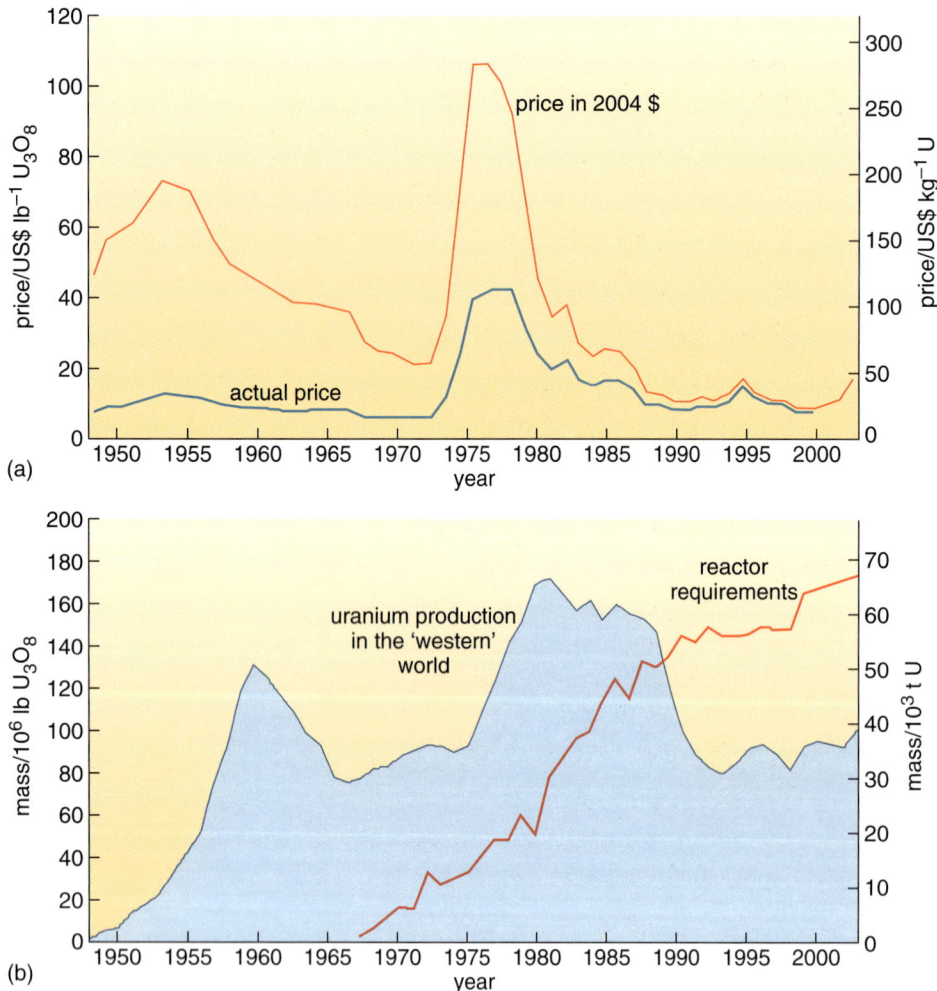

Figure 4.16 Changes in uranium mining economics from 1947 to 2004. (a) Uranium price (actual and that based on 2004 US$). Note that uranium prices are commercially quoted quaintly in US$ lb^{-1} for 'yellowcake' (U_3O_8). (Cost in US$ kg^{-1} U = 2.60 × cost in US$ lb^{-1} U_3O_8.). (b) Uranium mine production and nuclear reactor requirements.

Prior to the early 1970s the US government was the dominant customer for uranium in the Western World (partly for nuclear weapons), and effectively controlled prices. As demand for nuclear power increased during the 1970s, commercial demand for uranium grew. At the same time US government policy created an artificially high demand. Prices rose to a peak of around US$40 lb^{-1} in the late 1970s (equivalent, in terms of the decreased buying power of the US dollar, to US$105 lb^{-1} in 2004). That stimulated exploration and enabled exploitation of relatively low ore grades. Although demand continues to rise, prices and production dropped steadily through the 1980s and into the 1990s when prices flattened to around US$10–15 lb^{-1}. There were a number of reasons for this seemingly extraordinary situation. The US government first relaxed its contractual policies, then, with the Cold War coming to an end, stockpiles of uranium for military use were no longer needed and were released to the market. In addition, growth of nuclear power installations had slowed as environmental concerns grew. Thus demand for mined uranium fell markedly whilst military and other stockpiles were reduced.

Globally, some 440 nuclear reactors produced about 350 GW of electricity in 2003 (Table 4.1). This required an annual supply of about 67 000 t of uranium. Mined uranium amounts to only about 36 000 t globally. The remainder comes from recycling spent fuel, from stockpiles created while prices were low and from the use of ex-military weapons-grade enriched uranium (one tonne of which makes at least 30 t of reactor-grade uranium). Reduction of stockpiles and restrictions on the amount of military uranium released will eventually result in a resurgence of mined uranium to satisfy demand. Prices of uranium during 2003 started to rise in response to this (Figure 4.16a). Renewed interest in nuclear power as a means to check carbon dioxide emissions will undoubtedly spur further rises in the price of uranium. By mid-2005 it had risen to over US$29 lb^{-1}, but still much lower than in the early 1980s, taking into account the decreasing buying power of the dollar. If there is growth in nuclear reactor development, demand for uranium may well increase further, further driving up the world price. Price rises may eventually make more marginal deposits economic again. However, satisfying a rise in demand would not be immediate: lead times for any mine development can be as much as 15 years or more because of protracted planning for mine development and compliance with environmental legislation.

Is there enough uranium to satisfy likely future demands? Quantities believed to exist in an economically recoverable form are listed in Table 4.4.

Table 4.4 Uranium reserves estimated in 2003 by the International Atomic Energy Agency (IAEA) at a price of US$30 lb^{-1} U_3O_8. *Note*: Strictly speaking, the amounts are 'reasonably assured resources'.

	Reserves/10^3 t uranium	Proportion of global reserves/%
Australia	989	28
Kazakhstan	622	18
Canada	439	12
South Africa	298	8
Namibia	213	6
Global total	3537	100

What is the lifetime of the global reserves shown in Table 4.4, i.e. their R/P ratio (Section 2.4.5), assuming a 2003 production of 3.6×10^4 t?

Reserves/annual production = 3.5×10^6 t/3.6×10^4 t yr^{-1} = 98 years.

Global reserves would thus last for almost a century at constant 2003 production. However, if the whole of the current reactor demand for uranium (6.7×10^4 t yr^{-1} in 2003) had to be mined, the reserves would last only half that time. For the foreseeable future there are substantial reserves of uranium to supply global nuclear power needs, even if demand rose considerably. If prices rose, even greater quantities would be available, as this would encourage exploration. The IAEA has estimated that there might be as much as 14.4×10^6 t uranium in conventional resources (Table 4.2), enough for 200 years. If fast-breeder technology (Section 4.1.2) became commercially viable, however, requirements for newly mined uranium would actually decrease: fast breeders use mostly spent fuel from burner reactors, and the world is awash with this otherwise waste material.

But all this is highly speculative. The future is uncertain because every country that produces and/or consumes nuclear fuels will make its own political judgements about the competing technical, economic and environmental factors that affect nuclear power generation.

Question 4.3

Look at Figure 4.16 and answer the following:

(a) Significant uranium production began in the 1950s, yet requirements for power didn't take off until the 1970s. How do you explain this apparent disparity?

(b) How long was the time lag between price rises and increased production during the mid- to late-1970s and how do you explain it? *Note*: Refer to the price curve in terms of US$ in 2004.

(c) What was the reason for the fall in price in the 1980s?

(d) Explain the rise in price in the early 2000s.

Question 4.4

Decide, giving reasons, whether each of the following statements about uranium ore deposits and their formation is true or false.

(a) The charge and size of uranium ions are such that uranium does not fit readily into the crystal structures of common rock-forming minerals.

(b) Mineralized fault zones, particularly those at and beneath erosional unconformities, provide some of the world's richest uranium deposits.

(c) Uranium deposits in near-surface sandstones require higher grades for economic mining than underground vein-type ores because the former are more expensive to mine than the latter.

(d) Roll-front type uranium ores in sandstones form at an oxidation–reduction boundary where uranium carried in solution in groundwater is precipitated on conversion into insoluble uranium minerals.

(e) Yellowcake is an ammonium uranyl salt produced after ion-exchange treatment to purify the leachates of uranium ores.

(f) Uranium minerals may be precipitated from surface waters in the anaerobic reducing environments of lagoons and coastal swamps, so that coal (Section 2.1.3) and petroleum source rocks (Section 3.2.1) have high uranium contents.

4.3 Side-effects of the nuclear power industry

Nuclear power generation involves concentrated fissionable fuels which, after fission, leave significant quantities of fission-product isotopes, some of which are highly radioactive. Much of the criticism levelled against the industry falls under four main headings to which we have alluded in preceding sections:

1   the operational safety of nuclear reactors;

2   the biological effects of abnormal radiation levels arising from fuel transport, processing and reprocessing;

3   disposal of radioactive waste;

4   the increased potential for proliferation of nuclear weapons.

The implications of item 4 are beyond the scope of this book, but you should note that so much uranium and plutonium is easily available nowadays that those wishing to purchase materials for nuclear weapons would have little difficulty in doing so. For example, the smuggling of plutonium or enriched uranium from the former Soviet Union became public knowledge in the mid-1990s. Only small amounts are required to produce the high concentrations of uranium-235 or plutonium-239 needed for a nuclear fission bomb, and it is almost impossible to guarantee effective international control over nuclear weapons — despite the existence of the UN's Non-Proliferation Treaty. Such was the concern, that the development of nuclear technologies in Iraq was cited by UK and US Administrations as one reason for going to war in 2003 (although no evidence for such activity was uncovered). With Iran resuming its nuclear development programme and North Korea engaged in similar activity, diplomatic tensions between these countries and the US seemed likely to increase at the time of writing (late 2005).

4.3.1 Reactor safety: the Chernobyl incident

By far the worst nuclear reactor accident took place on 26 April 1986 when one of four 1 GW reactors at Chernobyl in the Ukraine released a radioactive cloud over Europe (Figure 4.17). The build-up to this accident has been related to a series of complex chemical reactions induced by operator errors during preparation for tests on the reactor — a kind of PWR but with a graphite moderator, normally operating at 700 °C. It appears that the tests required operation at 20–25% of full power but that as the power was being reduced, an order was issued to delay the tests because of unexpected electricity demand. This series of events led to the control rods being raised, leaving in the reactor core less than the minimum number specified in the operating instructions. Once the test did proceed, the reactor was extremely difficult to control in this state and it should have shut down automatically. But again the operators intervened, retaining manual control, and as the power output fell, the reactor core started to heat up rapidly. Once it was realized that an accident was imminent, the control rods could not be replaced fast enough. One opinion is that instability in the riverine sediments beneath the reactor complex had led to structural distortion in the plant and it was this that led to the control rods becoming jammed. Both the temperature and power output of the plant rose critically, leading to a massive explosion as molten uranium reacted with the cooling water. The explosion dislodged the 2000-tonne reactor cap, a fire started that took ten days to bring

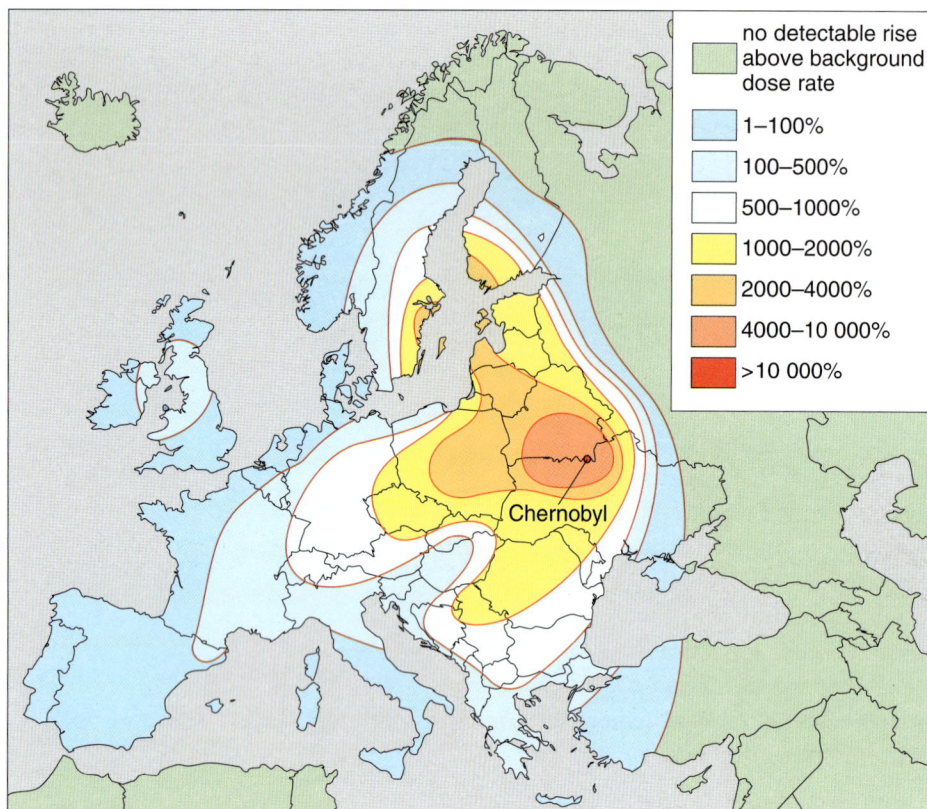

Figure 4.17 The dispersion of radioactivity across Europe on 3 May 1986, one week after the Chernobyl accident. Contoured areas represent percentage increases over background radiation dose rates as shown.

under control, and massive amounts of radioactive fission products were carried in a gas plume high into the atmosphere where they were dispersed by the wind.

The local effects of radiation were severe, with 31 fatalities and 200 incidences of radiation sickness among those working to bring the site under control. Some 130 000 people were evacuated and there were widespread restrictions on the use of fresh foods. After a week, radioactivity had spread across much of Europe (Figure 4.17). Parts of the UK were subjected to heavy rainfall when the dispersion cloud was overhead, leading to contamination of agricultural land and moorland, particularly by radioactive caesium. Enhanced caesium levels, though below the danger threshold, were still being found in sheep in the mid-1990s in parts of north-west Britain. But the major legacy of this accident will be felt in the Ukraine and in other areas close to Chernobyl where high radiation doses have increased the risk of cancer deaths among millions of people. To be fair, the statistical probability of death by cancer has probably been raised by just a few per cent among the great majority of these people, many of whom will live for several decades. So it is unlikely that the true death toll from Chernobyl will ever be known, though estimates in the tens of thousands are not unreasonable.

Chernobyl highlighted important lessons about reactor safety procedures, about the release and dispersion of radioactivity when a reactor gets out of control, and about long-term clean-up operations which continued into the early 21st century. In particular, risk assessment studies have been improved, and in the UK, for example, where it is claimed that a Chernobyl-like accident could not occur, efforts have been redoubled to assess the risk of seismic disturbance at reactor sites. Such assessments recognize that however well we control the chance of failure due to human error, nuclear reactors can never be totally immune from natural hazards. The Indian Ocean tsunami of 26 December 2004, for example, flooded the construction site of the prototype Kalpakkam fast breeder reactor in India — located only 150 m from the sea. With over 400 reactors operating globally, many of which are located in coastal areas, these are risks that we may have to live with.

4.3.2 Nuclear fuel transport and processing

Most radioactive materials connected with the UK nuclear power industry are transported by rail in massive transport flasks which have been shown in tests to survive high-speed impacts without fracturing. In fact, it is the highly radioactive spent nuclear fuels being transported to reprocessing sites that constitute the greatest danger.

Power stations unavoidably discharge radioactive gases to the atmosphere, but these escapes are sufficiently dilute not to be regarded as clinically hazardous, although the existence of a link between such emissions and the presence of cancer 'hotspots' is frequently claimed. Spent fuel rods are a far more serious problem.

● What are the main chemical changes that occur during the active lifetime of a fuel rod in a burner reactor?

○ The total uranium content decreases as fission products are generated, mainly from uranium-235 (Equation 4.2). Thus the uranium becomes depleted (Section 4.1.3) and, in addition, small amounts of plutonium are formed (Figure 4.6).

In practice fuel rods are removed before all the fissile uranium has been used up because some of the accumulating fission products impede the fission reaction efficiency (as noted in Section 4.1.2). The fission products are highly radioactive, and the result is an overall radioactivity typically about eight orders of magnitude higher than in the fresh fuel — hence the potential danger that spent fuel rods present if exposed to the environment. Many countries with a nuclear power capability simply store the spent fuel rods on site with other materials contaminated by radioactive isotopes. However, several countries, notably France and the UK, have developed nuclear waste reprocessing facilities, to which spent nuclear fuel rods are transported. At Sellafield (Cumbria), the state-owned company British Nuclear Fuels Ltd (BNFL) reprocesses depleted spent fuel to recover uranium and plutonium isotopes which can be converted into a new generation of fuel rods (Box 4.5 overleaf).

As with nuclear power stations themselves, nuclear fuel transport and processing present risks of accidental harm to human populations and the environment as a whole. Table 4.5 puts into perspective global radioactivity emissions, illustrating that under *normal* operating conditions the nuclear industry as a whole contributes a very small fraction above *natural background radiation* (mainly cosmic radiation and that emitted by natural radioactivity in rocks). The problem is that radiation leaks around nuclear facilities may be highly concentrated, making the data in Table 4.5 deceptive. For example, there is evidence of above-average incidences of childhood leukaemia at Sellafield. Exposure of the children's fathers to radiation at the pre-conception stage is one suggested explanation, but this has

Box 4.5 The Sellafield reprocessing facility (THORP)

Formerly known as Windscale, the plant at Sellafield was created in the 1950s to produce plutonium for nuclear weapons. As a secure centre of nuclear engineering expertise, Sellafield became the obvious place to store and reprocess spent uranium fuel rods from UK reactors, initially to retrieve plutonium for the fast breeder programme at Dounreay. In reality, little of this reclaimed plutonium has been used. Although some has been traded with the US, stockpiles of both plutonium and radioactive waste in the UK have grown since the early 1960s.

With expansion of nuclear reactors globally it became clear that large quantities of spent oxide fuels would be produced. After a 100-day public enquiry, plans were approved in 1977 for a new, commercial reprocessing plant at Sellafield, known as THORP (THermal Oxide Reprocessing Plant). The principal object of THORP is to concentrate the most radioactive fission products into a small volume of borosilicate glass for disposal.

THORP was designed to cater for needs beyond those of the UK nuclear programme. Advance payments by contracted customers for storage and reprocessing funded its total cost — £2.8 billion over a 15-year building and development programme. The main customers were Japanese and German nuclear generating companies which, throughout the 1980s, sent their spent fuel to Sellafield. Not surprisingly, this led to accusations that the UK was becoming a nuclear dustbin.

The other objective of THORP was to recycle uranium in the face of an anticipated shortfall in supplies.

- How reasonable was this concern at the time when THORP was first mooted in the mid-1970s?

- Eminently so, reactor requirements for uranium were rising rapidly during the 1970s (Figure 4.16) and at the time of the public inquiry.

By the early 1990s the uncertain future of the nuclear power industry, and increases in reserves and fuel stockpiles combined to make the reprocessing of spent fuel an uneconomic proposition. With the shutting down of the UK's fast breeder programme, one of the principal reasons for THORP's existence — to produce plutonium for fast breeders — no longer existed. Moreover, a report from the pressure group Greenpeace, commenting that adopting long-term storage rather than a 'reprocessing and disposal' strategy would reduce nuclear costs, influenced Scottish Nuclear to relinquish its option on reprocessing by BNFL. Similar questions about the long-term economics of reprocessing spent fuel at THORP were raised in Germany and Japan, two of the largest potential customers. Japan in particular already had a significant stockpile of plutonium (about 5 t in storage plus another 5 t at THORP and at reprocessing facilities in France) in anticipation of reprocessing, and was planning to build its own facility. There were also strong representations from members of the US Congress that the project should be halted, on environmental grounds and because of the increased potential danger of nuclear proliferation.

The UK Government finally gave approval for THORP to begin reprocessing in March 1994. Yet there was no guarantee that it would be profitable, given the global surpluses of uranium and plutonium, and continued opposition by environmental groups. Local release of radioactive krypton-85 gas from the plant and other leaks including plutonium dust continued to raise concern. By 2004 the THORP facility, dogged by technical problems, had reprocessed only 5000 of an anticipated 7000 t of spent nuclear fuel. Sellafield provides an example of how long lead times can result in major projects becoming less profitable (and even less socially acceptable) than anticipated when they were first proposed.

been disputed, and other studies have found occurrences of childhood leukaemia at other sites far from nuclear facilities. Another accusation levelled at Sellafield is the discharge and widespread dispersion of liquid wastes containing low levels of radioactivity into the Irish Sea through pipelines 2.5 km long. The early wastes

Table 4.5 Average annual radiation exposures globally from a variety of sources and incidents, relative to average natural background at an arbitrary level of 100.

Source	Level
natural background	100
Chernobyl (1986–87 only)	1.25
weapons testing (1986, 10% of 1963)	0.4
phosphate industry*	0.25
Sellafield fire (1957 only)	0.01
nuclear industry	0.008
Three-Mile Island (1979 only)	0.0002

* Phosphates contain low natural abundances of uranium, but are rich in potassium whose unstable isotope potassium-40 does emit alpha particles; they are widely distributed as fertilizers, giving rise to widespread low levels of radiation.

discharged contained some particulate plutonium which, following the intervention of a Royal Commission on Environmental Protection, is now removed before discharge. Radioactive discharges are now at a minute fraction of their original levels. But this opens up the whole question of radioactive waste containment and disposal, which you will examine in the next section.

4.3.3 Radioactive waste disposal

Most fission products from nuclear reactors are solid at ordinary temperatures. They cluster around atomic mass numbers 90 and 140 (see, for example, Equation 4.2). From the point of view of waste disposal, the problem is that most of them are highly radioactive. The common radioactive isotopes produced in nuclear reactors are given in Table 4.6. The shorter the half-life of these fission products, generally, the more intense their radioactivity.

Table 4.6 Common radioactive isotopes produced by fission of uranium-235.

Isotope	Half-life
iodine-131 ($^{131}_{53}$I)	8 days
krypton-85 ($^{85}_{36}$Kr)	10.7 days
strontium-90 ($^{90}_{38}$Sr)	28 years
caesium-137 ($^{137}_{55}$Cs)	30 years
cerium-142 ($^{142}_{58}$Ce)	5×10^5 years
zirconium-93 ($^{93}_{40}$Zr)	1.5×10^6 years
caesium-135 ($^{135}_{55}$Cs)	2×10^6 years

Besides these fission products is a range of *actinide isotopes* that do not occur naturally. These are similar in mass and atomic number to uranium, and produced from uranium by neutron absorption and electron emission, plutonium being the most widely known of them. They too are highly radioactive and some are extremely toxic chemically.

From the quantities of each radioactive product and its half-life, a curve that shows overall radioactivity decay can be constructed, which represents the gradual conversion of these materials to non-radioactive isotopes.

Figure 4.18 shows changes in the heat output — a measure of the level of radioactivity — from radioactive decay of the fission products and actinide isotopes produced in spent fuel from a 1 GW advanced gas-cooled reactor. Note that the axes of the graph are logarithmic. After 100–600 years there is a large fall in the heat production, and therefore radioactivity, from decaying *actinide* elements (from ~10 to 0.001 kW kg^{-1}). Thereafter, longer-lived *fission products* (cerium-142, zirconium-93 and caesium-135) dominate heat output, but at levels that decline from about 1 kW kg^{-1}. Yet it is widely agreed that this waste should not be allowed to leak into the biosphere for at least a thousand years. Ideally, it should be isolated for between 10^4 and 10^5 years. Moreover, it is vital to cool the waste in well-shielded storage for the first few decades until some of the short-lived isotopes have decayed. Spent fuel rods that contain products of nuclear reactions are much more dangerous than new fuel rods, which contain only uranium isotopes with very long half-lives.

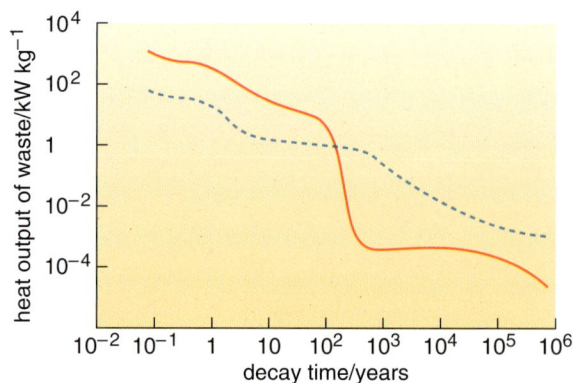

Figure 4.18 Change with time of the heat output per kilogram of spent fuel from a 1 GW advanced gas-cooled reactor, by actinide isotopes (solid line) and radioactive fission products (dashed line).

Radioactive wastes produced by the nuclear industry, together with those from other sources, are subdivided for disposal purposes into those with low, intermediate and high levels of radioactivity.

Low-level wastes (LLW) include several materials. *Gases* with half-lives of a few years at most are mainly isotopes of hydrogen, argon, krypton, xenon and radon, which are vented to and diluted by the atmosphere. *Liquids*, produced during waste treatment, are discharged into the sea or rivers. *Solids* that include worn-out equipment, crushed glassware, protective clothing, air filters, etc. are burnt in incinerators or buried at purpose-built sites, such as Drigg in Cumbria (Figure 4.8). Some 40 000 m^3 of low-level wastes are produced annually in the UK, but their disposal is not regarded as a major environmental problem. For example, it is often pointed out that some natural substances, such as brazil nuts and coffee beans, are sufficiently radioactive to be classified as low-level waste.

Intermediate-level wastes (ILW) have higher activities and so require more elaborate storage. They include solid and liquid materials from power stations, such as fuel cladding cans, and wastes from the radioisotope industry (e.g. hospital radiography departments) and defence establishments. In total, some 5000 m^3 are produced annually in the UK.

High-level wastes (HLW) are the concentrated products of nuclear fuel reprocessing, containing over 95% of the total radioactivity from the nuclear industry's waste products. At present HLW is stored in a low-density liquid solution, of which there is about 1500 m^3 at Sellafield, where it is contained by double-skinned, stainless steel tanks surrounded by concrete and cooled by water circulating through sets of stainless-steel coils. It is this waste that may ultimately be incorporated in glass (Box 4.5), for containment (Figure 4.19) prior to disposal.

Figure 4.19 Containers for high-level nuclear waste that are designed to resist groundwater penetration and to be suitable for long-term burial (100–1000 years) in deep boreholes.

High-level wastes do not occupy large volumes; Figure 4.6a shows that 4800 t of nuclear fuel for a typical 1 GW burner reactor end up as 4650 t of spent fuel after 30 years, having produced 150 t of HLW. At an average density of 10 t m^{-3}, the total volume of HLW produced after 30 years is 150 t/10 t m^{-3} = 15 m^3, the space occupied by a few filing cabinets. Even though these fission products remain extremely radioactive for centuries (Figure 4.18), storage and disposal of HLW might seem relatively unproblematic. The central issue is *where* they are to be kept. Several proposals have been made, one being that they could be lowered into deep-ocean trenches at destructive plate margins, to be covered naturally by sediment and eventually subducted into the mantle. But such areas are seismically active, and at subduction rates of centimetres per year it would take tens of thousand years for them to be subducted to a depth of 1 km; clearly a risky proposition. A more rational means of disposal would be in abandoned deep mines or specially constructed boreholes.

4.3.4 Geological criteria for safe radioactive waste disposal

Even in the best of circumstances, containers such as the one shown in Figure 4.19 will survive for only 100–1000 years, although the glass itself may inhibit the migration of radioactive isotopes for a further 1000 years. So, in view of the long decay times (Figure 4.18), the ideal geological site for waste disposal should also act as an impermeable barrier to any leakage. On land, the prime geological contenders for waste containment in the UK are unfractured clay-rich rocks, salt deposits, and hard crystalline igneous or metamorphic rocks. Salt and some clay deposits have the advantage of being self-sealing, owing to their high plasticity. Clay minerals also have a high capacity for ion adsorption (i.e. ions adhere to their surfaces). Crystalline rocks are generally dry, stable against tectonic movement, extensive and virtually inert to any heating effect from waste canisters.

A very important aspect of nuclear waste disposal is the choice of what form the disposed material should take. THORP is designed to produce borosilicate glass, but all glasses decompose over time and are relatively reactive. It would be better to create artificial minerals, such as apatite or zircon, that would mop up just as much material and are stable unless they are melted at very high temperatures (>700 °C).

Question 4.5

Taking three groups of impermeable rocks in turn — clay-rich sedimentary rocks, salt deposits (see Argles, 2005) and crystalline rocks — and using Figure 4.20, suggest which might be the best areas for radioactive waste disposal to be located in the UK.

In 1997, the proposal by the UK nuclear waste company Nirex to build an underground repository for HLW at Sellafield failed to obtain government approval after a five-month public enquiry. After rejecting the application the then Secretary of State for the Environment, John Gummer, justified the refusal by saying that he was 'concerned about the scientific uncertainties and technical deficiencies in the proposals presented by Nirex [and] about the process of site selection and the broader issue of the scope and adequacy of the environmental statement'.

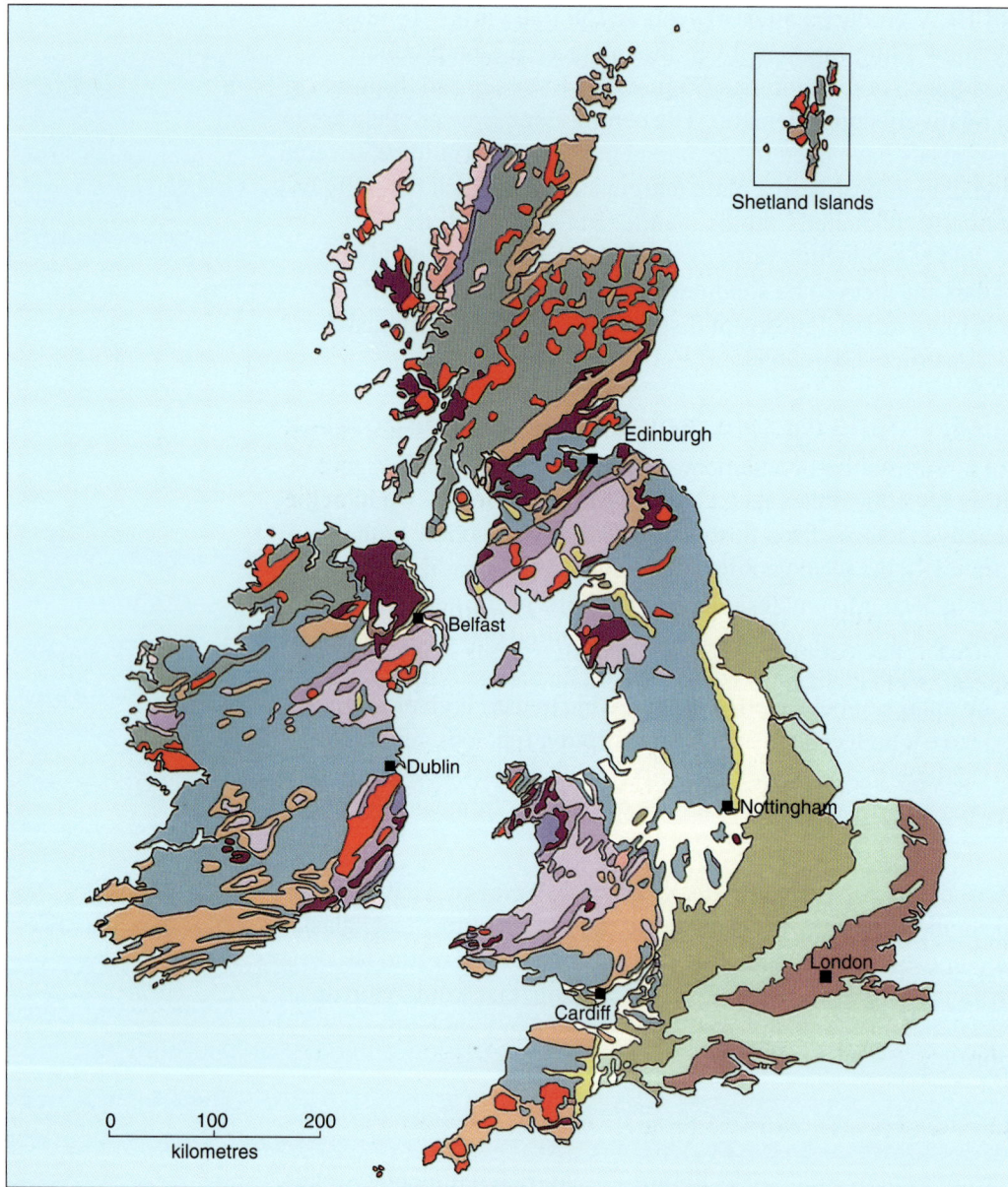

SEDIMENTARY ROCKS

CENOZOIC

		age/Ma
Tertiary and marine early Quaternary mainly clays and sands; Quaternary glacial deposits not shown	up to 65

MESOZOIC

Cretaceous mainly chalk, clays and sands — 65–142

Jurassic mainly limestones and clays — 142–206

Triassic mudstones, sandstones and conglomerates — 206–248

PALAEOZOIC

Permian mainly limestones, mudstones and sandstones — 248–290

Carboniferous limestones, sandstones, shales and coal seams — 290–354

Devonian sandstones, shales, conglomerates; (Old Red Sandstone) slates and limestones — 354–417

PALAEOZOIC continued

age/Ma

Silurian mainly shales, mudstones, some limestones — 417–443

Ordovician mainly shales and limestones; limestone in Scotland — 443–495

Cambrian mainly shales, slate and sandstones; limestone in Scotland — 495–545

UPPER PROTEROZOIC

Late Precambrian mainly sandstones, conglomerates and siltstones — 545–1000

HIGHLY METAMORPHOSED ROCKS

Late Precambrian, Cambrian and Ordovician mainly schists and gneisses — 443–1000

Mid Precambrian mainly gneisses — 1500–3000

IGNEOUS ROCKS

Intrusive mainly granite, gabbro and dolerite

Volcanic mainly basalt, rhyolite, andesite and volcanic ashes

Figure 4.20 Simplified geological map of Britain (BD/IPR/7-14 British Geological Survey. © NERC all rights reserved.)

Shortly afterwards the newly elected UK Labour Government was confronted with the need for a new technical approach and policy on nuclear waste management. As a result, a consultation was launched in 2001 due for completion in 2006.

Meanwhile the problem of waste disposal is equally acute in many other countries, several of which have technically aware environmental pressure groups. In 2002, the US Senate voted to approve using Yucca Mountain in a remote corner of Nevada as a repository for US waste. It is located in the United States' former nuclear weapons testing area. Although the site has been under investigation since the 1980s, there remain many political and regulatory hurdles before it can be approved for long-term waste storage.

Sweden has a repository for LLW and ILW, having adopted a no-expense-spared approach after widespread public consultation. Possible sites for a similar long-term HLW repository are being investigated but such a repository could not begin to operate until the mid-2010s at the earliest. France has similar plans and Finland's facility should be operational after 2020.

Even if there is no further expansion of nuclear power globally, filling up of temporary stores based at nuclear plants or at reprocessing plants like Sellafield clearly poses a long-term challenge that the world will have to face. The continuing debate about the UK nuclear industry illustrates well how 'Green' groups can successfully stimulate debate and even modify government policy. We leave you to decide whether or not their opposition to nuclear power and reprocessing of nuclear fuels is justified.

Question 4.6

(a) Why are spent fuel rods from nuclear reactors potentially so much more serious a hazard than new fuel rods if exposed accidentally to the environment?

(b) What factors have caused the economics of nuclear fuel reprocessing in the UK to change since 1986?

(c) What is the purpose of plans to vitrify high-level radioactive wastes?

4.4 Summary of Chapter 4

1   Nuclear power generation results from fission of uranium isotopes when bombarded by neutrons. Conventional burner reactors require relatively scarce uranium-235, whereas fast breeder reactors (which have not yet been developed on any significant scale) would exploit more abundant uranium-238.

2   In the early 21st century over 400 nuclear — mainly burner — reactors produced 16% of global electricity demand.

3   The UK played a leading role in nuclear power developments during the 1950s and 60s with its Magnox programme. During the 1970s new, more efficient reactor designs led to the building of AGRs in the UK, and PWRs and other analogous reactors in other countries. Progress was slowed in the late 1970s as concern grew over the operational safety of PWR nuclear reactors.

4   Including the costs of decommissioning reactors and managing radioactive waste, nuclear power is probably more expensive than power from fossil fuels. Nevertheless, arguments that nuclear power generation produces virtually no carbon dioxide and sulphur dioxide gases have begun to weigh in favour of its further development as fears of global warming grow.

5   The properties that determine the mobility and concentration of uranium to form ore deposits are: (a) it is an incompatible element by virtue of the high charge and large size of uranium ions, so it becomes concentrated late in the evolution of granitic magmas; (b) uranium-bearing ions are much more highly soluble in water under oxidizing conditions than under reducing conditions, thus controlling uranium transport and deposition in groundwater and hydrothermal fluids.

6   Uranium occurs in many geological settings, including disseminated magmatic and unconformity related hydrothermal deposits, sandstone-hosted and quartz-pebble conglomerate deposits.

7   Uranium is mined in surface and underground mines, and by *in situ* leaching. The higher costs of underground mining are offset by higher grade uranium ore. Uranium is extracted chemically from ore to yield uranium trioxide, or yellowcake, subsequently processed and enriched into reactor fuel. Safety precautions from the effects of radioactivity of uranium and its daughter products, especially radon, are essential in mining, processing and disposal of mining wastes.

8   Uranium demand has been affected by political factors and military requirements as well as the commercial demands of the energy industry. The low uranium prices of the late 1980s and 1990s led to many low-grade deposits becoming uneconomic. A reduction of stockpiles and the possibility of a brighter future for nuclear energy led to a rise in prices in the early 21st century.

9   Current estimates of reserves of uranium amount to about 3.5 million tonnes, sufficient to maintain global nuclear power supplies for many years even if demand ($\sim 6.6 \times 10^4$ t annually in the early 21st century) increases significantly.

10   Public concerns about nuclear reactor safety were exacerbated in 1986 by a major accident at the Chernobyl reactor in the Ukraine, after which widespread atmospheric dispersion spread radioactive contamination over most of Europe.

11   The intensity of nuclear-waste radiation is greater from short-lived radioactive fission products than from uranium and plutonium with long half-lives. The reprocessing of spent nuclear fuel rods therefore constitutes a potential hazard. However, by 2005, this has been tentatively linked only to increased incidences of childhood leukaemia. The small, but finite risk of accidents at facilities such as the controversial THORP plant on the Sellafield site could be much more serious.

12   Plans to dispose of radioactive waste by burial in the UK have had a chequered history, and plans to bury high-level wastes (HLW) in the UK and in other countries have either been postponed indefinitely or are awaiting approval.

ALTERNATIVE ENERGY 5

Energy from sources other than fossil and nuclear fuels is to a large extent free of the concerns about environmental effects and renewability that characterize those two sources. Each alternative source supplies energy continually, whether or not we use it. One of the alternative sources to consider is geothermal energy derived from the interior heat of the Earth (Section 5.1). Another is solar energy in forms that can be harvested at the Earth's surface. Direct conversion of solar energy is considered in Section 5.2. Yet the Sun also creates energy resources indirectly: from its concentration in chemical form by plants through photosynthesis; from wind; the surface flow of water; and ocean wave movement (Sections 5.3 to 5.6). The third alternative source stems from gravitational forces associated with the Sun and Moon — tidal energy (Section 5.7). You should note that the length of each of the sections in this chapter is not necessarily proportional to the potential of each alternative type of energy source. Rather, the length reflects the complexity of the science — especially the geoscience — that lies behind the origin of the energy source and its distribution; some are simpler to understand than others.

Many of these alternative sources have been used in simple ways for millennia, e.g. wind and water mills, sails, wood burning — but only in the last two centuries has their potential begun to be exploited on an industrial scale. Except for geothermal energy, all have their origins in energy generated outside the Earth, yet the potential of each is limited by its total supply set against its rate of use. You will see that each is likely to be renewable in the sense that the available rates of supply of each exceed those at which they are used. The main concern is whether or not such alternatives can supplant fossil- and nuclear-fuel use to power social needs fast enough to avoid the likelihood of future global warming and other kinds of pollution, which we discuss in Chapter 6.

5.1 Geothermal energy

Although energy from the Earth's interior that flows though the surface is on average very low — about a thousand times less than the solar energy that falls on the surface — it is sufficiently abundant worldwide to make it locally worth exploiting. The top 3 km of the Earth's crust stores an estimated 4.3×10^7 EJ of thermal energy by virtue of the temperature of rocks and their thermal capacity. Because global consumption of energy during 2002 was 451 EJ (Chapter 1), heat stored within the Earth might seem vastly more than sufficient for all humanity's needs. Stored energy is added to by heat *flowing* from the deeper Earth, but this heat is eventually lost by escape from the Earth's surface. In theory it would be possible to tap the *stored* heat, but in practice the principal geothermal potential is from heat that *flows* through the crust. As you will see, that is a far smaller potential resource.

Where geothermally heated water rises to the surface in hot springs it has sometimes been used *directly* for heating, recreational and horticultural purposes since Roman times. Geothermal energy was used *indirectly* for the first time when geothermally generated electricity was produced in 1904 at Larderello, Italy. In 2003 about 8 GW of geothermal electrical power capacity had been installed globally (Table 5.1), providing about 0.2 EJ of energy each year — a tiny fraction of global electricity consumption. Technologically less taxing, direct use of geothermal heat contributed about the same to global energy supply annually (Table 5.2).

The Earth's interior becomes hotter with increasing depth for two reasons: heat is generated by the *decay* of the long-lived, natural radioactive isotopes (uranium-238 and uranium-235, thorium-232 and potassium-40; see Section 4.1) and heat flows from the hot interior by means of convection and conduction. The manner in which geothermal heat flows through the Earth is reflected by

Table 5.1 All installed geothermal electricity generating capacities in 1995 and 2003.

Country	Installed capacity 1995/MW	Installed capacity 2003/MW	Average annual change in capacity (1995–2003)/%
USA	2816	2020	−3.5
Philippines	1227	1931	7.1
Mexico	753	953	3.4
Indonesia	310	807	20.0
Italy	632	791	3.1
Japan	414	561	4.5
New Zealand	286	421	5.9
Iceland	50	200	37.5
Costa Rica	55	161	24.4
El Salvador	105	161	6.6
Kenya	45	121	21.1
Nicaragua	70	78	1.4
Russia	11	73	70.5
Guatemala	—	29	3.6
China	29	28	−0.3
Turkey	20	20	0
Portugal	5	16	27.5
France	4	15	32.1
Ethiopia	—	7	0.9
Papua New Guinea	—	6	0.8
Austria	—	1.3	0.2
Thailand	0.3	0.3	0
Germany	—	0.2	0.03
Australia	0.2	0.2	0
Argentina	0.67	—	−0.08
Global total	6833	8402	2.9

Table 5.2 Annual direct, non-electric uses of geothermal energy in 2000. The top five countries using direct geothermal resources account for 60% of direct geothermal power use globally.

Country/Region	Annual energy use/TJ
China	37 908
Japan	26 933
USA	20 302
Iceland	20 170
Turkey	15 756
Areas apart from the above countries	
Rest of Europe and Russian Federation	47 885
Rest of Asia and Pacific	13 807
Middle East	3485
Africa	1812
Rest of North America	1023
South and Central America	920
Global total	190 699

the *geothermal gradient* (the rate at which temperature increases with depth). Figure 5.1 shows how the geothermal gradient varies beneath the surface of a continent; note how temperature increases much less with depth in the deep mantle than it does in the lithosphere.

Although they are not molten, rocks that form the deep mantle are hot enough to behave in a *ductile* manner. Parts of the mantle that are hotter than their surroundings rise slowly because they are slightly less dense than cooler mantle, thereby physically transporting their heat content towards the surface. This process, called **convection**, is an efficient means of heat transport. As a result, the geothermal gradient in the deep mantle involves less increase in temperature with depth than occurs within the lithosphere (Figure 5.1).

Down to around 150 km below the surface, the Earth's lithosphere behaves in a more rigid fashion — it is this rigidity that creates the stability of tectonic plates. Except at plate boundaries and hotspots (Sheldon, 2005) convection does not transfer heat in the lithosphere. Heat is transferred instead by **conduction**, which is a far less efficient process than convection, so the lithosphere acts as an insulating blanket (trapping the heat rising from below). Heat transfer by conduction explains why temperature increases more rapidly with depth in the lithosphere than in the deeper mantle (Figure 5.1).

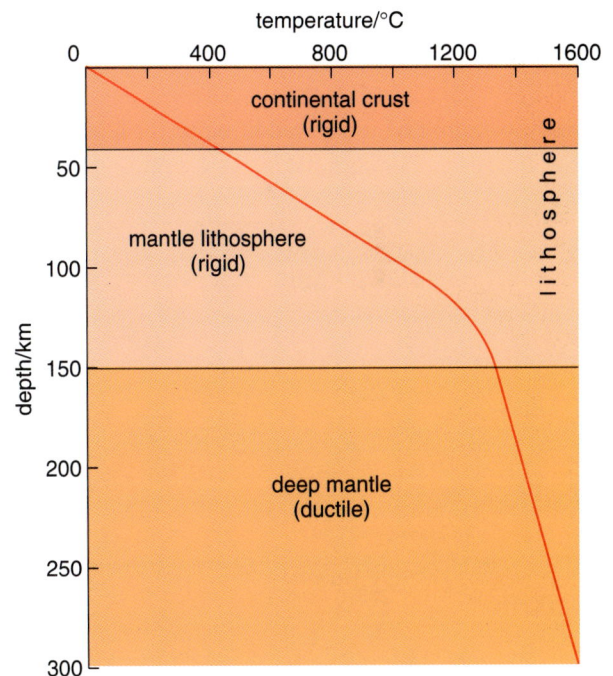

Figure 5.1 The geothermal gradient beneath a continent, showing how temperature increases more rapidly with depth in the lithosphere than it does in the deep mantle.

Harvesting geothermal energy depends on the relationship between heat flow through the lithosphere and the local geothermal gradient (Box 5.1). The most important practical factors involved in assessing an area's potential are:

- the presence of magmas close to the surface in volcanically active regions;
- high heat flow in near-surface solid rocks;
- the ability to transfer geothermal heat to power plants and other energy consuming technologies efficiently, usually as hot water or steam.

The most favourable areas for geothermal exploitation might seem to be those in volcanically active regions, at destructive and constructive plate margins. The distribution of geothermal electricity generation plants shown in Figure 5.2 confirms this. What is less obvious is that the geothermal potential of some areas that are remote from volcanically active regions is also worth exploiting. The geothermal energy exploited directly by two of the major users, China and Turkey (Table 5.2), is not related to active volcanism, but to abnormally high heat flow.

The local potential of geothermal energy is fundamentally a function of the *enthalpy* of the area, i.e. the *total* energy content of the geothermal system that lies below it. The concept of enthalpy was introduced in Box 1.2 — Equation 1.4 is repeated here for convenience.

$$H = U + PV \tag{5.1}$$

As you will see shortly, both the *heat content* (U) and the *pressure* (P) of steam in a geothermal system play a role in converting geothermal energy into electricity.

Enthalpy cannot be measured directly for a geothermal energy source. However, it can be estimated from the heat flow through the crust, which is measurable at

Figure 5.2 The main sites of geothermal electricity production in 2003 in relation to plate boundaries.

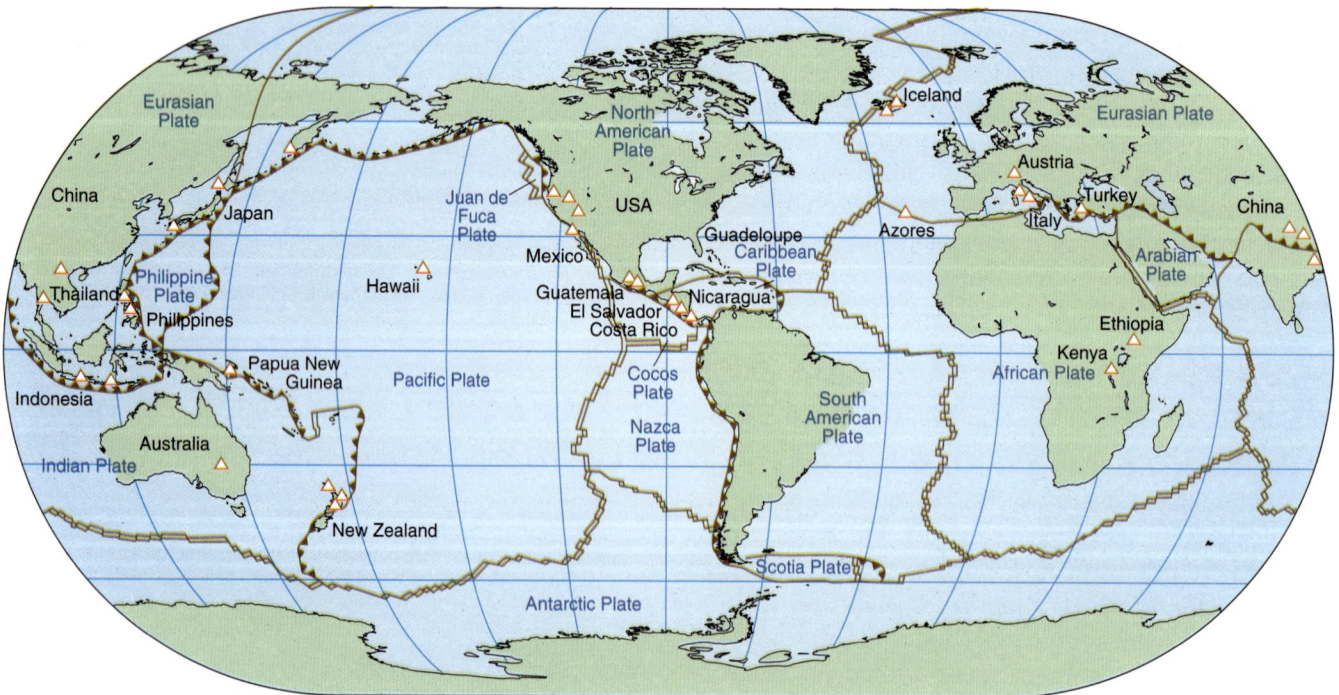

key //constructive plate boundary ⌇destructive plate boundary /transform fault plate boundary △geothermal plant

the surface. Heat flow is expressed as the geothermal power associated with an area of the surface, measured in mW m^{-2}. Figure 5.3 shows that large areas of the Earth's surface have much the same heat flow. However, the temperature at depth in the crust depends on how *conductive* different rocks are, and this in turn depends on their physical properties. This subsurface temperature is crucial in evaluating whether or not an area constitutes a geothermal resource (Box 5.1).

Figure 5.3 shows that on a global scale, the main areas where heat flow is much greater than the global average of 60 mW m^{-2} are close to constructive plate margins beneath the oceans. In volcanically active continental areas that lie above subduction zones, as around the Pacific, *regional* heat flow can be above the average, but only up to 120 mW m^{-2}. However, values on the continents can be as high as 300 mW m^{-2} where magma is being generated *locally*. Such continental occurrences and volcanically active oceanic islands are the most favourable for geothermal exploitation, for obvious reasons.

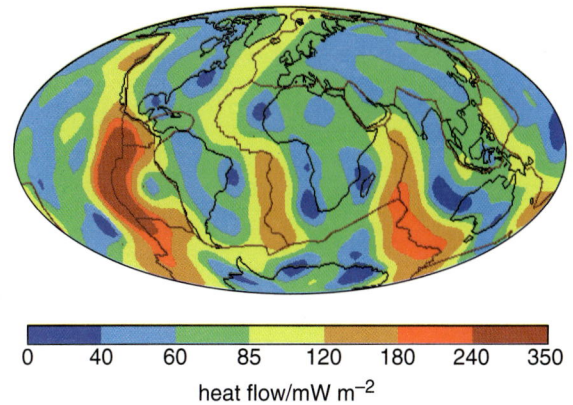

0   40   60   85   120   180   240   350

heat flow/mW m^{-2}

Figure 5.3 World map showing variation in surface heat flow in mW m^{-2} in relation to continental and oceanic crust, and major plate boundaries. The colours relate to the scale in mW m^{-2}. Note that there are areas in continents where heat flow does rise to very high values, but they are too small to show on this map.

Box 5.1 Heat conduction

The origin of a geothermal resource relates to the following equation for conductive heat flow:

$$q = \frac{k\Delta T}{z} \tag{5.2}$$

where q is *heat flow* (in W m^{-2}); k is *thermal conductivity* (in W m^{-1} K^{-1}); and ΔT is the temperature difference (in kelvin or K = °C + 273 °C) through a depth z. So, the geothermal gradient ($\Delta T/z$) has units of K km^{-1}. Thermal conductivity expresses the ease with which a material transmits heat. Thus a metal pan has a high thermal conductivity whereas an oven glove is a poor conductor of heat. All rocks are poor conductors of heat in the everyday sense, but some rocks, such as sandstones and granites, are better conductors of heat than others, such as shales and many metamorphic rocks.

● If heat flow q in Equation 5.2 is constant will the geothermal gradient ($\Delta T/z$) be greater where the thermal conductivity (k) is high (good conductor) or where it is low (poor conductor)?

○ For a higher value of $\Delta T/z$, at constant q a *low* thermal conductivity is required.

For example, if heat flow (q) is 100 mW m^{-2} (100 × 10^{-3} W m^{-2}), as it could be in some volcanically active areas, and the geothermal gradient is 100 K km^{-1}, i.e. the temperature difference ΔT between the surface and at 1 km depth is 100 K, then thermal conductivity k can be calculated from Equation 5.2:

$$100\times10^{-3} \text{ W m}^{-2} = \frac{100 \text{ K}}{1 \text{ km}} \times k = \frac{100 \text{ K}}{10^3 \text{ m}} \times k$$

Rearranging gives:

$$k = \frac{100\times10^{-3} \text{ W m}^{-2} \times 10^3 \text{ m}}{100 \text{ K}} = 1 \text{ W m}^{-1} \text{ K}^{-1}$$

If the geothermal gradient were *less*, say 50 K km^{-1}, then the same heat flow could only be produced if it passed through rocks with a thermal conductivity given by:

$$k = \frac{100\times10^{-3} \text{ W m}^{-2} \times 10^3 \text{ m}}{50 \text{ K}} = 2 \text{ W m}^{-1} \text{ K}^{-1}$$

i.e. if the geothermal gradient is halved, for the same heat flow the thermal conductivity must be doubled.

Specialists recognize three kinds of area with geothermal potential; those with high, medium and low enthalpy. High-enthalpy areas are those where subterranean water and steam are at temperatures greater than 180 °C, medium-enthalpy areas those where temperatures range between the boiling temperature of water (100 °C) and 180 °C, and low-enthalpy areas where temperatures are lower than 100 °C. A mass of pressurized steam has an energy content that includes the pressure–volume (PV) term in Equation 5.1, and also contains the heat involved in converting liquid water into gas (the *latent heat of vaporization*).

Question 5.1

A favourable site for geothermal development will have an above-average temperature at shallow depth, within the top few kilometres of the crust (i.e. there will be an above-average, or steeper, geothermal gradient). Leaving aside high-enthalpy volcanic areas answer the following.

(a) Which rock types might produce an above-average geothermal gradient by virtue of their low thermal conductivity?

(b) Which common rock types of the continental crust contain higher than average abundances of uranium and so might have the same effect by virtue of their locally high radioactive heat production? (Section 4.2.1 is helpful.)

Using geothermal energy effectively speeds up heat flow and locally increases the heat lost by the Earth; it is akin to mining. But geothermal heat is continually supplied, so the 'mining' depletes hot water and steam rather than the energy resource itself. However, if hot fluids removed through one borehole are replaced by cold water pumped in through another to recharge the geothermal field, the cold water heats up again. Carefully managed recharge enables the enthalpy of an area to be exploited continuously.

To allow their successful exploitation, geothermal fields usually need three characteristics:

1   An energy source to heat the fluids;
2   An aquifer (see Smith, 2005), either natural or artificially created (see Section 5.1.2), to act as a reservoir for fluids that can transport heat energy to the surface;
3   A cap rock with low permeability to seal in these geothermal fluids — in a similar way to the trapping of petroleum (Section 3.2).

● What type of rocks should make good cap rocks?

○ Mudstones and unfractured lavas are ideal.

5.1.1 High- to medium-enthalpy steam fields

When the geothermal gradient heats water above the temperature at which it boils at atmospheric pressure, at a depth accessible to drilling, conditions can favour using natural geothermal steam to generate electricity. Typically, the pressure can be several tens to hundreds of times that of the atmosphere. Even at 200 °C, high pressure can ensure that much of the fluid in a geothermally heated aquifer remains in the liquid state. Figure 5.4 shows why that seemingly odd situation occurs; the physical states of water (solid, liquid and vapour) transform to one another at different temperatures and pressures. The **critical point** of water is the temperature above which liquid and vapour are no longer valid concepts, irrespective of pressure. Instead, water exists at higher temperatures as a **supercritical fluid**.

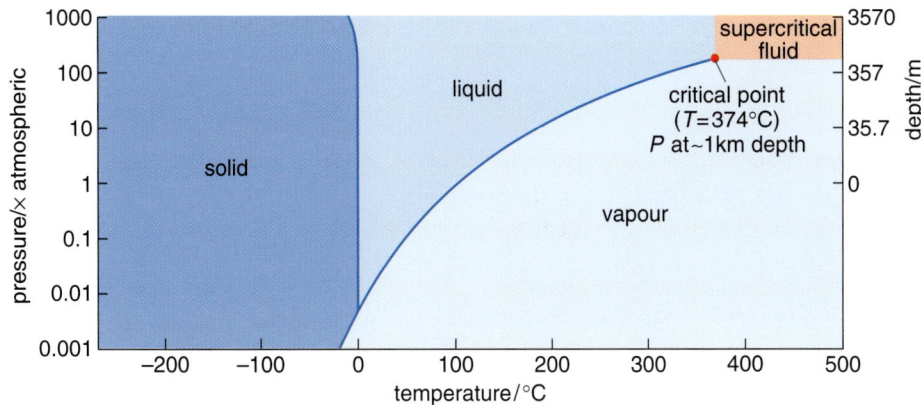

Figure 5.4 How the different states of water depend on temperature and pressure (the vertical scale is logarithmic). Above a temperature of 374 °C (the 'critical point' of water) there is no distinction between the liquid and vapour states of water, and it exists as a 'supercritical fluid'. (*Note*: A pressure of 1000 times that of the atmosphere at sea-level is equivalent to a depth of about 3.57 km in the crust. Pressures less than one atmosphere do not apply to the Earth's crust, but to water in the atmosphere.)

- In an area with a geothermal gradient of 60 K km^{-1} will water trapped in an aquifer exist as a liquid or steam at a depth of 2 km?

- A temperature of 120 °C that occurs between 0.357 to 3.57 km deep plots in the liquid field of Figure 5.4.

- What would happen to that liquid water if a well connected the aquifer to the surface?

- Pressure will drop, and as it rises toward the surface the water will become steam.

Effectively, deep groundwater that is above the boiling temperature at atmospheric pressure is *superheated*. Once pressure is released it '**flashes**' to steam, and jets to the surface. Many potential geothermal areas naturally connect to the surface along faults, so that steam is generated at depth and drives hot water to the surface, where it appears as hot springs and sometimes geysers that emit high-pressure steam and boiling water. To exploit this power-generating potential requires that such a water source be effectively trapped so that it cannot leak away naturally.

In Figure 5.5, traps below the Lardarello and Wairakei geothermal fields are at shallow enough depths for high-pressure steam to be present naturally in the aquifer (see Figure 5.4). Deeper systems contain superheated water that 'flashes' to steam when penetrated by wells. The fluids that issue from

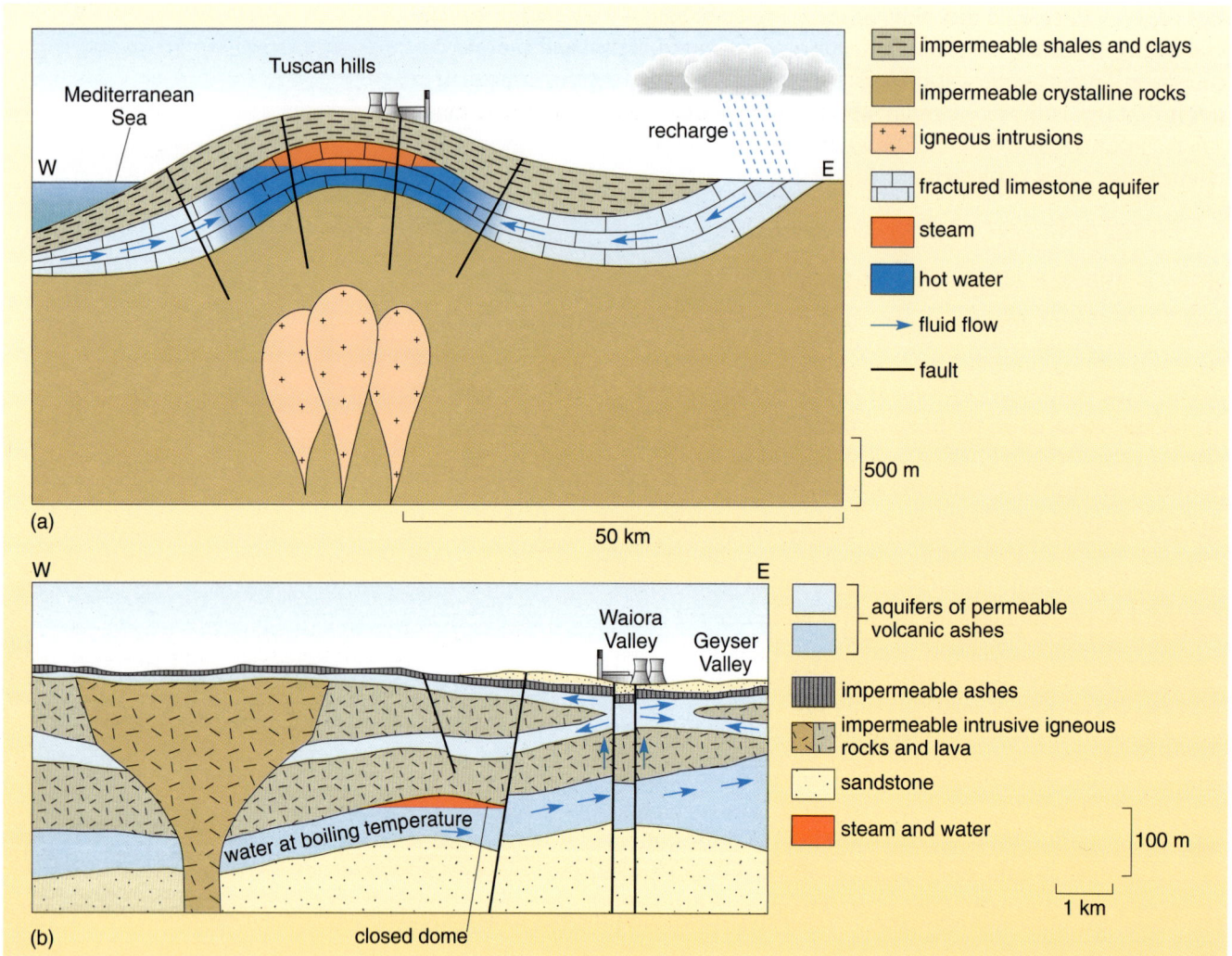

Figure 5.5 legend (a):
- impermeable shales and clays
- impermeable crystalline rocks
- igneous intrusions
- fractured limestone aquifer
- steam
- hot water
- fluid flow
- fault

Labels (a): Mediterranean Sea, Tuscan hills, recharge, W, E, 500 m, 50 km, (a)

Figure 5.5 legend (b):
- aquifers of permeable volcanic ashes
- impermeable ashes
- impermeable intrusive igneous rocks and lava
- sandstone
- steam and water

Labels (b): W, E, Waiora Valley, Geyser Valley, water at boiling temperature, closed dome, 100 m, 1 km, (b)

geothermal wells range from hot water and wet steam (close to 100 °C and easily condensed) to dry steam far above boiling temperature. Different kinds of fluid demand different electricity-generating technologies, so that each geothermal field is unique and is exploited on its own merits.

High- to medium-enthalpy geothermal resources are conventionally divided into water- or vapour-dominated fields. Vapour-dominated systems where the steam is at high temperature and dry tend to be the most economically viable geothermal resources, because the PV term in the enthalpy equation is high. Vapour-dominated fields also suffer less from problems of corrosion from the mineral-laden waters that are typical of deep aquifers. When such hot water passes through pipelines, not only is it highly corrosive, but solids dissolved in it precipitate in the pipes as temperature decreases, thereby clogging them in the manner of a 'furred' kettle.

5.1.2   Hot dry rock (HDR) fields

Heat flow through some parts of the continental crust can be well above normal locally because the underlying rocks contain abnormally high concentrations of uranium, thorium and potassium, which generate considerable heat. To add

Figure 5.5 Cross sections of high-enthalpy geothermal fields in volcanic areas: (a) The Lardarello field in Tuscany, Italy. Note the favourable dome-like structure over the active igneous intrusions, where a limestone aquifer forms a trap beneath shale and clay cap rocks. (b) The Wairakei field in New Zealand has several aquifers sealed by impermeable lavas, volcanic ash and intrusive igneous rocks. In both cases some hot water and steam escapes naturally to the surface along faults.

significantly to surface heat flow and thereby create high-temperature anomalies at shallow depths requires a large volume of such radioactive rocks. This condition is satisfied by some, but not all, *granitic* igneous intrusions, whose original magma became charged with heat-producing elements because of geochemical fractionation (Sheldon, 2005).

- Would such intrusions contain enough water to constitute geothermal resources?

- They are crystalline rocks and therefore have negligible porosity and permeability, so they are not natural geothermal resources.

Notwithstanding their lack of fluids to transfer heat toward the surface, thermal anomalies in such intrusions do constitute significant thermal resources, if they are shallow enough to be reached by drilling. **Hot dry rock (HDR)** systems require such rocks to be artificially fractured at depth so that water can be pumped into them, heated above boiling temperature and then returned to the surface to flash to steam in an electricity generating plant. Figure 5.6 shows in schematic form how hot, dry rock can become a high-enthalpy geothermal resource. A well that penetrates the intrusion is used to inject water at very high pressures into zones of natural fracturing in the rock. Water injected through the well not only heats up but enhances any fractures and creates more, by a process of *hydrofracturing*. This creates paths along which water can move, effectively creating an artificial heat exchange zone that feeds superheated water to another well that takes it back to the surface. The practical difficulties of extracting heat from such deep rocks can be formidable. Drilling through crystalline rocks is far more costly than through sedimentary strata. For an HDR field to be viable, water has to circulate through large volumes of hydrofractured rock. The HDR approach is being pursued because of its potential in areas well away from active plate margins. Steam generation that cools 1 km^3 of a fractured hot dry rock by only 1 °C will provide the same amount of energy as 7×10^4 t of coal.

A typical artificial HDR field would be at a depth of 3–6 km. During the 1970s and 1980s granitic areas of the USA, the UK, France and Germany became the focus of considerable research. The US experiment was conducted in an area with abnormally high heat flow (60 K km^{-1}) in New Mexico and proved that HDR-heated water was sufficient to power a 60 kW turbine for one month. The UK project was based at Rosemanowes in Cornwall and focused on the crucial techniques needed to hydrofracture crystalline rock.

Cornwall is underlain by several large, highly radioactive granite intrusions, parts of which are exposed at the surface (Figure 5.7). This creates a large zone where surface heat flow is higher than in surrounding areas, beneath which there

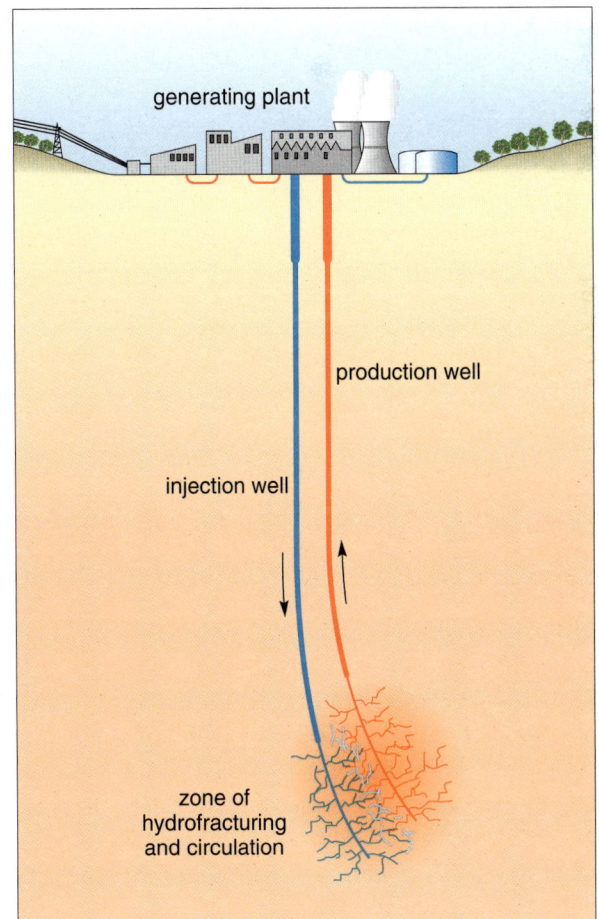

Figure 5.6 A hot dry rock circulation system using artificially enhanced fracturing that creates a large-volume heat exchange system linking injection and production wells by many connected fluid pathways.

are no granitic rocks. Rosemanowes (R on Figure 5.7) is not located over the area of highest heat flow associated with the Lands End granite, but where an abandoned quarry in another granite outcrop became available for the experiment. To be viable for geothermal power generation, the experimental HDR field would have to achieve at least 150 °C at depth.

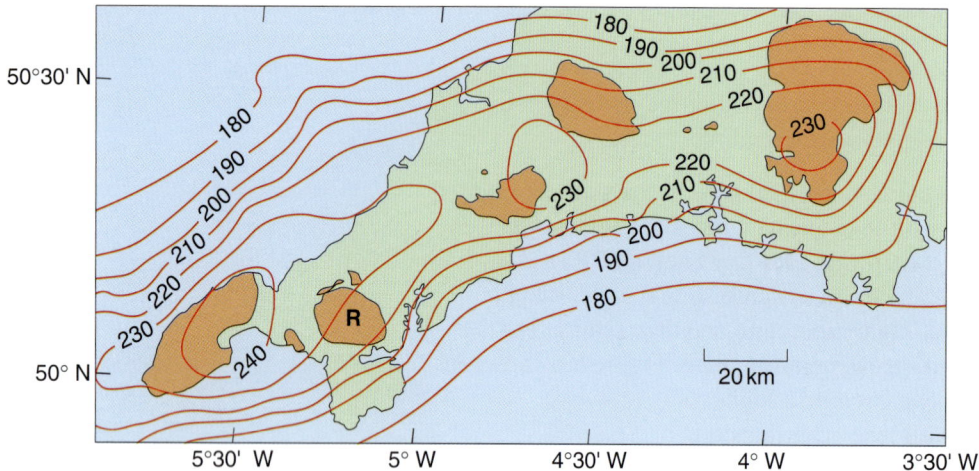

Figure 5.7 Surface occurrences of granite in Cornwall, which are connected at depth to form a much larger mass. Also shown are contours of the temperature in °C that are expected 6 km beneath the surface. The Rosemanowes HDR experiment is at R.

Question 5.2

If the geothermal gradient at Rosemanowes is 37 K km^{-1}, and the surface temperature averages 10 °C, what is the minimum depth to which drilling must penetrate to be viable for geothermal power generation?

The Rosemanowes experiment showed that it was more difficult to get a closed circulation path than had been hoped. At high pressures (100 to 1000 times atmospheric pressure) runaway hydrofracturing of the rocks occurred and the water leaked away; water losses are sustainable only if they remain less than 10%. The Rosemanowes site was abandoned. The findings at Rosemanowes indicated a need for real-time monitoring of the progress of hydrofracturing and paved the way for more elaborate projects, such as the EU funded programme at Soultz-sous-Forêts in France, near the upper Rhine Valley. Soultz was chosen because its higher geothermal gradient allowed high temperatures to be exploited at shallower depths than at Rosemanowes. Initially two wells were drilled, to 3.5 and 5 km, with ancillary wells containing geophones to monitor microseismic activity as the fractures formed — this approach can also locate the newly developed fractures. Careful injection of water allowed the two main wells to be connected efficiently with no loss of fluid. A four-month test showed that 10 MW of energy could be extracted from a fluid flow rate of only 25 kg s^{-1}. The project was a success and confirmed HDR technology as a major potential source of clean, renewable energy, provided it produced electricity at an economically competitive price.

5.1.3 Locating high-enthalpy geothermal fields

The search for potentially useful geothermal fields focuses initially on locating rocks that have been chemically altered by natural geothermal fluids, as well as looking for obvious surface features of geothermal activity such as geysers and hot springs. Measurements of fluid flow through the field allow estimation of its likely economic potential.

When a promising resource has been located exploration wells are drilled. However, given the high pressures and temperatures typical of a geothermal resource, special precautions need to be taken. For example, once the drill stem penetrates into a zone saturated with superheated water, the liquid will flash to steam as the pressure drops. This is more dangerous than simply venting the 'head' of pressure in an oil or gas field (Section 3.3.4). The well head itself is capped with valve gear to regulate the flow of fluids, and the whole assembly is then 'plumbed' to an electricity generating plant.

Exploration for HDR fields involves checking exposed granites for their content of radioactive heat-producing elements, and assessing subsurface heat production by measuring surface heat flow. Outlining the buried granite intrusions that are targets for HDR development depends on modelling their extent, depth and volume from gravity surveys; granites have a lower density relative to their surrounding rocks, and produce gravitational 'lows'.

5.1.4 Geothermal power plants

A typical geothermal power plant has a power output around 30–50 MW compared to several GW from conventional oil- or coal-fired plants. Therefore, between 20–33 geothermal stations are needed to replace a 1 GW conventional power station. However, once installed, geothermal operating costs are low.

Electricity generation using steam from geothermal fields is essentially the same as in fossil-fuel and nuclear power plants (Section 1.6.3), with various modifications to suit the enthalpy of the field. In high-enthalpy fields, such as Larderello in Italy and the Geysers Steam Field in California, the reservoir contains large volumes of superheated water at pressures up to 30 times atmospheric pressure, similar to that used in a conventional thermal power plant. When the fluid flashes to vapour the steam escapes at speeds up to 300 m s^{-1} (comparable with a jet engine at full throttle) and is suitable for electricity generation using the simplest turbines (Figure 5.8a) from which steam is vented to the atmosphere. The efficiency can be further increased if the steam is condensed behind the turbine (Figure 5.8b) as this increases the pressure gradient and so increases the speed of the steam. The 'spent' water will still be very hot,

Figure 5.8 Types of geothermal turbine. (a) Simple turbine used in dry steam fields that vents 'spent' steam to the atmosphere. (b) Turbine using a steam condenser behind the turbine to increase the pressure difference and therefore the steam speed through it.

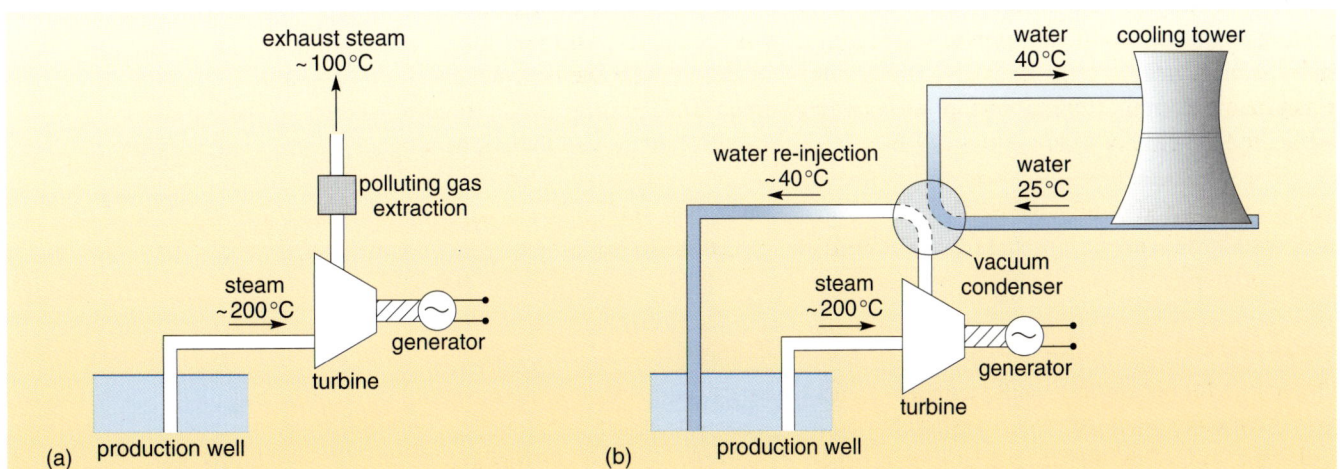

and can be re-injected into the geothermal aquifer system to maintain the volume and pressure of the geothermal fluid.

Lower enthalpy fields produce fluid that flashes to steam close to the well-head or within the plant itself at commensurately lower pressures and speeds, thereby reducing power output and efficiency. To maintain efficiency requires more complex turbines and heat-management schemes. At still lower enthalpy, conventional steam turbines become less efficient.

To increase the contribution of geothermal energy to electrical power supplies, development needs to focus along two lines. One option is developing lower temperature fields. Instead of using steam to drive turbines, hot water vaporizes a more volatile fluid, typically a light hydrocarbon. The other aims at harnessing the energy of fields in which fluids are at very high temperatures and pressures, including HDR fields. However, drilling technology has a depth limit, due to decreasing strength of the drill stem and decomposition of lubricants for the drill's cutting bit at very high temperatures.

5.1.5 Direct heating using geothermal energy

In the same way that waste heat from conventional power stations can be used for direct heating of buildings, and in industrial production and horticulture, low-grade geothermal energy has considerable potential. Many existing developments, such as those in Iceland, use spent fluids from geothermal electricity generation. Areas of natural hot springs are an obvious target, but it is also relatively simple to exploit normal heat flow using either natural groundwater or a variety of heat-exchange systems.

Deep sedimentary basins associated with average heat flow can have useful geothermal potential, if they contain sediments with low thermal conductivity and therefore above-average geothermal gradients (Box 5.1). The Paris Basin beneath northern France contains several limestone and sandstone aquifers (Figure 5.9) in such a setting. At depths between 1 and 2 km, groundwater temperature is above 60 °C. More than 50 wells tap this low-enthalpy resource to use the hot water to heat domestic and industrial buildings. The corrosive, saline groundwater transfers heat to fresh water through heat exchangers, which then circulates through housing complexes at around 55 °C, before returning to the heat exchanger. Such a system is little different from the circuit of a conventional gas or electrical heating boiler, except for its scale. The cooled geothermal fluid is then re-injected into the aquifer. One advantage of the Paris Basin is that the deep groundwater is semi-artesian (Smith, 2005), so that less pumping is needed. Each heating circuit typically supplies between 3 and 5 MW of power over its

Figure 5.9 Cross-section of the Paris Basin, showing the depth to the 60 °C isotherm — temperatures above 60 °C shown by darker shading.

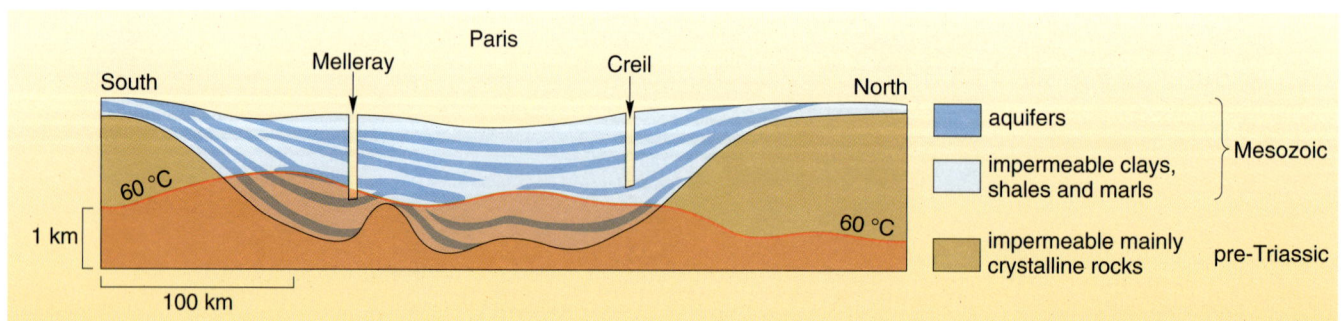

designed lifetime (30–50 years). Although this scheme is only marginally economic — during the oil glut of the early 1990s, development work all but shut down, the scheme saves the equivalent of 2×10^5 t of coal per year.

A similar scheme that supplies 1 MW of power was installed in Southampton in the 1980s to provide central heating and hot water for the city's Civic Centre, the surrounding area and a local swimming pool. A single well penetrates a Triassic sandstone at a depth of almost 2 km beneath the city, in which groundwater is at 70 °C. Semi-artesian conditions bring the water to a depth of about 650 m, thereby reducing the need for pumping. The spent water is discharged into the sea, the aquifer being naturally recharged where it is exposed at the land surface.

Between 1992 and 2000 the global capacity of direct geothermal heating increased fourfold from 4 GW to 16 GW. Compared with heating that uses fossil fuels, supplied very efficiently either by gas pipelines or electricity grids, direct geothermal heating is economic only when fossil fuel prices are high. The capital outlay on wells and distribution systems is the main financial burden, making direct geothermal heating unattractive for small-scale development. However, there is an approach that may become viable for a great many more individual consumers or small housing schemes.

In soil, even at shallow depths, geothermal heat flow and heat retained from summer solar warming ensures that subsurface temperatures are higher than winter air temperatures. This very-low enthalpy source can be exploited using a heat exchanger, known as a **ground-source heat pump** that uses upward and downward pipes about 100–150 m long set into the ground (Figure 5.10). A heat transfer fluid, usually water, circulates within the loop to transfer heat from the ground to the properties above. When demand is low during summer the

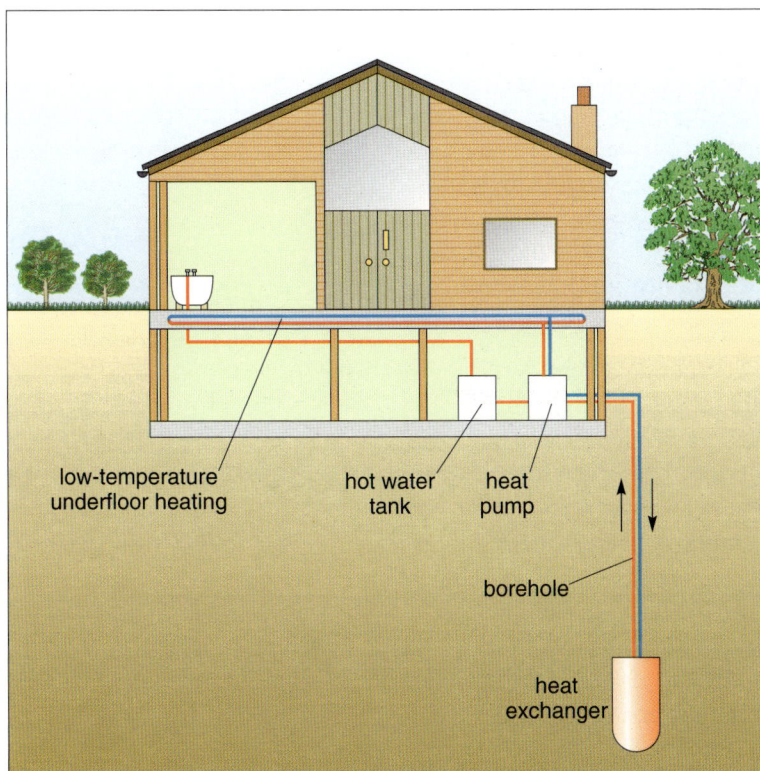

low-temperature
underfloor heating

hot water
tank

heat
pump

borehole

heat
exchanger

Figure 5.10 Ground-source heat pump (not to scale). Some designs use a system of shallower pipes, rather than a borehole.

underlying soil and rock reheat, ready once again for winter use. A variant on this approach is to use heat pumps — effectively reversible air conditioners — to pump excess heat downwards in summer and upwards in winter; at least some of the high energy cost of air conditioning is recycled.

Ground-source heat pumps can be deployed almost anywhere, and there has been a recent explosion in their popularity. The approach was developed simultaneously in the USA and in Europe, particularly Switzerland and Sweden. Installations in Switzerland are growing by 10% a year, and at an even higher rate in the USA, where there are already over half a million units. If 1000 residential units at the latitude of the UK were to have such geothermal heat pumps for winter heating and summer cooling, peak demand for electricity could be reduced by 1.7 MW, or gas demand by 0.7 TJ per year. The UK is a late-comer to this technology, but a new health centre in the Isles of Scilly now uses a 25 MW reversible ground-source heat pump to supply hot water, heating and cooling.

5.1.6 The pros and cons, and future of geothermal energy

Geothermal energy is renewable but the fluids emit gases such as CO_2, H_2S, SO_2, H_2, CH_4 and N_2 when used for electricity generation. However, geothermal power plants are usually sited in areas of natural geothermal activity, where such emissions occur anyway. Other potential pollutants are various ions dissolved in the geothermal fluids, but these are almost always returned to the reservoir when the spent fluids are re-injected.

As regards safety, accidents are rare, although in 1991 a well failure at a geothermal plant on the flanks of a volcano in Guatemala vented hundreds of tons of rock, mud and steam. Protracted geothermal exploitation can induce ground subsidence, similar to that which occurs when groundwater is extracted, although this is minimized by re-injecting spent fluids. Small earthquakes sometimes result from large-scale exploitation. However, most high-enthalpy geothermal areas are naturally prone to seismicity.

The main disadvantage of geothermal power generation is that suitable high-enthalpy areas are geographically very restricted, many being in areas of low population density (or under the sea). Conversely, the very low-enthalpy potential of normal heat flow is universal, and potentially useful for heating and even air conditioning, given the necessary investment.

In 2003, global electricity generating capacity from geothermal energy was 8.4 GW (Table 5.1), equivalent to just eight power stations using fossil fuels. Existing technology is potentially capable of providing global electricity generating capacity up to 72 GW, rising to 138 GW with HDR exploitation. Yet that falls well short of the 2004 globally *installed* nuclear power capacity of 362 GW (Section 4), and would only provide about 9% of current world electrical generating capacity (1.5 TW). Moreover, as with all energy resources, there is an absolute upper limit to geothermal power.

● The Earth's surface area is about 5×10^{14} m^2, and its average surface heat flow is 87 mW m^{-2}. What is the total geothermal power of the Earth compared with present total power use in TW (Section 1.2.2)?

● Total geothermal power = 5×10^{14} m^2 × 87 mW m^{-2} = 4.35×10^{16} mW = 4.35×10^{13} W = 43.5 TW. The total power use world-wide in 2002 was 14.3 TW, or about one-third of total geothermal power.

However, only 13.5 TW of this *total* global heat flow passes through the continental crust, where it can be used most conveniently. Converted into annual emission of energy, continental heat flow totals 425 EJ globally — about the same as global annual energy use (451 EJ in 2002). Only a tiny amount of this continental heat flow gives rise to high-enthalpy geothermal conditions, so is clear that heat flow is destined to play only a minor role in future electricity generation. Using low-enthalpy sources for heating (Section 5.1.5), however, has far greater potential.

The vast amount of heat (4.3×10^7 EJ) *stored* in the top 3 km of the Earth's crust, might suggest that geothermal energy could have a much larger role to play. However, exploiting heat stored by rocks would be akin to mining, once use exceeded the annual heat flow; it is then a non-renewable resource, however vast.

Table 5.3 shows estimates by the International Geothermal Association of the ultimate geothermal potential, assuming advances in geothermal technology, on a continent-by-continent basis (including some use of stored geothermal heat). The total electricity-generating potential of 2.24×10^4 TWh yr^{-1} is 45% more than total world electricity production in 2001 (1.55×10^4 TWh). Note also that low-temperature resources are more than three times current primary energy use globally. To achieve such output from current levels clearly requires enormous growth in the geothermal industry.

Table 5.3 Ultimate potential development of geothermal energy.

Geographic region	High-temperature resources suitable for electricity generation/TWh yr^{-1}	Low-temperature resources suitable for non-electrical uses/ EJ yr^{-1}
Europe	3700 (13 EJ yr^{-1})	>370
Asia	5900 (21 EJ yr^{-1})	>320
Africa	2400 (9 EJ yr^{-1})	>240
North America	2700 (10 EJ yr^{-1})	>120
South and Central America	5600 (20 EJ yr^{-1})	>240
Pacific	2100 (8 EJ yr^{-1})	>110
global potential	22 400 (80 EJ yr^{-1})	>1400

Figure 1.6 showed the proportion of energy delivered as electricity in the UK; around 12% of the total energy 'delivered' to users, the rest being by burning coal and petroleum products. It is about 1 EJ yr^{-1}, i.e. about five times current *global* electricity production from geothermal sources and around 1.25% of the total in Table 5.3. The UK is not well-known for its active volcanoes and geysers, and although it may have some HDR potential, the UK's geothermal electrical energy cannot supplant more conventional supplies. However, the UK could probably extract sufficient low-enthalpy energy to cater for a significant proportion of space heating.

In the short term, even twenty- to forty-fold growth in geothermal capacity will not result in it being a significant contributor to global energy needs by 2020. Installing capacity takes time (note the average annual changes in capacity in Table 5.1) and of course money, whatever the benefits. You should recall that geothermal power plants are generally much smaller than fossil-fuel and nuclear generators; tens of MW, rather than the tens of GW for the biggest 'conventional' plants. To replace a single 1 GW fossil fuel power station requires between 20 to 33 geothermal stations (Section 5.1.4). The seeming benefit of local and 'environmentally friendly' geothermal power generation, in the economic context of their construction, delays its adoption at national to international scales. Direct-use geothermal projects suffer from slow growth also, because of economic factors. Even though installation of domestic ground-source heat pumps is growing at 10% per annum in a few countries, that growth rate would need to be sustained globally for a great deal longer than 20 years to make any real difference to energy use patterns. That issue characterizes all alternative energy sources (Section 5.8).

5.2 Solar energy

The Sun will radiate energy until it ceases thermonuclear fusion, in around 5 billion years. As you saw in Section 1.3.1, the solar power that enters the Earth's system is 1.1×10^5 TW (0.3×10^5 TW to atmospheric heating and 0.8×10^5 TW absorbed at the surface — Figure 1.7). This is equivalent to a global energy supply of 3.5×10^6 EJ yr^{-1}. About 33% heats the atmosphere and contributes to setting winds and waves in motion. Of that reaching the Earth's surface, 70% falls on the sea, setting in motion ocean currents and a large proportion of the circulation of water vapour in the atmosphere because of evaporation from the ocean surface. The remainder falls on the land. Figure 1.7 shows the complexity of primary redistribution of solar energy through interlinked surface systems:

- the carbon cycle based on photosynthesis;
- atmospheric circulation and the water cycle;
- winds and ocean waves; and the ocean current system.

Each of them is a potential source of useable energy, as covered in later sections: this section discusses the *direct* use of solar energy.

In every case, with the exception of the energy available from surface water flow (Section 5.5), humanity comes nowhere near exploiting the Sun's potential to supply useable energy; in fact, we really do not know the practical limits. Whatever those are, they will not disappear as a resource — all are renewable. Compare this with the solar energy stored chemically by the degraded products of photosynthesis in fossil fuels. Although carbon burial adds continually to that resource, its pace of renewal (between 1 to 10 GW — Figure 1.7) is about 2000 times slower than we use it. Fossil fuels are non-renewable and declining extremely quickly in terms of human history.

In British summers direct sunlight (**insolation**) on a clear day can reach power densities of up to 1 kW m^{-2} when the Sun is at its highest in the sky (the daytime average is about 450 W m^{-2}), and supplies around 6.5 kWh m^{-2} per day (450 W m^{-2} × about 14 hours of daylight in June) (Figure 5.11a). This is enough to heat a bathful of water to a comfortable temperature. Summer insolation in southern Europe (~30° N on Figure 5.11a) reaches 7.5 kWh m^{-2} per day (by virtue

(a)

(b)

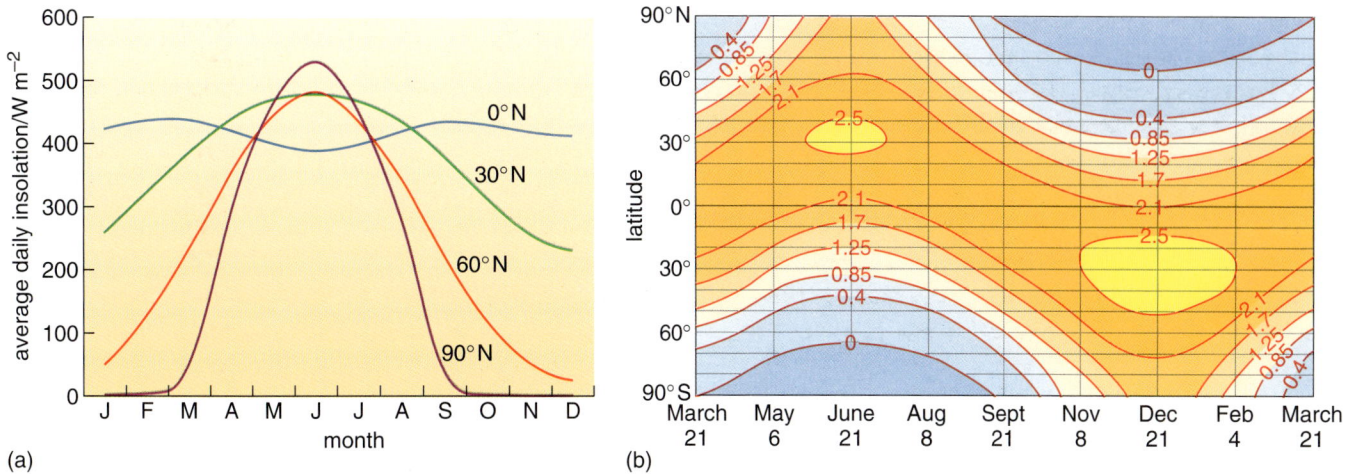

Figure 5.11 (a) Seasonal variation of the daily average incoming solar power (W m⁻²) at the Earth's surface, under *cloud-free* conditions. Variations in solar power are shown for four northern latitudes, and allow for variations in the length of the day and that of the Sun's elevation above the horizon. (b) Seasonal variation with latitude of daily incoming solar energy (in 10^7 J m⁻²) — assuming cloud-free conditions. The axes are latitude and time of year (this is not a map), and the contours are for solar energy received per day under cloud-free conditions. (*Note*: There is a difference between Northern and Southern Hemispheres because during the Southern Hemisphere summer the Earth is closer to the Sun than it is during the Northern Hemisphere summer.)

of a higher Sun angle). Winter insolation is much lower. In January, northern Europe (~60° N on Figure 5.11a) has an average insolation around 10% of its summer value and around only 20% of the January value in southern Europe. In the tropics (0° N on Figure 5.11a), insolation shows far less seasonal fluctuation. Solar energy collection in northern Europe is therefore most effective in the summer, in southern Europe there is useful winter potential as well, and at low latitudes it can be used throughout the year. High latitudes have a problem with solar energy; it is most abundant when it is least needed, i.e. in summer. Although variations in cloudiness reduce insolation, diffuse solar energy still reaches the surface at a power density that is often useable.

Curvature of the Earth's surface and the inclination of its rotational axis relative to the Sun result in insolation declining from a maximum in the tropics towards higher latitudes (Figure 5.11b). There is also considerable variation with the seasons. Satellite data give a more realistic view of the amount of available solar energy across the planet (Figure 5.12). One surprising feature is that summer months in Greenland and Antarctica have the highest *actual* insolation on the planet. That is partly because the Sun shines for 24 hours each summer day, but

Figure 5.12 Average available insolation (i.e. depending on cloud cover) at the Earth's surface for (a) January (1984–1991); (b) July (1983–1990) from NASA satellite data. The 'rainbow' colour range (blue–green–yellow–red–magenta–white) represents average power densities through the day (0–>350 W m⁻²). Values of available solar insolation are at their highest over Antarctica in January and over Greenland in July. High values also occur along the subtropics during Northern and Southern Hemisphere summers.

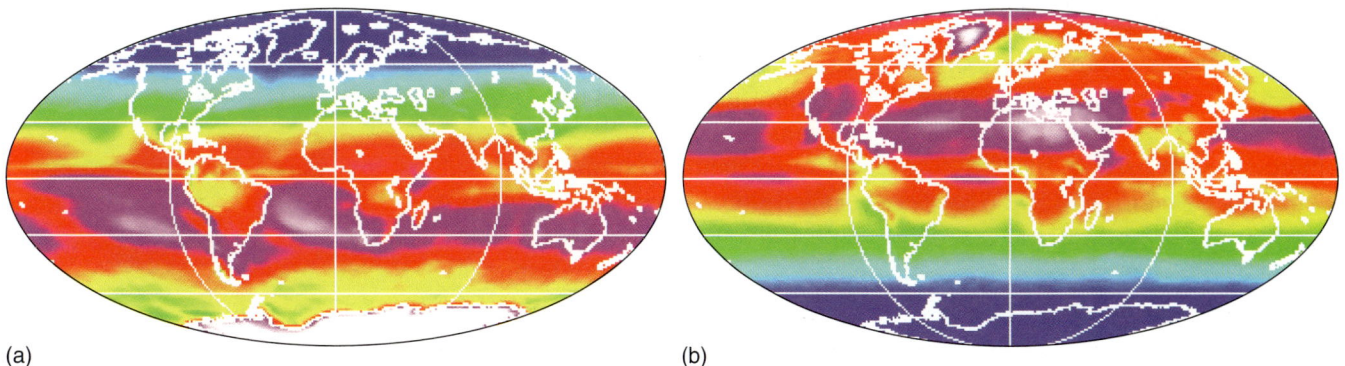

(a)

(b)

also because ice caps are usually free of cloud during summer months — high-pressure air masses exclude moist air from the oceans. (*Note*: Much is reflected away by ice and snow.) Bar that oddity, available solar energy varies roughly as expected, but always lower than theory predicts (Figure 5.11) because of cloud cover.

Direct solar energy use is of two basic kinds; using it as a source of heat (Section 5.2.1) or converting it into electricity using photovoltaic technology (Section 5.2.2).

5.2.1 Solar thermal energy

Solar heating of trapped air, water and solids has been used for centuries, but modern architectural design can enhance all three effects for space heating, hot water supply and heat storage. Such **passive solar heating** relies on short-wave radiation being absorbed by materials so that they heat up and then slowly re-emit long-wave radiation. The most obvious example is inside a greenhouse, where solar radiation that passes through the glass heats the inside air to temperatures well above those outside. Glass is more transparent to short-wave radiation than to long wavelengths, so a glazed room absorbs and retains heat, especially if the windows face the Sun. By absorbing solar radiation, house bricks, beach sand and rocks can become uncomfortably hot, even in the UK. The darker solids are, the more short-wave radiation they absorb — black cars become hotter inside than do white ones on sunny days.

The most common design using water as a heat absorber is a flat glass plate set above a waterproof black surface (Figure 5.13a). Water flows slowly between the two, becomes heated and can either be stored in tanks or pumped through radiators for space heating. Greater efficiency is achieved by the use of fluids that are more volatile than water. They are vaporized in the gap and then pass through a condenser, where the gas to liquid transformation releases the latent heat of vaporization imparted to the gas by solar energy; a form of heat pump. Some solids require more energy to raise their temperature than others; they have a high *specific heat capacity*. If this property is combined with that of slow release of energy — because of high density and low thermal conductivity — such solids are ideal for night-storage heaters. Conventionally, such storage devices are heated by electricity, but can be used in walls to absorb solar energy. The more massive they are, the more energy they absorb, and the longer they are capable of heating spaces. Used in well-insulated buildings, solar energy can be the sole means of space- and water-heating.

Passive solar heating systems first appeared during the 19th century, particularly in the southern states of America, where water filled tanks were warmed behind glazed walls. As cheap and more convenient oil and gas came to dominate domestic heating during the 20th century this bulky approach all but disappeared. Since the late 1960s, growing environmental consciousness has brought renewed interest in solar heating. By 2001, 57 million square metres of simple solar collecting panels (Figure 5.13a) had been installed globally, mainly in industrialized countries. That they are not even more widely used reflects their high price, driven by high demand relative to a restricted number of suppliers. Despite current high costs of installation, in the long-term almost totally free energy input makes passive solar heating a good investment.

(a)

(b)

(c)

Figure 5.13 Types of solar collector. (a) Passive solar energy collectors mounted on the roof of an eco-hotel in Essaouira, Morocco. The hot water obtained from this system is also used in the local *hammam*, or Turkish bath. (b) A solar cooker based on mirrored plates arranged in the rough form of a parabolic mirror. The point of focus is the base of the pot, and little heating occurs away from that, so such cookers are safe to use. (c) Experimental solar furnace in the French Pyrenees.

A more technically advanced exploitation of insolation uses mirrors to focus solar radiation. **Active solar heating** is increasingly used in poor countries for cooking (Figure 5.13b), thereby reducing the demand for fuelwoods and the risk of cancers from smoke inhalation. Similar, roof-mounted devices heat water for domestic use or radiators to much higher temperatures than does passive solar heating. On a larger scale, arrays of large reflectors boil water to power turbines and generate electricity, or create extremely high temperatures in solar furnaces (Figure 5.13c). The mirrors are motorized and controlled so that they track the movement of the Sun.

5.2.2 Photovoltaic conversion of solar energy

You have probably used direct conversion of solar energy to electricity though a solar cell that powers a pocket calculator or a solar 'trickle' charger to top up car batteries. Both exploit the **photovoltaic (PV) effect**, first described in 1839 by Edmond Becquerel. He observed that the voltage of an early 'wet cell' battery increased when its silver plates were exposed to light. In 1877 Cambridge physicists discovered that selenium crystals created an electrical current when exposed to light. A New York electrician, Charles Edgar Fritts, devised the first working PV cells in 1883, using selenium plates covered in gold wires, but they converted less than 1% of the incident solar energy into electricity. Since then

PV technology has been mainly concerned with improving efficiency, so that today's solar cells convert up to 25% of incident solar power to electricity.

In the 1950s the Bell Telephone Company's laboratory in New Jersey, USA experimented on the effect of light on **semiconductors**, that had been invented as miniature analogues of the valves used in radios. These first transistors were made of silicon that had been 'doped' with minute amounts of other elements, including boron or phosphorus. The impurities increase the electrical conductivity of poorly conductive silicon to that intermediate between non-conductors and metals — hence the name, semiconductor.

The outer electron shell of a silicon atom is only half filled, leaving *four* unfilled electron sites. In crystalline silicon the atoms share their electrons. Consequently, pure crystalline silicon is a poor conductor of electricity because the outer electrons are barely able to move. 'Doping' the silicon with elements whose atoms have different numbers of outer electrons modifies this reluctance to conduct electricity.

Phosphorus, with *five* electrons in its outer shell, is one example of a doping element. For each phosphorus atom that bonds with an adjacent silicon atom there is an excess electron. So silicon doped with phosphorus has a slight overall *negative* charge; it is an *n-type* semiconductor. When a photon of light meets a doped silicon–phosphorus pair the weakly bonded, 'excess' phosphorus electron breaks free and moves. Passage of electrons through the silicon lattice creates a weak electrical current and an electrical potential.

Silicon doped with boron is left with *unfilled* electron shells, since boron has *three* outer electrons, leaving the boron-doped silicon with a slight *positive* charge; this is a *p-type* semiconductor that is able to *accept* electrons. Layering together both n- and p-type semiconductors in wafers (Figure 5.14a) enhances the PV effect. Negatively charged electrons from the illuminated n-type cross the gap between the wafers to restore electrical equilibrium. However, despite the high electrical potential across the gap, large positively charged silicon atoms cannot cross it, and so only electrons can flow. The resulting current is conducted to its point of use through metal ribbons embedded in the PV wafer, which inevitably block some incoming photons. Silicon is highly reflective, so an antireflection coating is applied to the cell surface to maximize photon capture. Doped wafers and electrical conductors are generally mounted in a vacuum between glass sheets as **solar PV cells** for assembly in tiled arrays (e.g. Figure 5.14b).

Solar PV cells based on doped crystalline silicon have a theoretical maximum efficiency of 28%, because only a narrow range of short wavelength radiation from the Sun has the appropriate energy to displace electrons. In practice, they convert about 15% of the incoming solar energy in that wavelength range to electrical current. Crystalline silicon is durable but expensive to manufacture. Fine-grained, doped silicon formed into wafers is cheaper, but its efficiency is around 5 to 7%, and it eventually degrades. Current research centres on low-cost semiconducting materials, one promising candidate being fine-grained titanium dioxide — the white pigment used in paints, and therefore cheaply available. Combined with dyes to increase the sensitivity of TiO_2 to sunlight, it has a practical PV efficiency of 7 to 12%, and degrades very slowly. Titanium dioxide occurs as a natural mineral (*rutile* — TiO_2), concentrated in some beach sands because of its

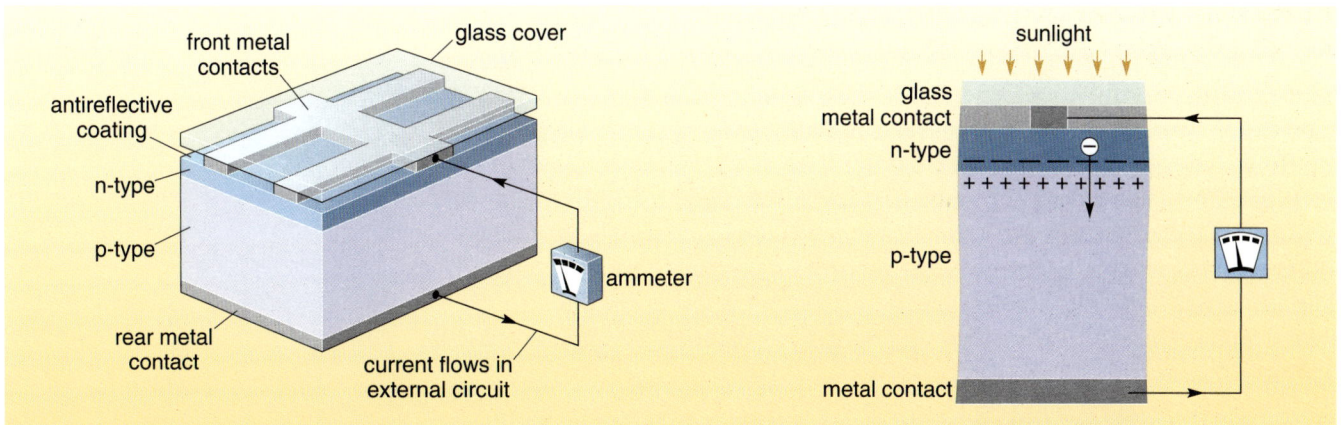

(a)

(b)

Figure 5.14 (a) Simplified photovoltaic cell, showing its bonded, multilayered structure. The 'working parts' consist of bonded wafers of n- and p-type semiconductors (enlarged). The moving electrons produced in them by the PV effect have to be conducted away. Metal contacts are bonded onto the top and bottom surfaces of the pair of semiconductors, using metal ribbons that are as narrow as possible so that as much sunlight as possible can pass through. Another layer above the top conductor lessens reflection of light, and the whole cell is sealed by a glass cover. The bonding has to be as near perfect as possible, free of air bubbles and extremely durable. (b) Solar PV cells used by Berber householders in the Atlas Mountains, Morocco. They show clearly as rectangular white frames about 1 m × 0.3 m in size, as near the woman preparing mint tea, contrasting with the general irregularities of traditional rural life in Morocco.

high density. Titanium dioxide embedded in thin plastic films may prove to be a much cheaper alternative to silicon cells. Research into PV materials is understandably a 'hot topic', and among novel possibilities are organic compounds that behave as PV semiconductors, and could be formed into flexible and strong solar cells.

- What is the main practical problem that might inhibit the widespread deployment of solar cells? (Think of your household electricity supply, and refer back to Section 1.3.1.)

- *Direct* current (DC) is produced, whereas all appliances linked to the 'mains' use *alternating* current (AC) generated by conventional turbines. Photovoltaic cells need DC–AC *inverters* to supply alternating current to the grid system (see Chapter 1).

Another drawback of photovoltaic solar power is the area needed to give a useful supply; a 1 m^2 solar panel operates at an average power of about 30 W, so three are needed to light one 100 W light bulb. To put this in perspective, to supply the entire electricity needs of New York City with solar power would require solar cells covering an area of 192 km^2; equivalent to about 25% of the city's area.

5.2.3 The future of direct solar energy use

The immense global energy flux from the Sun makes it the prime candidate for future sustainable energy production. Both solar thermal energy and solar PV conversion involve technologies that can be deployed on personal through community to regional scales, using both simple and advanced technologies. You have probably seen solar PV panels that power automatic roadside weather stations and other low-drain communications systems. The panels require low maintenance and usually charge batteries to allow them to remain operational during the night. In poor countries where the energy infrastructure is rudimentary or absent, PV systems hold out great potential. An important use is for daytime pumping of water from wells.

In its 1997 renewable energy plan, the EU set a target of half a million village-scale direct solar systems to be deployed in developing countries and a similar number in European houses by 2010. The United Nations has asked world governments to deploy 4.5 GW of solar PV electricity generating capacity in developing countries by 2012. Both Japanese and German governmental subsidies have boosted both photovoltaic production and deployment, and a similar 'hard-sell' stems from commercial sources in the US. As a result, power capacity of solar PV rose from a global 50 MW in 1995 to over 2 GW in 2002, and is estimated to be growing at a rate of 40% per year. The main hindrance to greater deployment is simply that of cost; at between US$ 0.2 to 0.5 per kWh, solar PV electricity was almost ten times as expensive in 2005 as that from the cheapest fossil-fuel source, natural gas. To progress, the technology requires continued reduction in the cost of the solar cells themselves — but the enormous reduction in cost of silicon-based computer hardware since the 1970s is cause for optimism.

Solar PV could theoretically supplement grid-power during daylight hours to reduce generating costs and environmental emissions. However, at this scale serious disadvantages emerge. The daily intensity of sunlight varies dramatically because of cloud cover. Moreover, solar power is greater in the summer while the demand for electricity is lower, except in areas with high use of air conditioners. Provided photovoltaic conversion contributes no more than 10–20% of the total amount of electricity in the grid, its integration seems feasible. This is because electricity grid systems are designed to cope with large variations in demand, and they can cope equally well with fluctuations from different forms of supply.

Should future solar PV power rise above 20% of the total electricity supply, then existing grid systems built to be dominated by coal, oil and nuclear generation would have to be modified. This is because conventional power plants are slow to start up and shut down; they are **slow-response systems**. Solar conversion, along with other alternative sources whose power source fluctuates uncontrollably (e.g. wind and waves), is a **fast-response system**. A distribution grid with solar PV power as a major component would need to be supplemented by *controllable* fast response systems, principally hydroelectric and gas-turbine generators. Another solution would be short-term electrical storage installations, but they are both costly and inefficient. A means of 'having one's green cake and eating it', however, would be to use electricity from solar PV to generate hydrogen by electrolysis of water. Hydrogen gas is combustible, storable and moveable, and so avoids most of the problems associated with electrical storage. The hydrogen could even be converted back into electricity using fuel cells. We pick up on the theme of a clean, efficient, *hydrogen economy*, in Chapter 6.

Despite these caveats, the potential of solar PV is enormous. If photovoltaic conversion with 10% efficiency was installed over an area of 500 000 km^2 (about 1.3% of the area of tropical deserts) humanity's present energy requirements would be met. That outlook is probably far off. Of the electricity generated from all alternative energy sources in the early 21st century, solar PV contributes only about 0.02%, with solar thermal generation a little more significant at 0.06%.

5.3 Biomass conversion of solar energy

Photosynthesis in the geological past was responsible for all fossil fuel reserves, but its products are buried about 2000 times more slowly than we use them at present. The total carbon content of all biomass growing on land is estimated to be 5.6×10^{14} kg and, as Figure 1.10 shows, about one-fifth of this mass is renewed each year. Figure 5.15 shows how modern plant biomass is distributed across the continents. Clearly, biological conversion of solar energy to a chemical form in combustible materials presents a potentially vast resource. Using biomass as a fuel need not add to carbon dioxide in the atmosphere, because its products, CO_2 and water vapour, are exactly the same as those of natural aerobic decay. If plant photosynthesis is carefully managed to renew carbon consumed by using vegetable matter as an energy source, the natural balance of the carbon cycle can be sustained. About 6% of the total biomass that grows annually — containing about 7.5×10^{12} kg of carbon content — would meet all humanity's current demands for energy. However, such a perspective has nightmarish overtones because its production would involve about 11% of the global land area (a much greater proportion of fertile areas — see Figure 5.15). Nonetheless, 'high-tech' harvesting of solar energy via photosynthesis is a potential alternative to fossil-fuel use. There are five main approaches to extracting biomass energy: solid fuel combustion; gasification; pyrolysis; digestion and fermentation.

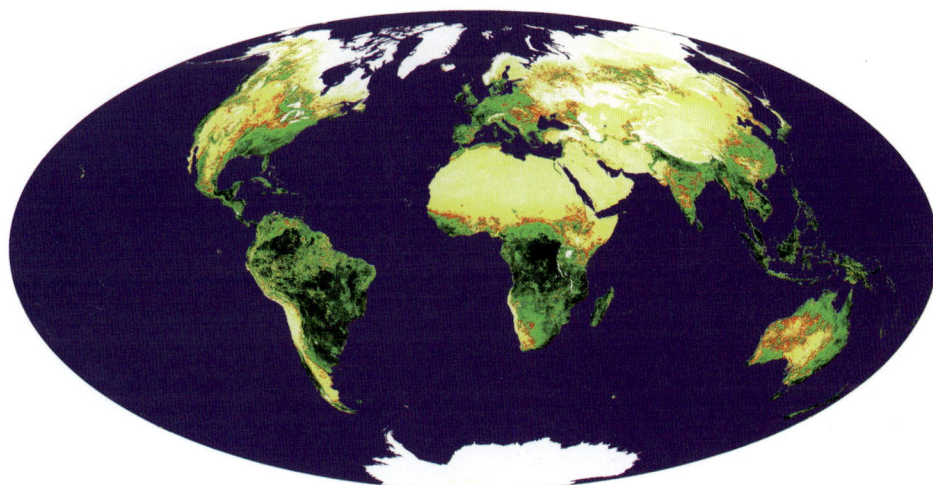

Figure 5.15 Estimated variations in plant biomass on the continents, based on NASA satellite data. Pale areas have low vegetation cover; red and orange indicates areas where vegetation varies widely from wet to dry seasons; the green areas have permanently dense vegetation cover, the darker the green, the greater the plant biomass.

Solid fuel combustion is self-explanatory. Although the main combustible is wood, animal dung and waste products — even carcasses — are also used. Although only at the stage of research and public discussion in the UK, power generation from waste combustion and conversion of combustible waste into fuel briquettes are growing industries in other parts of Europe. Some of the world's poorest countries obtain the majority of their energy needs from burning wood, animal dung and other biomass, but in doing so contribute to deforestation and desertification, and in the case of animal waste reduce the amount of natural fertilizers used on the land.

Traditional ways of burning biomass are inefficient. Open fires allow large amounts of heat to escape, and a significant proportion of the biomass is not fully burnt. Up to 75% is released in unburned gases and particles as the fuel heats up; energy literally goes up in smoke. Deaths in Africa from smoke-induced diseases, particularly among women and children, are comparable with those from malaria and HIV/AIDS. However, there are simple technologies that increase burning efficiency, routinely used in modern multi-fuel stoves, and even in wood-burning power stations.

Pyrolysis involves heating biomass by very slow combustion in a low-oxygen environment, which is the ancient principle behind making charcoal (Figure 5.16a). Most of the water and other volatile contents are driven off to leave impure carbon. Almost the same energy content as the starting biomass is contained in half the mass, making the fuel easier to transport. Charcoal burns at a much higher temperature than the original biomass, and it is more flexible and useful in a range of manufacturing processes. Pyrolysis also produces a mixture of combustible gases — carbon monoxide, hydrogen and methane; a process known as **gasification**. By exploiting both processes the harmful impurities in solid fuel are removed by gasification, giving fewer pollution problems when charcoal is burned. By retaining the volatiles, rich in hydrogen and carbon monoxide, which would otherwise be lost, the industrial 'flash' pyrolysis process uses these gases as reducing agents to produce a hydrocarbon-rich liquid or 'bio-crude'. The charcoal that is bagged and sold for summer barbecues probably came from such a dual-purpose plant.

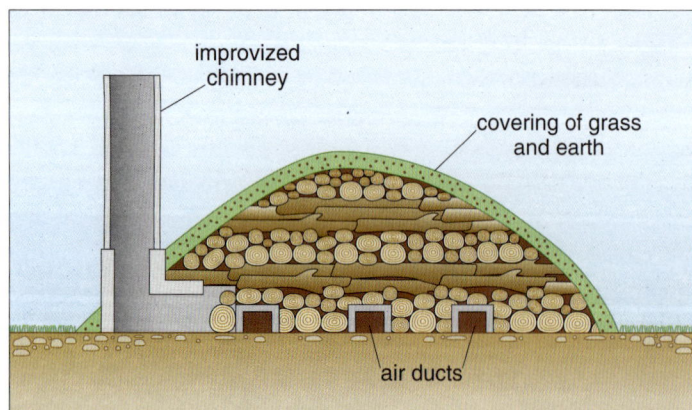

(a)

(b)

Figure 5.16 (a) Traditional charcoal burner. (b) Modern pyrolysis plant.

(a)

(b)

Figure 5.17 (a) Indian village biogas generator. (b) Industrial biogas plant.

Digestion of biomass involves its metabolism by anaerobic bacteria, which do not use oxygen and therefore do not completely oxidize carbohydrates to CO_2 and H_2O. Two groups of these bacteria yield hydrogen and methane, which are emitted by natural swamps. Semi-liquid biomass, usually human or animal waste, is collected in gas-tight digesting tanks that are 'seeded' with anaerobic bacteria. Their metabolic gases (known as **biogas**) are then collected and distributed as an energy source. Waste digestion is typically very efficient, with up to two-thirds of the energy content of sewage being recoverable. Indian scientists pioneered this simple technology in rural areas and the concept has caught on for industrial development (Figure 5.17). The same anaerobic digestion occurs in landfill where domestic rubbish is dumped. At many UK landfill sites, the powerful greenhouse gas methane is vented to the atmosphere, adding to global warming. Once landfill is capped with a layer of clay it is relatively simple to extract and pipe methane to mix it with domestic gas supplies.

Using bacterial yeasts to ferment the sugar content of plant matter into ethanol has been around for thousands of years. You may be astonished to learn that in Sweden aquavit (a caraway-flavoured spirit usually made from grain) is made from lichen! Generating fuel from biomass **fermentation** is merely a diversion of brewing and distilling skills to a more mundane application. Virtually any plant material can be fermented to produce ethanol, including sugar cane, fibre from fast growing trees and biomass waste. The leader in this technology is Brazil, where ethanol derived by fermentation of sugar cane is mixed with conventional diesel to make 'diesehol'. In fact a diesel engine will even burn chip fat, but diesehol is a great deal less polluting. By tailoring genetically engineered bacteria to the material to be fermented, yields and efficiencies can be greatly increased. There is growing interest in the simpler technology of converting oils from widespread crops, such as oilseed rape and maize, to produce diesel-like liquid fuels.

The main advantage of biomass conversion lies in reducing the emissions of carbon dioxide that are released by burning fossil fuels. Products from the various conversion technologies are suited to a variety of conventional uses;

vehicle fuels, means of cooking and space heating, and even electricity generation. However, to become substantial contributors to global energy consumption they require vastly increased exploitation of both agricultural and hitherto unused land. The greatest biological productivity is in the humid tropics, mainly areas of extreme poverty and intensive subsistence agriculture. Biofuels, except for those generated from human and animal waste products, are currently income-generating crops. Replacing the production of staple foodstuffs by other 'cash' crops in low-income areas has been implicated in worsening poverty, as a result of declining commodity prices (e.g. cotton, coffee and sugar), and removal of land and people from food production.

5.4 Wind energy

Wind energy was the fastest growing power source at the start of the 21st century, yet wind-driven mills and pumps, and nautical sails for transport were, along with waterwheels, the first mechanical devices to power industrial production. The advantages of harnessing wind energy are obvious; it is free, clean and widely available (but see later). Although a favoured source of 'green' energy, the increasing deployment of wind turbines where they are most efficient, on hilltops and coasts, together with their increasing size has led to outcries because of their impact on landscapes (Figure 5.18).

The USA and Canada, where the 'green' movement gained its early momentum in the 1970s and which have vast tracts of open land, once led in developing wind-power technology. However, countries in the EU took over that initiative at the start of the 21st century, with 23 GW of potential production installed in 2004

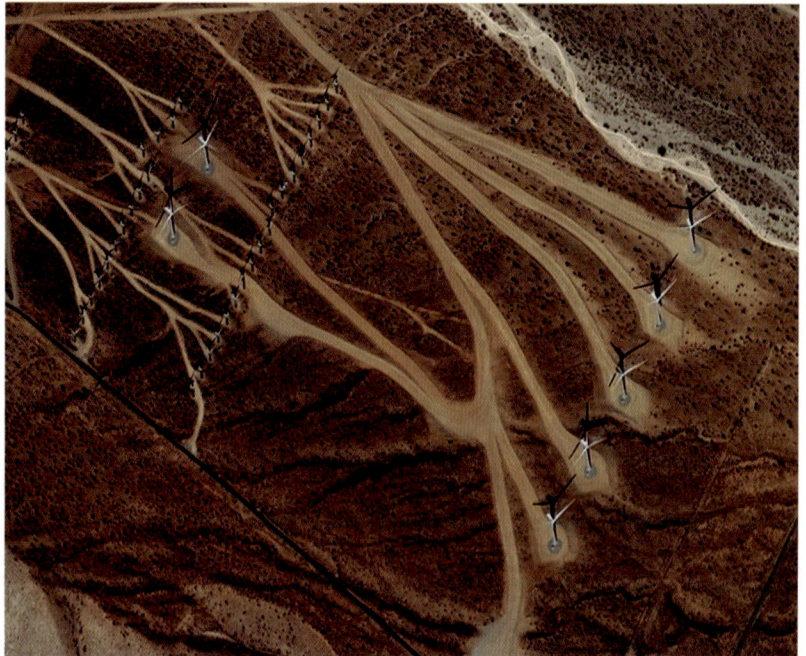

(a)   (b)

Figure 5.18 (a) Wind turbines at Tebay, Cumbria. (b) Aerial view of wind turbines in Wyoming, USA, showing the network of access tracks.

(12 GW in Germany alone), compared with only 4.6 GW in the USA. The UK, incidentally, has a mere 0.5 GW of installed wind power, despite the enormous fuss made by those for and against that technology.

The power available to a wind turbine depends on its size and the wind speed. The kinetic energy of wind is proportional to the mass (m) of air moving through a turbine and the square of its speed (v):

$$\text{kinetic energy} = \tfrac{1}{2}mv^2 \tag{5.3}$$

In one second, the mass, m, of air passing through a turbine is given by

$$m = \rho Av \tag{5.4}$$

where ρ is the density of air, A is the swept area of the turbine (i.e. the area swept out by the rotating blades) and v is the wind speed.

Therefore, the kinetic energy per second, or the power available to a turbine is

$$\text{power} = \tfrac{1}{2}(\rho Av)v^2 = \tfrac{1}{2}\rho Av^3 \tag{5.5}$$

So, wind power is proportional to the *cube* of wind speed. Consequently, the output of a wind turbine varies dramatically with wind speed.

The wind's kinetic energy forces the aerodynamic blades of a turbine to rotate, but air that has passed through the swept area becomes slower and more turbulent (more on this shortly). Because of this, only a proportion of the available wind energy, between 30 to 40%, can be converted into electricity (the efficiency of a wind turbine). Moreover, there are very few parts of the world with a constant and useful wind speed, so power generation can be considered only as a local resource (Figure 5.19a). Losses due to friction mean that a minimum wind speed of around 5 m s^{-1} is needed to start the blades rotating. With speeds over about 12 m s^{-1} damage to the turbine might result, so the turbine blades are progressively twisted (feathered) to present a smaller effective area to the wind. The net outcome is that the average power at a well-selected site is only around 30% of a turbine's rated capacity.

Question 5.3

Calculate the power output of a wind turbine that has an efficiency of 40%, and whose blades (about 5 m long) sweep an area of 300 m^2, for a wind speed of 10 m s^{-1}. The density of air is 1.2 kg m^{-3}.

The answer to Question 5.3 is not particularly impressive.

● How can the output be increased?

○ Increasing the length (r) of the blades, and hence the swept area A ($= 4\pi r^2$), increases the power.

With blades 10 m long the swept area is about 1200 m^2, and the power available from a turbine rises by four times, i.e. doubling the length of turbine blades quadruples the power output. This simple geometrical relationship lies behind the

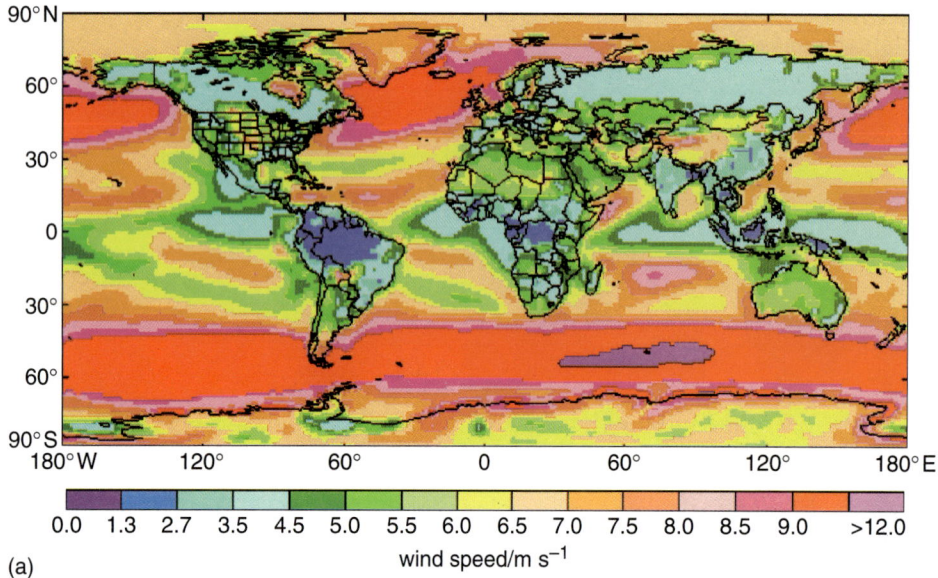

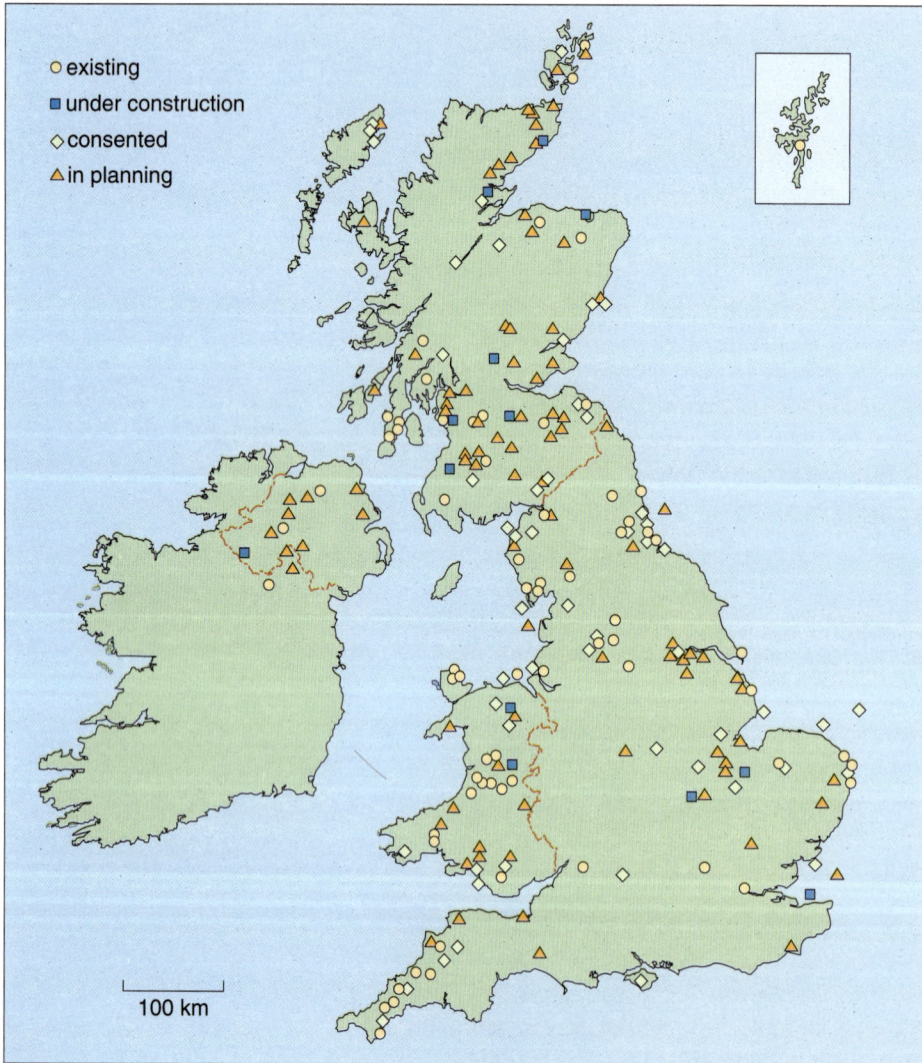

Figure 5.19 (a) Global estimates of average wind speed at 50 m above the Earth's surface. (b) Wind power developments in the UK (2004).

progressive growth in the size of wind turbines. In 2005, a Danish company built a wind turbine rotor blade 63 m long, using lightweight fibre composites. Using the same parameters as in Question 5.3, such a turbine would have a power output of around 25 MW at a wind speed of 10 m s^{-1}. It would stand between 150 to 200 m high.

⦿ Apart from avoiding awesome damage to any unwary vessel (it would be an offshore turbine), why do you think this vast structure would need to be considerably higher than the diameter of the rotating blades?

⦿ Because of surface drag, wind speed near the sea surface would be both slower and more turbulent than higher in the atmosphere. Such a huge turbine would encounter mechanical problems unless the wind speed was almost constant across its diameter.

These calculations show that to supplant a 1 GW fossil-fuel or nuclear power plant with wind power requires a great many of even the largest wind turbines — more than 40 in the case of the biggest. This has given rise to the well-known concept of 'wind farms', whether on land or offshore. Figure 5.19b shows the existing and planned wind farms in the UK, both onshore and offshore.

The economics that set the capital cost of turbines and their maintenance costs against income, together with considerable incentives in some countries, makes wind farms an attractive proposition for investment. The more so, as land which they occupy can yield other income too, as it seems that livestock become used to whirling blades. But there is another issue regarding wind farms.

Packing turbines closely on a single wind farm is not viable; turbines slow down the wind and create turbulence on their lee sides, which can extend many kilometres down wind. Designing the layouts of wind farms is critical, and many that you will see around the UK have relatively few turbines arranged in lines and offset to avoid downwind effects. Both offshore and onshore installations take up much more surface area per MW of output than a simple view might suggest; they are extremely large ventures. For instance, to supply the electricity demand of New York City using 1.6 MW turbines would require an area of 27.4 km^2; around 3.5% of the city's area. Hence the outcry at their impact on scenery.

As well as being steadier, wind energy can be up to five times greater at heights of 50 to 80 m — the height of the turbine axis on many modern wind turbines — than at the surface. Figure 5.19a shows how average wind speeds at 50 m above the surface vary globally. Note that average wind speeds over much of the western British Isles exceed 8 m s^{-1}. Globally, about 30% of the land surface has average wind speeds below the 5 m s^{-1} threshold for wind power generation.

Wind turbines have two basic configurations; horizontal axis wind turbines (HAWTs) include a vane so that the turbine blades are continuously pointed into the wind, whereas vertical axis wind turbines (VAWTs) can harness wind energy from any direction (Figure 5.20).

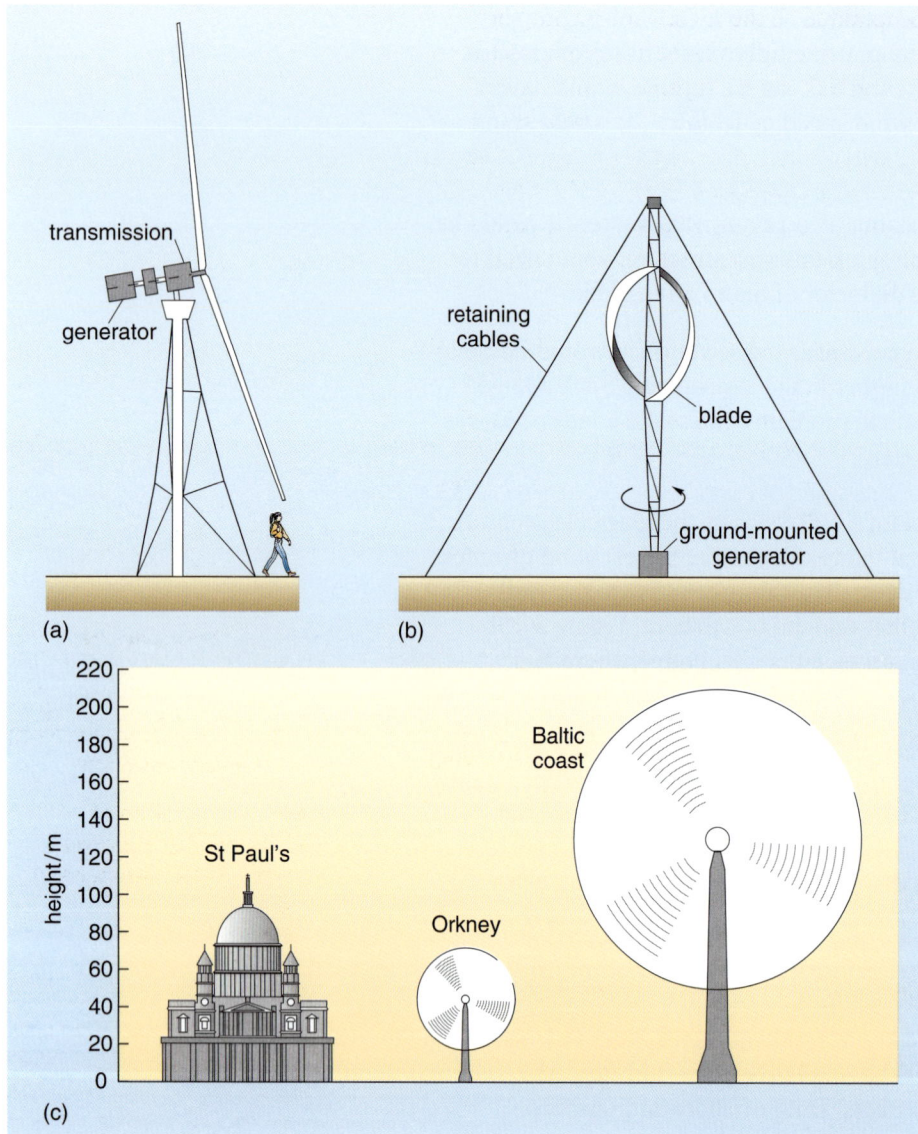

Figure 5.20 The two basic designs of wind turbines: (a) horizontal axis (HAWT), in which the vane is commonly incorporated in the design of the generating pod; (b) vertical axis (VAWT), in which the generator is on the ground. (c) The sizes of the largest onshore turbine, the 3 MW generator on Orkney, and the largest offshore turbine, a 25 MW generator planned for the German Baltic coast, compared with St Paul's Cathedral. *Note*: Power output is for a wind speed of 10 m s^{-1}.

5.4.1 The future of wind energy

A great advantage of using wind energy is that, unlike power generation from combustion of fossil fuels, it produces no gas emissions. Even a small 750 kW wind turbine operating with wind speeds just above that of turbine cut-off would reduce annual emissions to the atmosphere by 1200 t of carbon dioxide, 6.9 t of sulphur dioxide and 4.3 t of nitrogen dioxide, compared with the equivalent power output from coal-fired generators. Nevertheless, wind turbines and their infrastructures are substantial constructions with a short life span (around 30 years), so production of the materials used in their construction (e.g. concrete) would not be effect-free. However, wind energy is inexhaustible and free of fuel costs. Given advanced technology, wind energy could provide up to 20% of the energy needs of the United States. This is equivalent to the amount currently provided there by nuclear power.

Negative reactions to wind turbines can be somewhat subjective. They are noisy; a turbine designed to supplement power on a Milton Keynes housing estate has never operated, because residents complained of the noise. Yet by the very nature of the resource, many wind farms are usually placed far from dwellings. Moreover, modern turbines are much quieter than their predecessors, as they have to conform to more stringent noise pollution guidelines. Undoubtedly, wind farms that deploy many large turbines, either on high ground or in coastal areas, dramatically change the vista. That disturbs many people who live in the vicinity or who use the locality recreationally. Interestingly, an object of comparable height to a wind turbine, the Angel of the North on the outskirts of Gateshead, is the subject of local pride yet only serves a symbolic function.

The main practical problem with wind turbines is their variable output. Obviously, wind speed does not match the variation of electricity demand (Figure 5.21). Wind energy suffers from **intermittency of supply**. Whereas higher overall consumption of electricity in winter coincides with generally higher average wind speeds than in summer, demand varies by a factor of around two during normal days: plus the legendary power-demand surge for cups of tea at half time in televised soccer matches. Wind power by itself cannot cope with such socio-economic vagaries. The high capital investment in wind turbines requires that they maximize income, except when wind speed is too high or too low. That requires either complex switching between power sources or an energy storage system when wind-power output is high. One strategy is the use of excess wind power to pump water to high-elevation hydropower reservoirs (Section 5.5). Another is to ensure that the electricity grid accepts all power supplied by wind generation, while more controllable sources are deployed to match demand (see Section 6.4.2).

Wind energy is also geographically variable. In windy places, small-scale development combined with power storage devices could supply all the electricity needs of individual dwellings and small communities; an ideal environmental solution, but not an equitable one, which applies to other alternatives too.

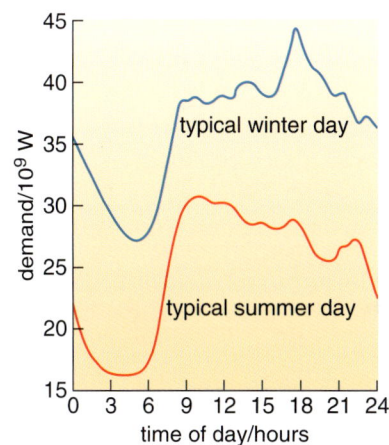

Figure 5.21 Variation of demand for electrical power during typical UK summer and winter days.

5.5 Hydropower

Hydroelectric energy is ultimately solar energy converted through evaporation of water, movement of air masses and precipitation to gravitational potential energy and then to the kinetic energy of water flowing down a slope. That energy was harnessed for centuries through the use of water wheels to drive mills, forges and textile works, before being supplanted by coal-fired steam energy. Electricity generation using water turbines, although first centred on constricted streams, has increasingly focused on dams and reservoirs that store gravitational potential energy until it is needed.

World capacity of hydroelectric generation has increased every year for over a century and by 2002 had reached about 740 GW, by far the largest alternative contributor to global electricity supplies. At the present time, large-scale hydroelectricity generation is increasing by about 50 TWh yr^{-1}, although there are fluctuations related to the precipitation in any particular year. This illustrates an important point about hydroelectricity; despite its ultimate reliance on the combination of solar and gravitational energy, the immediate mediator of output is the *water cycle*.

● What is the principal advantage of flowing water over moving air as a means of power generation?

○ Water is very much denser than air so that, for a given speed, the kinetic energy of a given volume is much greater than for an equivalent volume of air (recall Equations 5.4 and 5.5) — flowing water is a more concentrated form of energy.

The principle of hydroelectric power generation is simple; water flowing down steep gradients or from dams is channelled through pipes and transfers its kinetic energy to rotating turbines that drive electricity generators. Steep gradients ensure high flow speeds, and piping is necessary to maintain the pressure head. Some energy is lost to friction and turbulent flow in the pipes, in turbine rotation, and in conversion of mechanical energy to electricity in the generator. However, these losses are not great; indeed modern hydroelectric turbines have efficiencies exceeding 90%, so most of the potential energy of water stored in a reservoir can be converted into electricity.

The power developed depends on the product of the water discharge rate (Q, in $m^3\ s^{-1}$) and the **working head** (the vertical fall of water flowing into the hydroelectric plant, H, in metres) as follows:

$$N = Kg\rho QH \qquad\qquad (5.6)$$

where N is the power output in W (= $J\ s^{-1}$), g is the acceleration due to gravity ($9.8\ m\ s^{-2}$), ρ is the density of water ($10^3\ kg\ m^{-3}$) and K is the overall efficiency of the generating system. Clearly, installations that have a large working head require smaller discharge rates for equivalent power output than do those with a small working head. Hence mountainous countries, such as Norway, Switzerland and New Zealand, can rely greatly on hydropower, often using small water flow rates, because working heads exceeding 1000 m are often available. Today almost all of Norway's electricity is produced by hydroelectric means (40% of its entire energy use). However, since most of the best sites for large working head installations have already been developed in industrialized nations, attention is turning to improving the economics of using working heads of only 20–30 m on more slowly flowing rivers. Even working heads less than 10 m can give useful yields.

Question 5.4

Use Equation 5.6 to calculate the power output in MW from the River Severn's average discharge of $60\ m^3\ s^{-1}$ in its middle course. Assume a generating system efficiency of 80% and that a 5 m high barrage can be built safely.

Currently, the three main rivers that flow through the flat plains of India and Bangladesh (the Ganges, Brahmaputra and Meghna) discharge an average of $2.5 \times 10^4\ m^3\ s^{-1}$ to the Bay of Bengal. Their hydroelectric potential from single installations on each river is about 1 GW, using a 5 m head. Similar capacities characterize the lower courses of most of the world's major rivers. But this is no more than the capacity of moderately sized fossil- and nuclear-fuel generating plants, so despite their potential to act as flood regulators as well as power generators, major rivers have proved unattractive for hydropower development low in their courses. In their higher courses rivers have lower discharges, but where they flow through narrow valleys and gorges it is possible to build higher dams (with higher working heads). The Three Gorges Dam across the Yangtse River in China involves a dam 185 m high. With an average discharge of 1.3×10^4 m^3 s^{-1}, it has a generating capacity of 18.2 GW, and will be the largest single hydropower project on the planet when completed in 2009.

All electricity generation that exploits moving fluids requires a turbine to rotate the generator. Turbines driven by flowing water have different designs from those used with either pressurized steam or wind. The Pelton turbine (Figure 5.22a) is essentially the same as a water wheel, except it is submerged in the water flow. The water's force on cup-like buckets rotates a horizontal shaft connected to the generator. Turbines resembling boat propellers (Figure 5.22b) are placed in the piped water flow and rotate because of the aerodynamic shape of the vanes. Another basic design (Figure 5.22c) involves water flow entering the turbine radially from a spiral section of the supply pipe, and either horizontal or vertical shafts can drive the generator.

As well as hydroelectric projects designed for continuous supply, there are also projects where hydropower potential makes a minor contribution, compared with other kinds of generation, to a grid with a widely fluctuating demand (Figure 5.21). With these projects, the hydropower plant is not required to operate all day. Stopping and starting power generation can be very rapid; involving merely closing and opening the water inlets to cope with the fluctuating power demands. In such cases **pumped storage schemes** have been developed where the same water is recycled between two reservoirs at different topographic elevations. Electricity from other power stations, usually nuclear stations which are not easily closed down over short periods, is used during low-demand times, usually at night, to pump water to the upper reservoir. The process is reversed to generate electricity rapidly during periods of peak demand. The Dinorwic pumped storage scheme at

Figure 5.22 Basic water turbine designs: (a) waterwheel-like Pelton turbine; (b) propeller-like Kaplan turbine; (c) radial-flow Francis turbine.

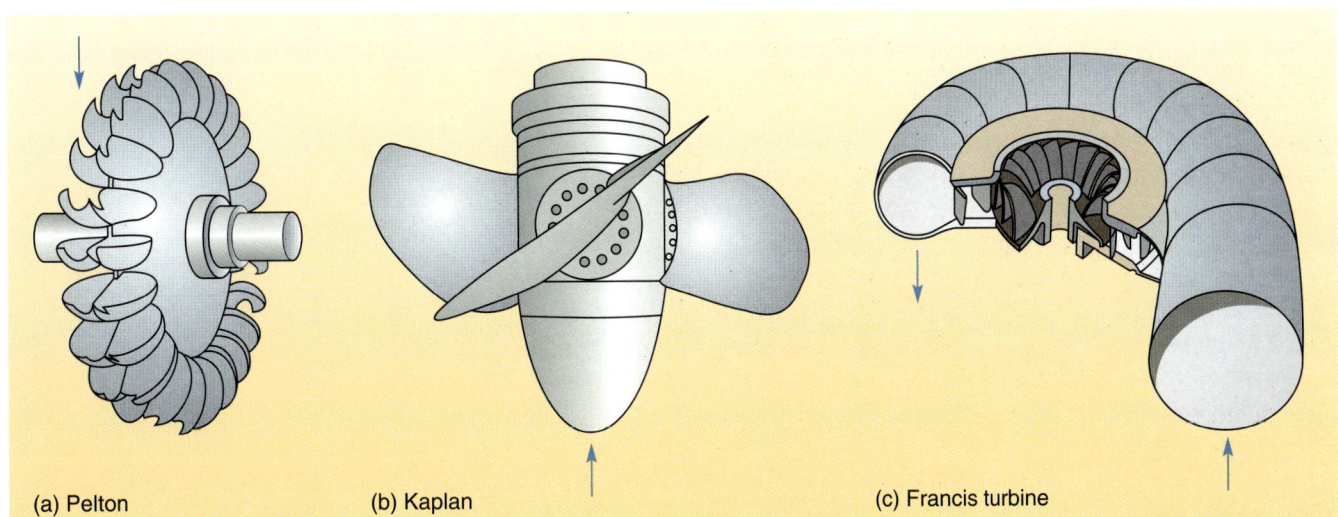

(a) Pelton   (b) Kaplan   (c) Francis turbine

Llanberis in North Wales (Figure 5.23), constructed in the 1970s to be linked with the nearby Trawsfynydd nuclear power station, is the largest such scheme in the UK; its maximum output of 1320 MW can be reached in 10 seconds during an emergency. The Trawsfynydd nuclear station developed structural problems in the late 1980s and was closed in 1993, so the Dinorwic scheme now relies on nighttime power from the UK National Grid.

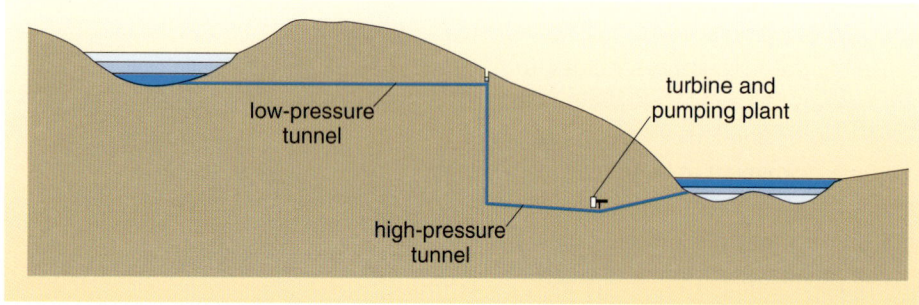

Figure 5.23 Cross-section of the Dinorwic pumped storage hydroelectric scheme at Llanberis, North Wales. The vertical shaft linking the low-pressure and high-pressure tunnels is 440 m long and 10 m wide. Different levels are shown in the reservoirs; when power demand is low, the upper reservoir is filled by pumping from the lower one. At times of peak demand flow from the upper reservoir generates power and fills the lower one.

Pumped storage schemes are net consumers of electricity because it takes more energy to pump water to the top reservoir than is returned when the same amount of water is released through the turbines. However, this method of *storing* electrical energy represents an enormous saving in generating costs over alternative methods of meeting peak demand, which involve operating coal-fired or oil-fired stations well below their economic capacity for most of the time.

5.5.1 The future of hydropower

Hydroelectric power generators produce no emissions — except indirectly during their construction — and the water used is renewable. Generation can be started and stopped almost instantaneously, simply by opening or closing the inlets to the turbines. The generating technology is well-established. Other positive aspects of hydroelectric schemes are river regulation, and provision of irrigation and drinking water supplies, fisheries and recreational facilities.

These advantages have to be tempered by several drawbacks. Topography and precipitation determine suitable sites, so that potential is highly variable from country to country. As with any dam projects (see Smith, 2005), large bodies of water destroy natural environment, may displace people, induce seismic and slope instability, and can encourage water-borne diseases. China's Three Gorges project will displace more than 1 million people, affects an area of outstanding scenic appeal and may induce landslides on the steep slopes that flank the reservoir. The Koyna hydropower project in central western India was associated with swarms of minor earthquakes, once the reservoir was filled. Large hydropower reservoirs in Africa, such as Kariba and Aswan, are havens for snails and the bilharzia parasites that they carry, and increase the incidence of disease-carrying mosquitoes.

Inevitably, reservoirs fill with sediment carried in suspension by the rivers that supply the water. This has two outcomes: storage capacity and thereby generating potential decreases; and downstream flood plains cease to be replenished by fertile alluvium deposited by rivers. Both have affected the huge Aswan hydropower scheme that harnesses the Nile in southern Egypt, Lake Nasser being rapidly filled with sediment and the Nile's flood plain and delta being starved of annual silt deposition because the Nile floods are controlled by the Aswan dam.

Large-scale hydropower schemes are among the largest construction ventures undertaken. They are enormously expensive, and place insupportable debt burdens on poor countries, which are unlikely to be relieved by income generation. Building them is a lengthy process, 17 years in the case of the Three Gorges project, and, once begun, construction is difficult to stop, since that would mean abandoning the high initial investment and construction contracts.

In general, the larger the dam, the bigger the investment, the more serious the social and environmental disruption, the longer the lead time, and the greater the danger that circumstances will change before the project is completed and make it redundant. Despite these negative features, flowing water that is free of cost continues to give hydroelectric schemes a high profile. So, what are the global prospects of exploiting this free power, and, given its long history, has hydropower kept pace with increasing energy use?

Figure 5.24a compares the increase in hydroelectricity generation in several regions of the world between 1965 and 2004. Figure 5.24b compares the growth in global electricity generation with that of hydroelectricity between 1990 and 2004.

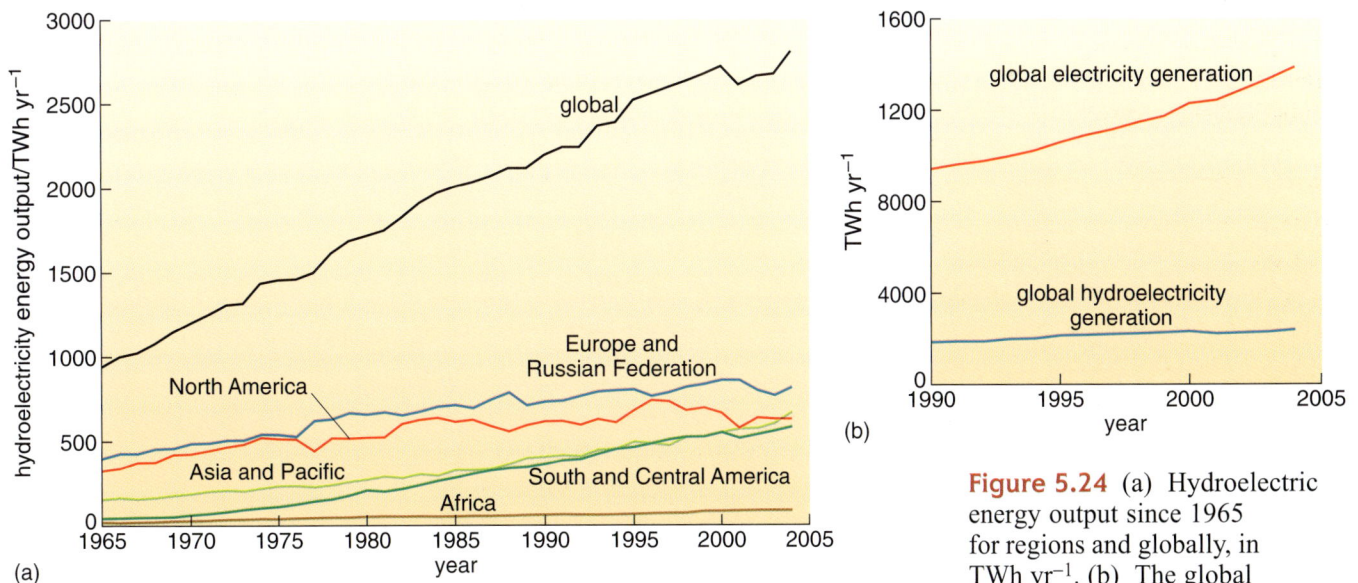

(a)

(b)

Figure 5.24 (a) Hydroelectric energy output since 1965 for regions and globally, in TWh yr^{-1}. (b) The global growth of all electricity production compared with that for hydroelectricity between 1990 and 2004.

● What can you deduce about the contribution of hydropower to global electricity production over time from Figure 5.24b?

○ Global hydropower production shows slower growth than electricity generation from all sources, so the overall contribution of hydropower is declining. Only in South and Central America is there a sign of more rapid growth.

Several factors help to explain the slow growth of hydroelectricity generation relative to demand. One is the high capital cost of large-scale hydropower plants, combined with lengthy periods for construction. Another is that suitable, large-potential sites are often remote from consumers, adding to the cost of distribution. There has also been increasing resistance, both from environmental groups to the impact of dams, and from developing world governments because of the burden of debt repayments for foreign loans. Outside North America and Europe, hydropower potential remains under-exploited (Figure 5.25). Some specialists put the theoretical global hydropower potential at 4.6 TW, whereas others reckon that the technically feasible limit is 1.5 TW, with a limit on economic grounds of about 0.9 TW. Whichever estimate is valid, the potential for hydropower development is not so far in excess of actual use (0.7 TW in 2003, Figure 5.25) as for other alternative energy sources. Despite this slow growth, there is growing interest in small-scale hydropower installations for community use. At the smallest scale (0.1 to 5 MW capacity), mere diversion of existing lake water and stream flow though pipes minimizes environmental impact.

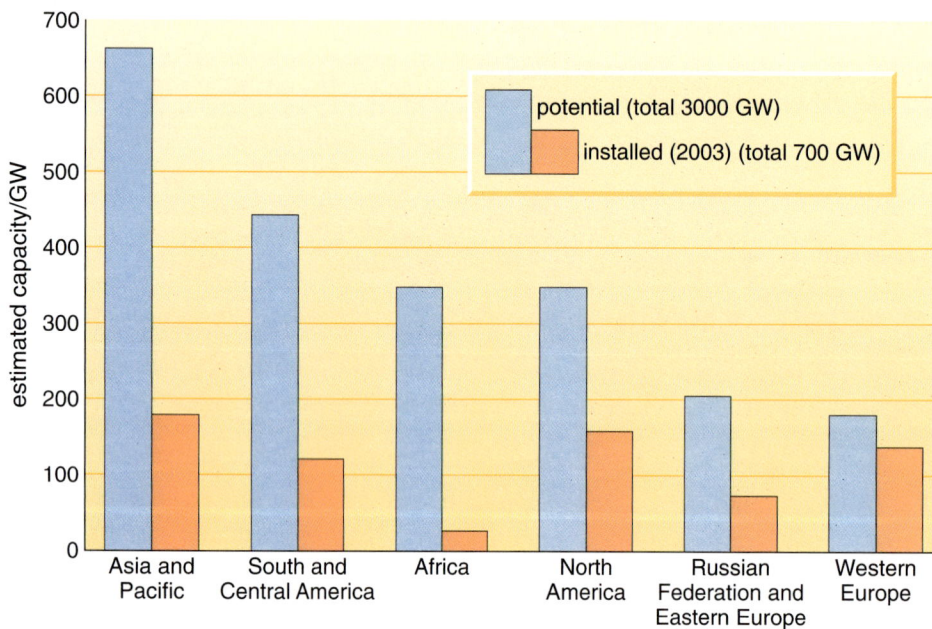

Figure 5.25 Estimated potential and installed (2003) world hydroelectric power capacity, by region. The inset shows global totals.

5.6 Wave energy

The energy carried by ocean waves derives from a proportion of the wind energy transferred to the ocean surface by frictional drag. So, ultimately it stems from the proportion of incoming solar energy that drives air movement. Just how much energy is carried by a single wave depends on the wind speed and the area of ocean surface that it crosses; wave height, wavelength, and therefore wave energy, are functions of the distance or *fetch* over which the wind blows. Not surprisingly the main countries with an interest in the development of wave energy are those with long ocean coastlines. In favourable locations wave energy along coastlines can average 40 MW km^{-1} and the global wave energy potential has been estimated at around 2000 TWh (7.2 EJ) per year. Large as that potential might appear, wave energy can never supplant more than a few per cent of world energy use (451 EJ in 2002).

● How does this compare with primary energy use in the UK in 2002 (Chapter 1)?

○ The UK used around 10 EJ in 2002, so even the entire world's wave-energy potential cannot satisfy the energy demand of a small, albeit highly industrialized country.

The wave-energy potential around the UK is shown by Figure 5.26. Take a moment to judge the locations around the UK of the greatest and least energy available from waves, as shown by Figure 5.26.

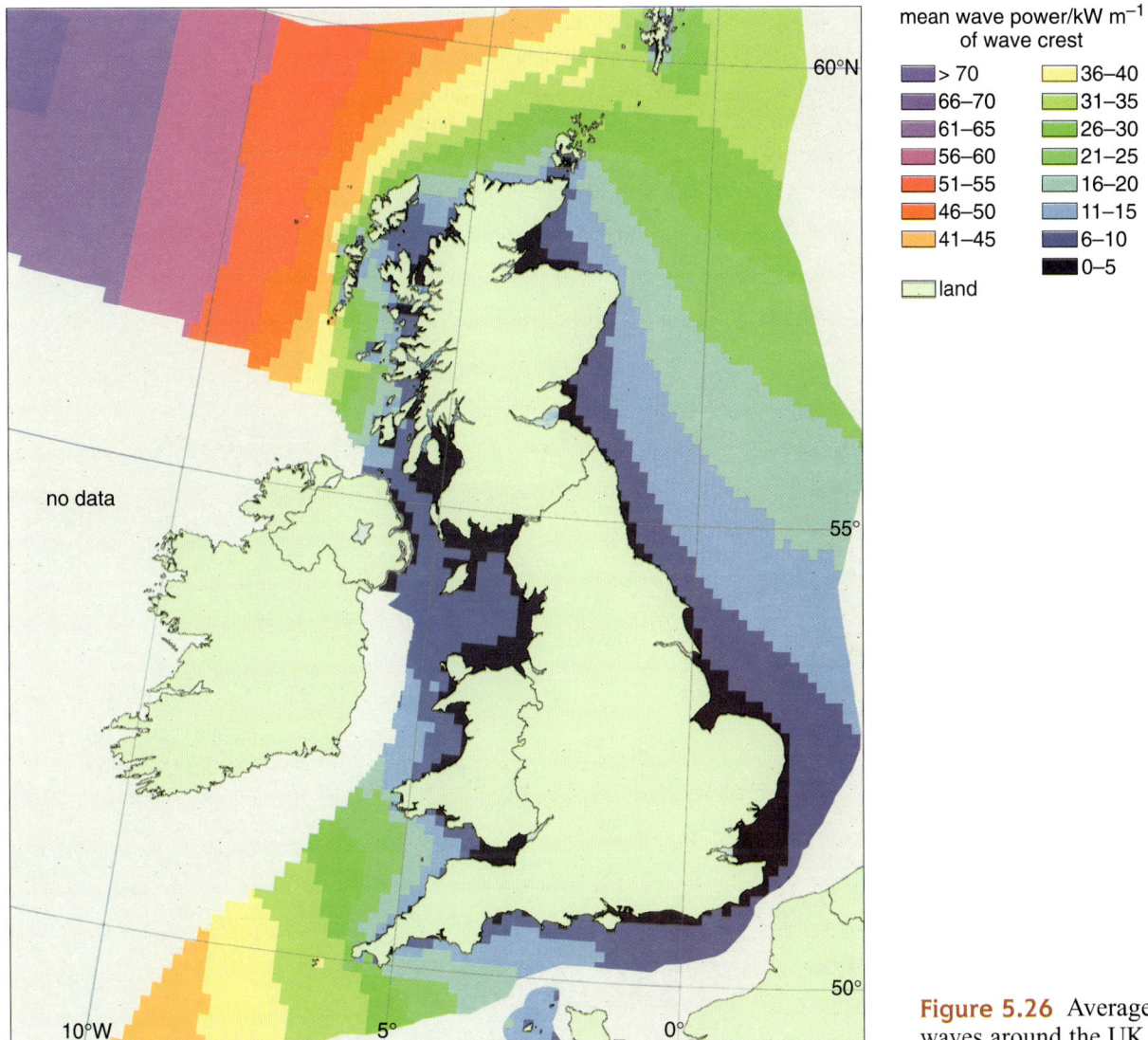

mean wave power/kW m^{-1} of wave crest

> 70	36–40
66–70	31–35
61–65	26–30
56–60	21–25
51–55	16–20
46–50	11–15
41–45	6–10
	0–5
land	

Figure 5.26 Average power of waves around the UK.

One thing is very clear; apart from along the western coasts of the Outer Hebrides, Northern Isles and the south-western approaches, coastal wave energy is often about a tenth of that available in the open North Atlantic. Despite the great excitement in some circles, both for and against (on environmental grounds) wave energy exploitation around the shores of the UK, much of the country has insignificant potential. The prevailing winds are westerlies, so Ireland and the Hebrides protect much of the UK from usefully energetic waves.

Wave energy may be exploited at the coast, in near-shore waters and in the open sea. Shoreline and near-shore wave-power generators are fixed to the seabed and can be installed and serviced easily. They supply electricity close to the point of use. Their operation exploits processes that can often be seen in sea caves and blowholes; surging of waves and the trapping and pressurization of air. Figure 5.27a to c show three coastal wave energy devices. The first focuses waves through a tapered channel, to increase their amplitude, so that seawater floods a raised lagoon and then runs through a turbine back out to sea again. The second uses an air tank above sea-level. As waves move up and down, the air is compressed and then drawn back to drive a turbine. The third uses the motion of a 'gate' that moves as waves break and recede, to drive a hydraulic pump to produce electricity. Such coastal wave-driven generators need to be exposed to wave action continually to produce continuous electricity, and therefore require either a small tidal range or water at the shore that is deeper than the tidal range. However, Figure 5.26 shows that near-shore waves have less energy than those further out to sea.

Offshore devices (e.g. Figure 5.27d) are typically located in water depths greater than 40 m at a considerable distance from land. There are several types. Some

Figure 5.27 Varieties of wave energy device. (a) Tapered channel (TAPCHAN) device. (b) Air compression device (LIMPET). (c) Hydraulic device driven by a flapping gate (Pendulor). (d) Pumping device driven by floats (Hose Pump).

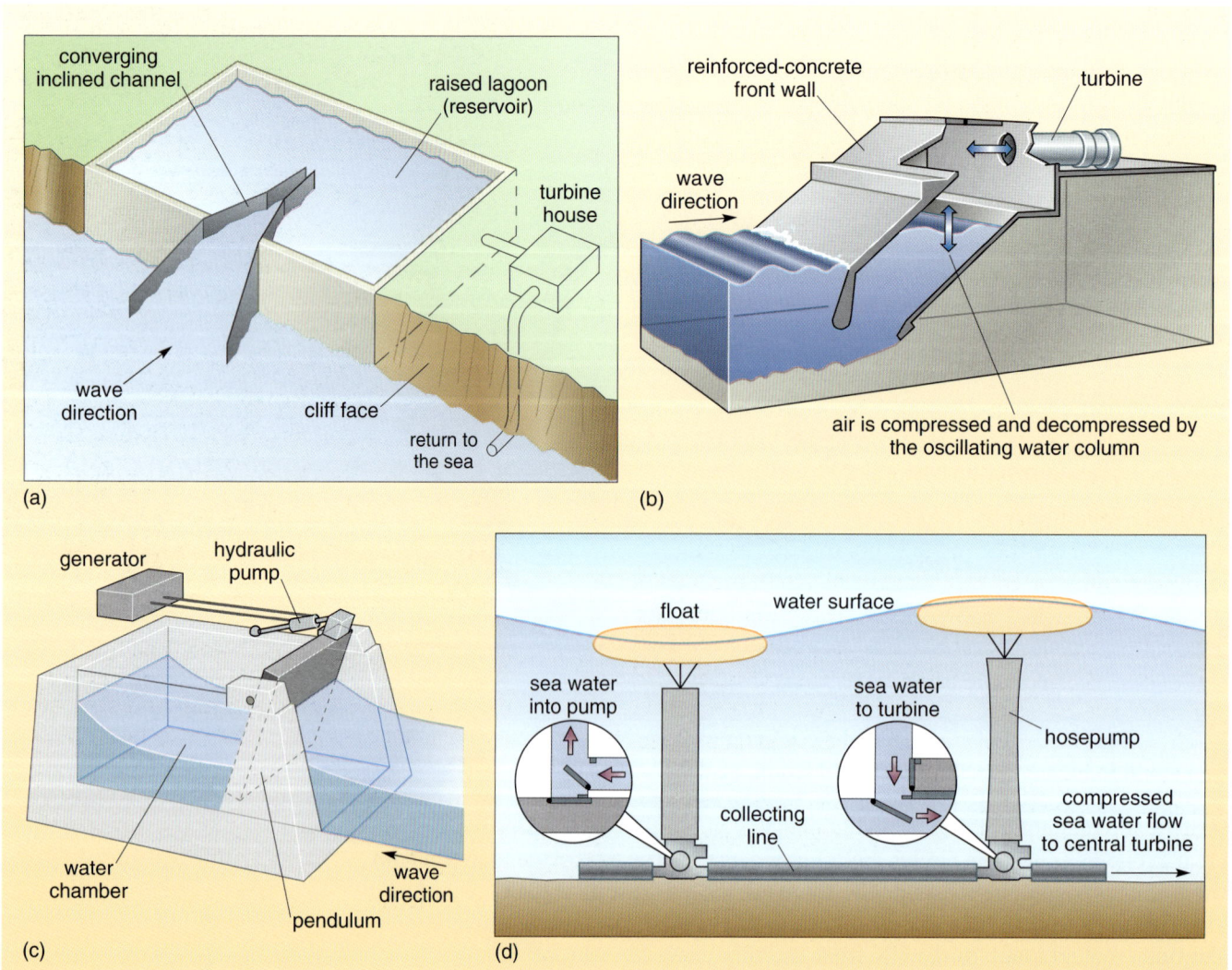

(a)

(b)

(c)

(d)

are part of a fixed platform and use the same principles as shoreline and near-shore generators, driving air or seawater back and forth through a turbine. Others are buoy-like and tethered to the seabed, using floats to drive hydraulic devices (e.g. Figure 5.27d) that rotate a generator, or mechanical and exploit the duck-like rocking motion of floating objects.

As you saw earlier, total ocean wave power falls far short of satisfying global energy demand — its potential is roughly the same as that of current global hydropower production (2800 TW h yr^{-1} in 2003, Figure 5.24). Moreover, the most efficient sites capable of sizeable supplies are very restricted. Even there, wave energy, like the wind, is rarely constant, except in the most inhospitable places, such as the circum-Antarctic region. Consequently, some form of energy storage is needed to ensure stable power supplies from waves. The future for wave-energy generators may therefore be limited, probably to remote locations and to power offshore installations.

5.7 Tidal energy

The rise and fall of ocean tides result from the combined gravitational pull on water by the Moon and, to a lesser extent, by the Sun, which exerts a force on water directed towards the two astronomical bodies. These gravitational effects combine with centrifugal forces that result from the Earth and the Moon orbiting each other to make the details of tidal changes complex. The physics of this second contribution to tides is beyond the scope of this book.

Tides are not completely synchronized with the position of the Moon and Sun relative to the Earth because of viscous drag. Figure 5.28 shows how the gravitational force exerted by the Sun and Moon together affects tidal range, according to phases of the Moon. Tides are a great deal more complicated as regards their driving forces than the effects that we notice. However, the forces are enormous and serve to generate the energy bound up with the ebb and flow of the tides.

Globally, total tidal power has been estimated to be 2.7 TW. Far out at sea, the tidal range is not as great as that experienced in the UK, and is generally no more than 0.5 to 1.0 m. Where coastlines face the open ocean, there are two tidal cycles per day, separated by approximately 12 hours 25 minutes. Complex coastlines, such as those around the UK, and where narrow seaways constrict tidal flow, result in modifications to both the timing of tides and their range. These can result in bizarre modifications to the tidal cycle; for instance, the Solent has four noticeable tides each day. Similarly, coastal tide ranges vary markedly around the land, rising to more than 15 m in a very few places, such as the Bay of Fundy in Newfoundland. The higher the tidal range the greater the potential for electricity generation; tidal range is akin to the working head of hydropower schemes.

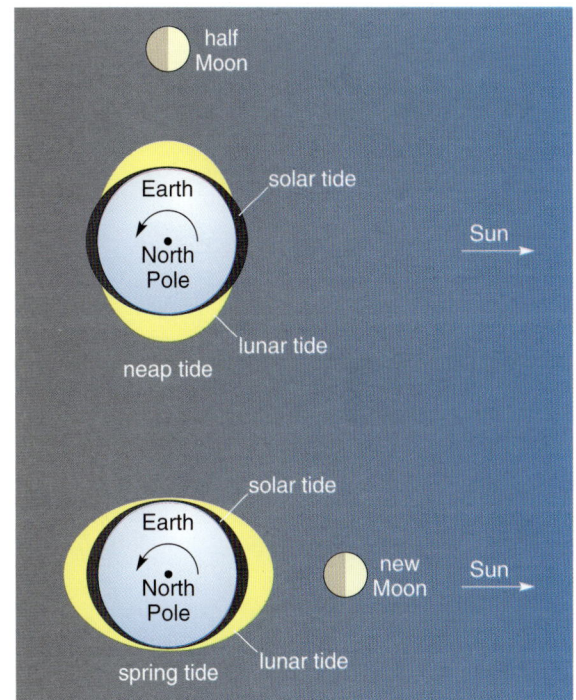

Figure 5.28 When Earth, Moon and Sun are aligned, at full and new Moons, the tides have their highest range (spring tides). At half moons the tidal range is at its lowest (neap tides).

Figure 5.29 expresses available tidal power in terms of kW m^{-2} (i.e. the power per unit area of the sea surface) around the coast of the UK. The greatest potential coincides with narrow tidal straits, such as the Pentland Firth, constricted estuaries, such as that of the Severn, and areas where major tidal flows meet, as in the southern North Sea and the English Channel. The highest energies are associated with the many narrow channels separating the islands of the Inner and Outer Hebrides, which are notorious for their tidal rips and whirlpools. Because of its complex tidal flow at the margin of the North Atlantic, the UK coastline focuses about half the total tidal energy potential of Europe. This is estimated to be in the region of 53 TWh yr^{-1}; potentially about 14% of total UK energy needs.

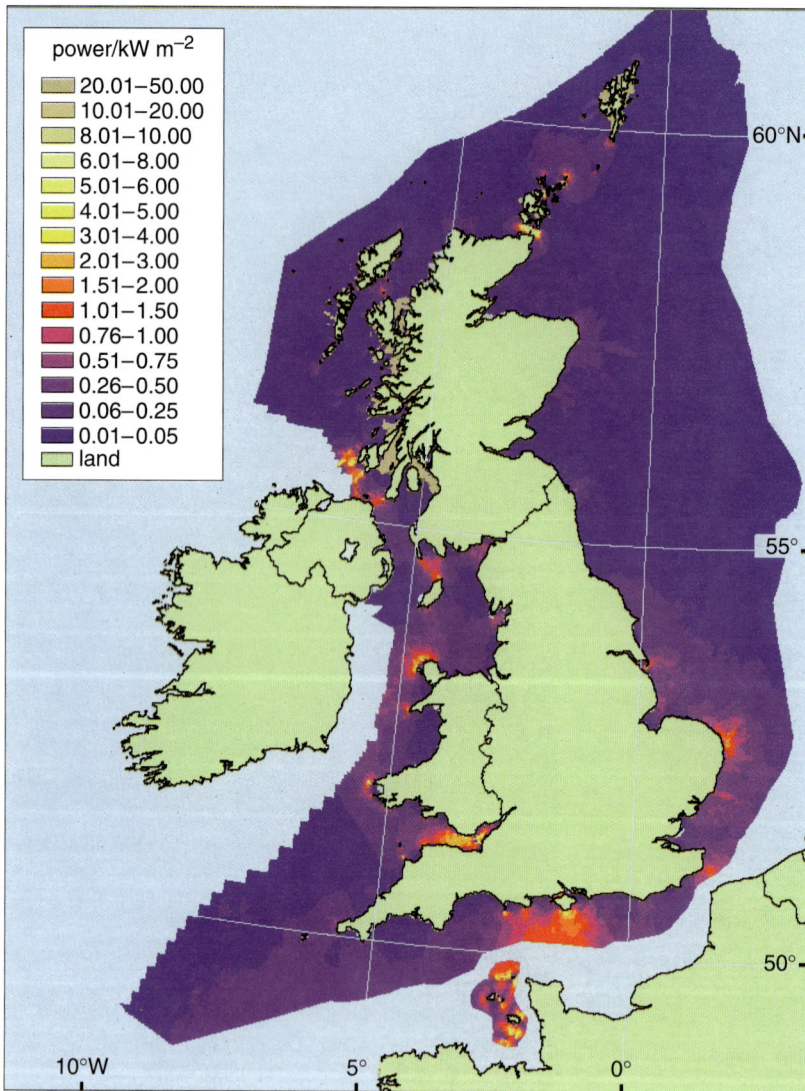

Figure 5.29 Potential tidal power per unit area around the UK coast.

Potential is one thing, but being able to harness it is quite another matter. The key factor, as well as tidal power per unit area, is the volume of tidal flow that can be constrained through turbines. Figure 5.29 shows one location that offers major possibilities, the Severn Estuary. In more detail, two other west-coast estuaries would also seem to be possible sites, the Solway and Morecambe Bay. The very high power potential along the intricate coastline of NW Scotland is associated with dangerous tidal currents, and presents construction problems that cannot be overcome by modern civil engineering practices.

5.7.1 Harnessing tidal energy

The principle of tidal power generation is very similar to that employed in conventional hydropower schemes, except the working head of water is no greater than about 10 to 15 m. The critical term in Equation 5.6 is the water discharge rate Q. Tidal barrages across estuaries can extract energy from the incoming (flood) and the outgoing (ebb) tides using turbines located in sluices through the barrage. Estuarine barrages can enclose vast amounts of water, including river flow into the estuary, thereby increasing power capacity.

The most common type of barrage uses just the ebb tide. The incoming flood tide is allowed to pass though sluices with the generators idling, to trap the water behind the barrage at high tide. This head of water then drives the turbines to generate power on the ebb tide. Flood tide generation is simply the opposite but in either case, given that there are two tides a day, power is produced in two bursts every day. Two-way operation is possible, using both ebb and flow, although power output is not doubled. This is because the tidal basin would normally empty and fill through its entire natural outlet, and when constrained through sluices the full cycle would not be complete before the tide turned. The major disadvantage of tidal barrages is the same as for hydroelectricity dams, they silt up. Moreover, by constraining tidal flow through narrow sluices instead of the full width of an estuary the natural processes of tide-dominated sedimentation and erosion are changed over large areas.

There are alternatives to estuarine barrages, albeit with much lower power capacity. One is to create artificial lagoons in embayed coasts or to cut off parts of estuaries to minimize environmental change. In especially powerful tidal flows that pose insurmountable challenges to construction, such as those around western and northern Scotland, tidal energy could be 'harvested' using small, floating turbines anchored to the sea bed.

As with wind and wave energy, that available from tides is geographically highly varied, and obviously restricted to maritime countries. Some, like the UK are well endowed, but most have far less potential, particularly islands surrounded by major oceans and areas with simple coastlines.

5.8 Alternative energy in perspective

Alternative energy sources are seen by many people as potential solutions to the many economic and environmental challenges posed by the current dominance of world energy supply by fossil and nuclear fuels. Just how realistic are these hopes?

The first thing to note is that around 18.4% of primary energy used globally in 2002 (Figure 1.4) was indeed supplied by alternatives that you have studied in this chapter. However, over 16% of that proportion was from hydropower and traditional uses of biomass, i.e. the inefficient and environmentally damaging burning of fuelwood and animal dung in the poorer parts of the world. Although several alternative sources, such as geothermal, solar and biomass energy have potential for heating and other non-electrical uses, the alternative energy industry's significant feature is the production of electricity by several different means.

Figure 5.30a shows global changes over three decades for primary energy sources used in electricity generation, in graph form. The two pie charts (Figure 5.30b) compare the proportions of the major energy sources being used for electricity generation in 1973 and 2003. Coal remained in much the same dominant role throughout those three decades. Significant changes were a marked decrease in the role of oil and increases in nuclear and natural gas. The only alternative source that had a significant role in electricity generation, hydropower, steadily declined in its relative contribution, by more than 5%. The 'other' category consists of the other alternatives discussed in this chapter. Although their role had tripled since 1973, it was still less than 2% in 2003. However rapidly a means of electricity generation is growing, if the growth is from a low starting level it will take a long time to reach a significant proportion of global production.

The pie chart in Figure 5.30c is even more revealing. It emphasizes the overwhelming dominance of alternative energy sources by hydropower, the first alternative energy source to be used for electricity generation and also the one

Figure 5.30 (a) The graph shows the increasing contribution of nuclear power to global electricity generation since its introduction in the early 1970s, linear growth in that from fossil fuels, and little change in hydropower. (b) The pie charts show in more detail the changes in contributions from major energy sources between 1973 and 2003, including a breakdown of the fossil fuel contribution shown in the graph in (a). (c) The percentage contribution to global electricity generation by different alternative energy sources in 2000.

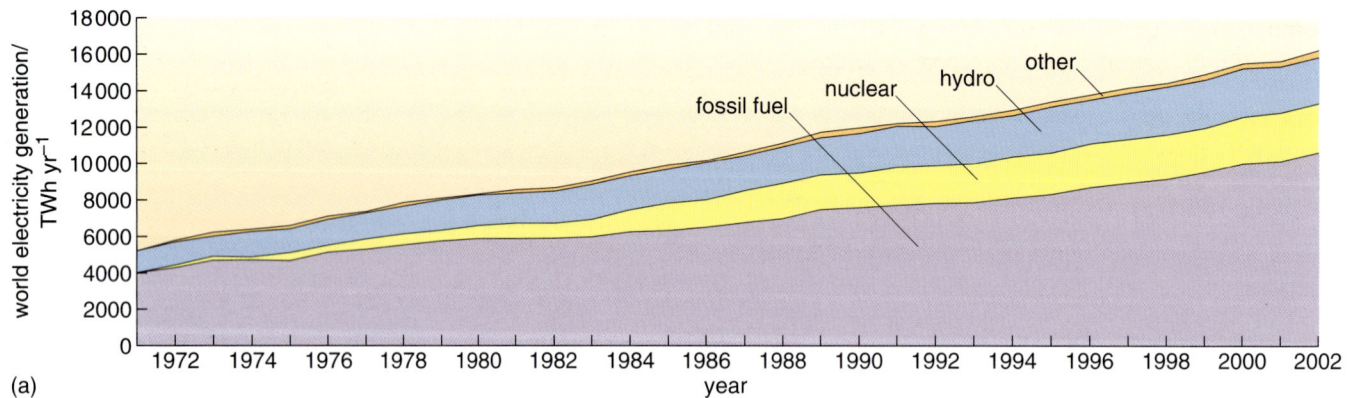

(a)

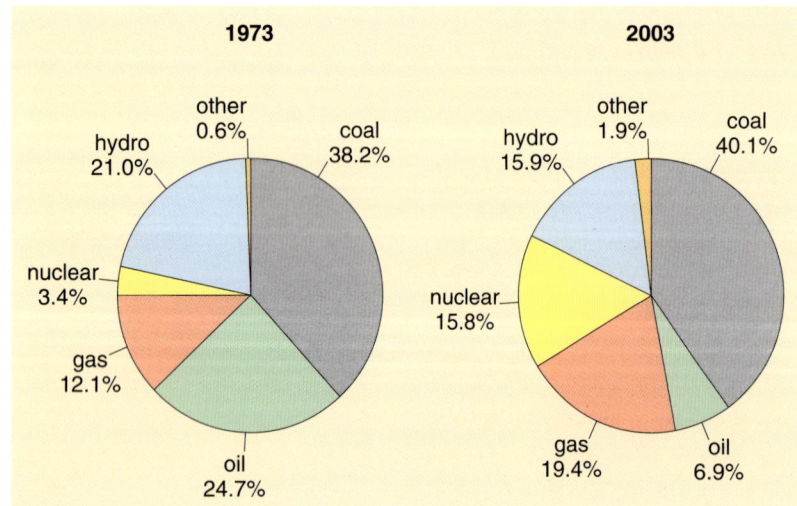

(b)

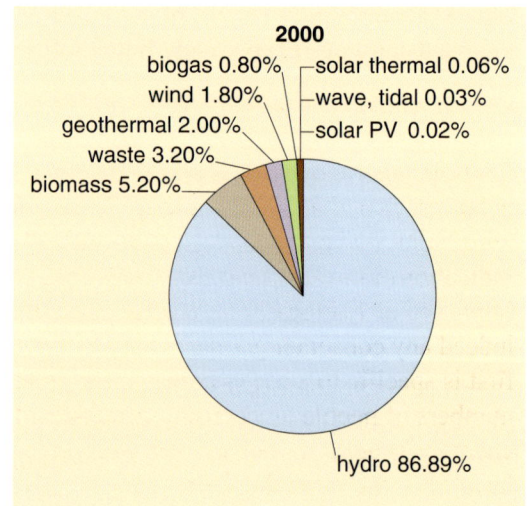

(c)

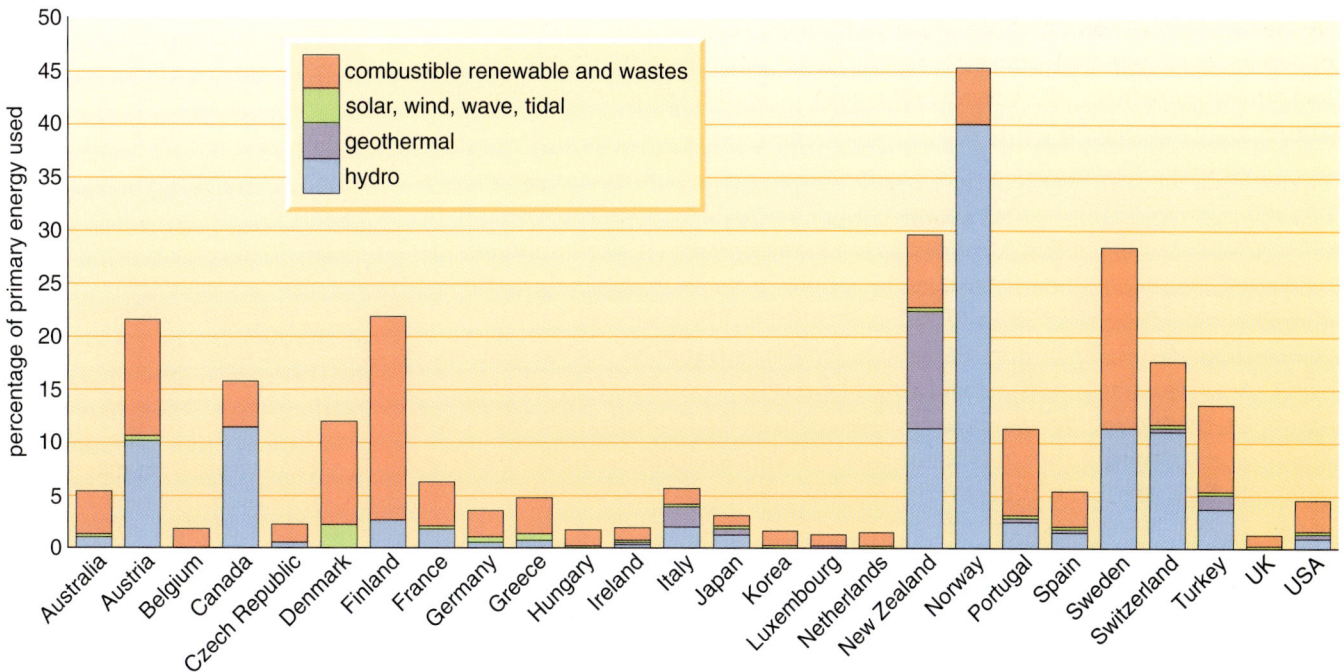

Figure 5.31 Percentage contributions of alternative energy sources to primary energy uses in countries which belong to the Organisation for Economic Cooperation and Development (OECD).

closest to reaching its limits. Those alternative sources that receive most attention in the media, particularly solar, wind, wave and tidal, are dwarfed by comparison. Figure 5.31 shows the disparities between various countries, in terms of their use of alternative energy sources.

Figures 5.30 and 5.31 give a vivid impression of the magnitude of the changes needed to make any impression on the dominance of fossil and nuclear fuel usage. To achieve those changes requires political and industrial commitment to invest in and construct alternative sources, and to phase-out 'conventional' energy sources. For example, an estimate of the contribution of solar PV electrical power to global electricity generation use in the early 21st century is 0.001%. Judge from that the magnitude of effort required for photovoltaic technology to have a significant effect on energy use patterns.

● If solar PV power currently contributes 0.001% of global electricity generation, by how many times would it have to increase to reach 20%?

○ It would have to increase 20 000 times.

For other alternatives, the outlook is somewhat more optimistic, yet their potential is far less than that of solar energy. Interestingly, in 1941 only 0.02% of US households had a television, yet by 2000, 98% had at least one. That is a good indication of how demand can drive rapid growth in supply, especially when improved technology helps reduce price. But the demand for television sets, and indeed any consumer goods, is very different from that for energy supply; the first is specific to a use, whereas the second is general. Although increasing numbers of people are becoming aware of the necessity for changes in how energy is produced, they are still a tiny minority. Whatever their sentiments, they have little choice in the form and sources of the energy that they use, as shown by Figure 5.30. Nonetheless, in 2004 there were high growth rates for some alternatives in some areas; China was developing geothermal resources at about 12% growth annually (doubling period of 5.8 years); 'high-tech' use of biomass was growing globally at 20% (doubling every 3.5 years); in the EU wind power was growing at about 25% (a 2.8-year doubling).

As you have seen, alternative energy sources are very different as regards the technologies required to extract power that exists naturally. They are just as diverse in the context of their geographic availability. Coal, oil, natural gas and nuclear fuels are commodities that can be shipped globally to wherever there is a market for them. For the industrialized parts of the world, much the same happens with electricity, through national and international grid distribution using cables. Apart from biomass (an expensive proposition considering its low intrinsic value), alternative energy sources cannot be exported or imported in their natural form! So, it is very difficult to estimate what the ultimate potential of each type of source for human use might be.

The 'baseline' from which to judge the relative potential of each alternative is the global power that they 'deliver' naturally, shown by Table 5.4. Each section in this chapter has summarized the technical and geographic challenges posed by each alternative source. It is left to you to judge the feasibility of implementing these changes against the claims for 'alternative' solutions to global energy challenges that are regularly made. It does seem, however, that solar and wind energy are likely to make the most rapid impact, judging by their industries' rates of growth in the early 21st century, and by their potential to deliver by far the greatest power. However, changing the pattern of energy supply is a matter neither of relative abundances nor environmental common sense. It boils down to cost, and some interesting information that compares alternative and conventional costs of installation and supply appears in Tables 5.5 and 5.6. Note that the cost of installing renewable energy plant is from 3 to 30 times higher per kW of power capacity than for conventional power stations. Note also that solar photovoltaic is 'priced out' of all markets at present, and both coal and natural gas will continue to have a competitive edge for at least the foreseeable future.

One aspect of cost for 'conventional' fuels is rarely mentioned, yet is a real cost nevertheless. It concerns the environmental effects of using different 'conventional' fuels. Such 'externalities' are notoriously difficult to assess, the supreme example being the threat of climate change from continued burning of

Table 5.4 Total power potentially available globally from alternative energy sources.

Alternative energy source	Total power/TW
solar (surface insolation)	8×10^4
wind	1×10^4
waves	1×10^3
biomass (terrestrial)	9×10^2
geothermal (heat flow on land)	13.5
hydropower	4.6
tidal	2.7

Table 5.5 Energy and investment costs for electric energy production from renewables, compared with those from nuclear and fossil fuels.

	Current energy cost/US\$ kWh^{-1} (2002)	Potential future energy cost/US\$ kWh^{-1}	Current investment cost of installation /US\$ kW^{-1}
biomass	0.05–0.15	0.04–0.10	900–3000
geothermal	0.02–0.10	0.01–0.08	800–3000
wind	0.05–0.13	0.03–0.10	1100–1700
solar (photovoltaic)	0.25–1.25	0.05–0.25	5000–10 000
solar (thermal electricity)	0.12–0.18	0.04–0.10	3000–4000
tidal	0.08–0.15	0.08–0.15	1700–2500
coal (US)	0.01–0.04		
natural gas (US)	0.02–0.05		
oil	0.06–0.08		
average fossil and nuclear fuel power installation			300

Table 5.6 Energy and investment costs for direct heat from renewables.

	Current energy cost/US\$ kWh^{-1} (2002)	Potential future energy cost/US\$ kWh^{-1}	Current investment cost of installation /US\$ kW^{-1}
biomass (including ethanol)	0.01–0.05	0.01–0.05	250–750
geothermal	0.005–0.05	0.005–0.05	200–2000
solar heating	0.03–0.20	0.02–0.10	500–1700

fossil fuels. Wherever its source, CO_2 and its climatic consequences spread globally. Environmental costs fall on us all, yet do not figure in the point-of-use cost in US\$ per kWh of fossil fuels in Table 5.5. Such costs are clearly less for alternative, more 'environmentally friendly' energy sources, but by how much? Chapter 6 revisits some of these interwoven issues.

5.9 Summary of Chapter 5

Chapter 5 covers a great diversity of energy sources and the scientific and technological issues that surround them. So, summaries are linked to each main section for ease of reference.

Section 5.1

1   Geothermal energy stems from the decay of naturally radioactive isotopes in the Earth's interior, which results in varying heat flow across the Earth's surface, depending on tectonic setting. In volcanic areas heat flow can locally be very high.

2   Geothermal electricity generation requires: heat (most effective in areas of high enthalpy); abundant heated water in an aquifer (or in artificial fractures in crystalline hot, dry rock); an impermeable cap-rock that prevents its escape.

3   Global heat flow, although renewable and present everywhere, is limited in its potential contribution to energy needs, being of the same order overall as global primary energy use in the early 21st century. More than two-thirds of it is through the ocean floor. Only small areas on land have sufficiently high heat flow to generate electricity. That will limit geothermal power to a small fraction of current electricity requirements.

4   Areas of low enthalpy can supply heat to buildings and other applications, thereby supplanting heating using fossil fuels.

Section 5.2

5   Solar power is an immense source of directly useable energy and ultimately creates other energy resources: biomass, wind, hydropower and wave energy.

6   Most of the Earth's surface receives sufficient solar energy to permit low-grade heating of water and buildings, although there are large variations with latitude and season. At low latitudes, simple mirror devices can concentrate solar energy sufficiently for cooking and even for driving steam turbines.

7   The energy of light shifts electrons in some semiconducting materials. This photovoltaic effect is capable of large-scale electricity generation. However, the present low efficiency of solar PV cells demands very large areas to supply electricity demands.

8   Direct use of solar energy is the only renewable means capable of ultimately supplanting current global energy supply from non-renewable sources, but at the expense of a land area of at least half a million km^2.

Section 5.3

9   Photosynthesis converts solar energy into a form that is directly useable through solid fuel combustion, gasification, pyrolysis, digestion and fermentation.

10 Biomass products have an advantage over direct solar power, being storable, transportable and useable with current technologies.

11 Biofuels produce carbon dioxide, but unlike that produced from fossil-fuel burning, it is an active part of the carbon cycle, adding nothing to global warming.

12   Satisfying global energy demand with biofuels would require about 6% of the biological productivity of the continents, with potentially massive impact on natural ecosystems and agriculture.

Section 5.4

13   Power output from wind turbines is proportional to the area swept by their blades, and to the cube of wind speed. The narrow range of useable wind speeds restricts the areas where wind energy can be exploited.

14   Wind power has great potential, but has three main drawbacks. Output depends on intermittent wind speeds, irregular distribution of suitable wind speeds, and occupancy of large areas of land.

Section 5.5

15   Hydropower was the earliest means of commercial electricity generation, and currently dominates alternative electricity supply. However, its global capacity for large-scale exploitation is less than six times that currently installed.

16   Growth of hydropower is slow and its contribution to global electricity supply is falling. Both are due to economic factors, the slow pace of large-scale project construction, the remoteness of high-potential sites, and increasing resistance to the social and environmental effects of large dams.

17   Small-scale hydropower projects for local electricity supplies may expand rapidly, particularly in remote areas, but are unlikely to add significantly to global supplies.

Section 5.6

18   Despite claims made for its potential, wave energy can never make any significant contribution to global energy supplies, although it may find a use in coastal communities.

19   The greatest potential from wave energy exists far from shore, but waves are wind-dependent and so supplies are bound to be irregular.

Section 5.7

20   Tidal power uses the same principles as hydropower, but involves smaller working heads and larger volumes of water.

21   Tidal power is predictable, but globally has low potential.

22   To be economic, electricity-generating tidal barrages require high tidal ranges, large-volume estuaries and safe conditions for large-scale offshore construction. Tidal power is highly restricted geographically.

Section 5.8

23   At the outset of the 21st century, alternative energy sources contributed almost a fifth of global primary energy supply, mostly from large-scale hydropower and traditional biofuel burning in poorer countries. Neither are environmentally 'friendly', or without serious damage to society.

24   The more promising alternative sources account individually for tiny proportions of current global energy supply. In order to significantly supplant fossil fuels, avoid global warming and foster economic sustainability in the next half century, they require very high global growth rates.

25 Unlike fossil and nuclear fuels, alternatives other than biofuels are not practicably portable. Nonetheless, most of them have no economic cost except for that involved in conversion to useful forms of energy.

26 Despite the publicity that surrounds all forms of alternative energy source, only direct and PV solar and wind power can make a significant contribution to global energy supplies.

FUTURE ENERGY DEMAND AND SUPPLY

6

Human demand for energy has only quite recently exceeded the relatively modest amounts available locally: wind and water power, wood or dung for heat. Since the mid-19th century, expansion in the large-scale exploitation of cheap, plentiful, concentrated energy sources — the fossil fuels — has outstripped global population growth (Figure 6.1). When you consider that the global annual consumption of primary energy increased more than ten-fold during the 20th century, the importance of planning future energy supply becomes clear. This final chapter examines how future global energy consumption might develop from its current level, and summarizes the main influences on that development, including economic, environmental and technological factors.

6.1 The present-day perspective

Notwithstanding the rapid increase in energy consumption depicted by Figure 6.1, the average per capita consumption of commercial energy sources has not increased much, if at all, since 1975 (Figure 6.2a), even in Europe and North America.

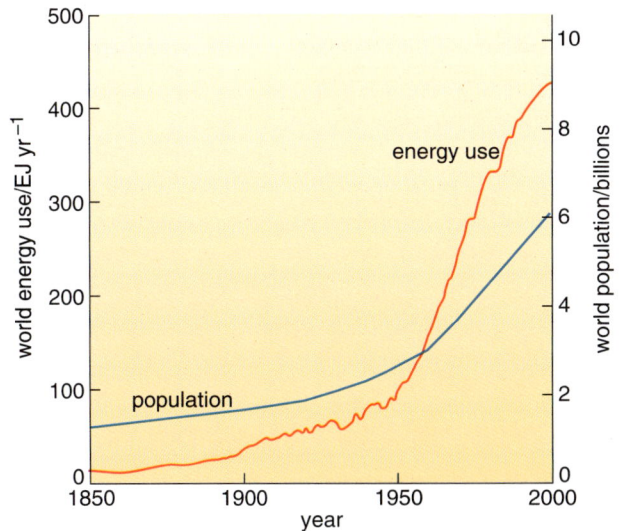

Figure 6.1 Population growth and global primary energy consumption from 1850 to 2000.

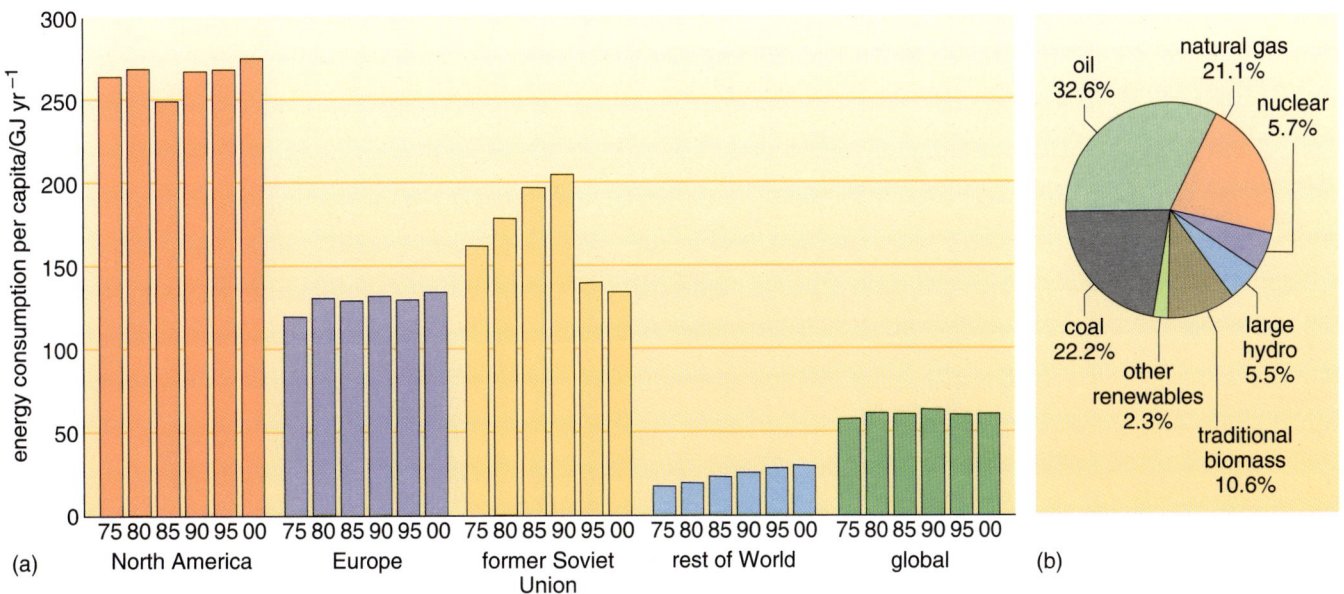

Figure 6.2 (a) Energy consumption per capita per year for different regions and for the world as a whole, highlighting the contrast in consumption between industrialized and developing regions. These figures are for commercially traded fuels only, which neglects the >10% contribution from 'traditional biomass' shown in (b). *Note*: The bars refer to individual years. (b) Percentage contributions of energy sources to global primary energy consumption in 2002. *Note*: The contributions from nuclear, hydro and other renewable power sources are the inputs that would be needed to produce the actual outputs at 38% plant efficiency (i.e. an approximation, to compare them with fossil fuel contributions). If the actual electricity outputs of these minor sources were used in this diagram, their contribution figures would be substantially lower.

However, the bulk of this annual consumption depletes energy sources that are non-renewable.

⬤ How can the static per capita figures for global energy consumption in Figure 6.2a be reconciled with the dramatic increase shown in Figure 6.1?

◯ The global population increase (Figure 6.1) drives a corresponding increase in energy consumption despite a per capita consumption that is approximately constant.

In 2002, fossil fuels provided >75% of global primary energy consumption — a further spur for long-term planning, since these finite resources are being rapidly depleted. As Figure 6.2b shows, renewable energy sources contributed only 18.4% towards global consumption in 2002, and of that the greatest proportion was from traditional burning of biomass. As reserves of fossil fuels inevitably dwindle, other sources of energy must be exploited more than at present, to ensure that current levels of consumption can be sustained (or increased).

The inequalities in consumption shown in Figure 6.2a are further complicated by unequal distribution of fossil fuel resources (especially oil) across the globe (Figure 6.3). As reserves dwindle, these inequalities are likely to foster increasing political tensions. It is notable that, for many of the major conflicts of the late 20th and early 21st centuries (e.g. Iran–Iraq war, both Gulf Wars), fossil fuel supply was a critical (if understated) factor. The same inequalities of consumption and reserves may also force the development of alternative energy sources at different rates in different regions.

How long will global fossil fuel reserves last at current consumption rates? *Proven* reserves suggest ~190 years of coal, ~70 years of gas, and ~40 years of oil, but these figures do not take account of discoveries of new reserves, or technological advances in extraction or production. However, the rate of new

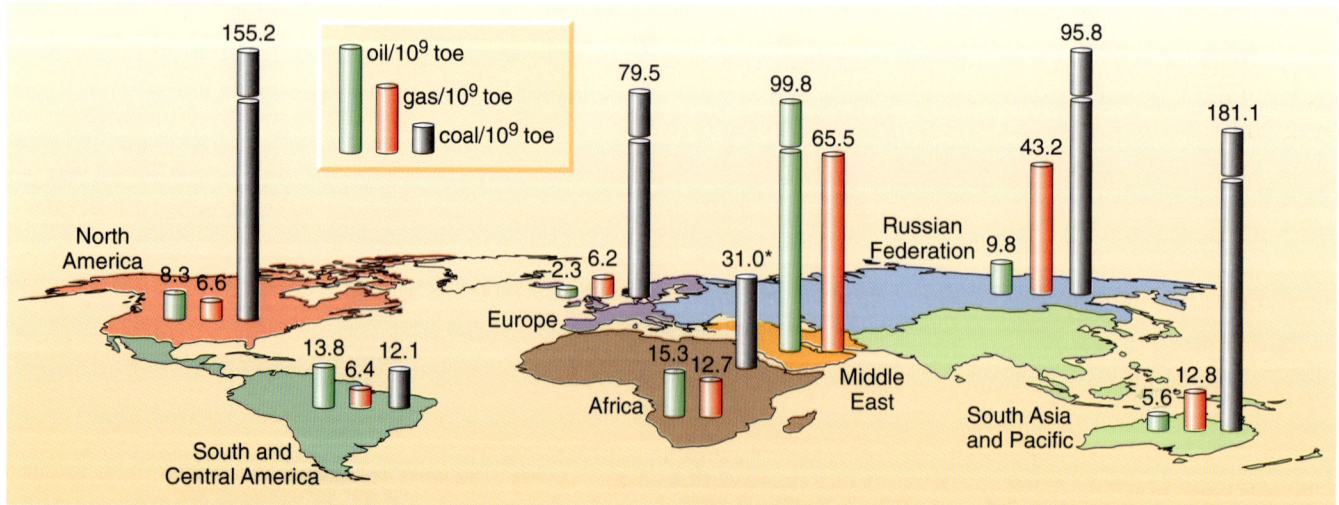

Figure 6.3   The unequal geographic distribution of proven reserves of fossil fuels. The reserves are expressed in billions of tonnes of oil equivalent (toe), so that those for each fossil fuel can be compared with the others in terms of their primary energy content (1 toe = 42 GJ). *Note*: Oil reserves are dominated by those in the Middle East (61.7%); natural gas by the Russian Federation and Middle East (67.3%); coal is more equitably distributed. * The coal reserves of Africa and the Middle East have been combined in this figure.

petroleum discoveries is declining relentlessly, while rates of energy consumption have risen steadily from 1900 onwards, suggesting that current consumption rates are not sustainable in the long term. The underlying truth is that fossil fuels are ultimately finite resources, and at anything approaching current global consumption rates they will be depleted as rapidly as they have risen to dominance. Were non-conventional sources of petroleum (Section 3.7), such as the huge oil sand deposits of western Canada, to become economic, that truth would still apply in the long run. As Figure 6.4 shows, production of conventional oil and gas is predicted to *peak* between 2005 and 2030, and the graphs are a stark illustration of the short lifetime of oil and gas use relative to human history.

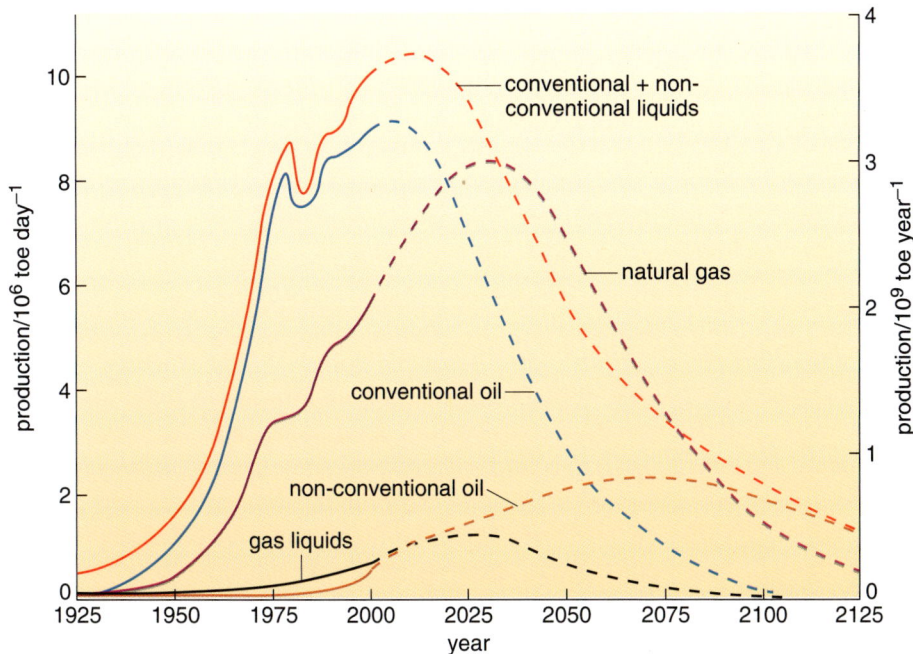

Figure 6.4 Global production of oil and gas. Solid curves (prior to 2000) represent historical data; dashed curves are predictions of future supply.

It is clear from this brief summary of the global energy situation in the early years of the 21st century that a shift in energy strategy is inevitable on geological, economic and political grounds. And, of course, there is also a profound environmental dimension to such decision making. The extent to which each contributing consideration will govern such a shift is hard to judge. So how can we make such long-term decisions from our short-term perspective? The next section examines this dilemma, and evaluates some initial attempts at forecasting the future of energy supply.

6.2 Forecasting: energy in the future

So far as human society is concerned, the only certain thing about the future is that nothing is certain. The case of the UK deep-mined coal industry (Box 6.1) illustrates how difficult it is to plan ahead, when lead times for changes in infrastructure are on a scale of years to decades, yet the timescale of political and economic change is commonly weeks or months (or less). These problems are most crucial for the energy industries, because without energy nothing functions.

Box 6.1 UK Coal: decline and fall

In an earlier version of this course, which first appeared in 1983, the authors commented:

> 'In the UK we are fortunate in having large enough coal reserves to produce most of our electrical power — an important factor that has delayed the need to expand rapidly the generation of power from nuclear or renewable resources.'

During the early 1980s, few in the UK could have foreseen the day in October 1992 when the Government announced the immediate closure of 31 coal mines out of a total of 50 still working at the time. Vast, proven reserves of coal were rendered worthless at a stroke, damaging the UK's principal long-term foundation for energy self-sufficiency. Yet the decision was not prompted by environmental concerns, despite a UK commitment at the Earth Summit in Rio de Janeiro a couple of months earlier to stabilize or reduce carbon dioxide emissions by reducing the use of fossil fuels. Nor were the closures determined by Britain's geology, except in so far as thin coal seams hundreds of metres deep were costly to extract compared to imported coal from thicker, near-surface seams.

Ironically, it could have been argued cogently that in the long run the virtual demise of the UK's deep-mined coal industry was a desirable outcome, considering the numerous detrimental environmental impacts associated with its extraction and combustion. Unsurprisingly, there was an outcry in the immediate aftermath of the decision due to the social impact of widespread redundancies on communities traditionally dependent on coal mining, and the widely held view that the sudden closures were a political retaliation for the year-long miners' strike of 1984–85.

The object of the closures was purely commercial: to cut costs to the recently privatized electricity generating companies (and hence consumers), and ultimately produce a smaller, more efficient UK coal industry which would be attractive for privatization. Cheap electricity was to be generated from cheap, imported coal and new power stations fired by North Sea gas (the 'Dash for Gas'). In fact, the writing had been on the wall for the UK's deep-mined coal industry even before the miners' strike. The decline since 1950 is graphically illustrated in Figure 6.5: from 170 deep mines in 1984, to 17 at privatization in 1994, and only 9 by the end of 2004 (although their productivity had doubled since the 1990s). For comparison, both France and Belgium had ceased coal mining entirely by 2004; 120 Belgian coal mines were shut down between 1957 and 1992. Coal industries in most other European countries, including the main producers Germany and Poland, are substantially subsidized.

The abrupt changes in the fortunes of the UK coal industry are a good example of how resource planning and policy depend more upon economic relationships and political considerations than upon geological availability.

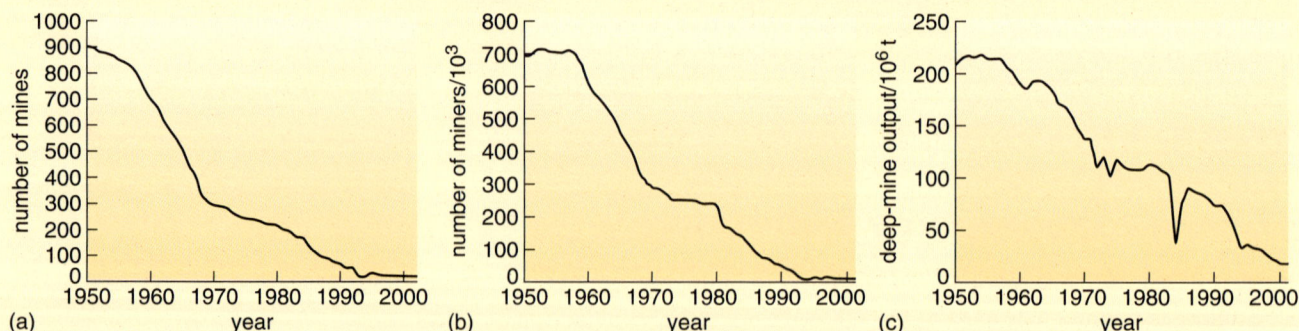

Figure 6.5 The decline in UK deep-mined coal from 1950 to 2002, as shown by three indicators: (a) number of deep mines; (b) number of miners employed; (c) output from deep mines. The slight increase in the number of mines at privatization (1994) reflects the fact that from 1950–1994 only National Coal Board mines were recorded in the statistics.

6.2.1 A lesson from the past

One famous example of forecasting was Marion King Hubbert's prediction of US oil production, published in 1956. Hubbert, a respected geophysicist, predicted that US oil production would peak in the early 1970s — an extraordinary statement at a time when oil production was rising steadily, with plenty of spare capacity. Hubbert recognized that the rate of consumption had exceeded the rate of discovery of new reserves, and the result would be a decline in production that mirrored the growth over the previous century. As Figure 6.6 shows, his forecast has proved remarkably accurate, even down to the total ultimate production (represented by the area under each curve). Hubbert's figure for this, 2.7×10^{10} t, is remarkably close to modern estimates (2.9–3.0×10^{10} t).

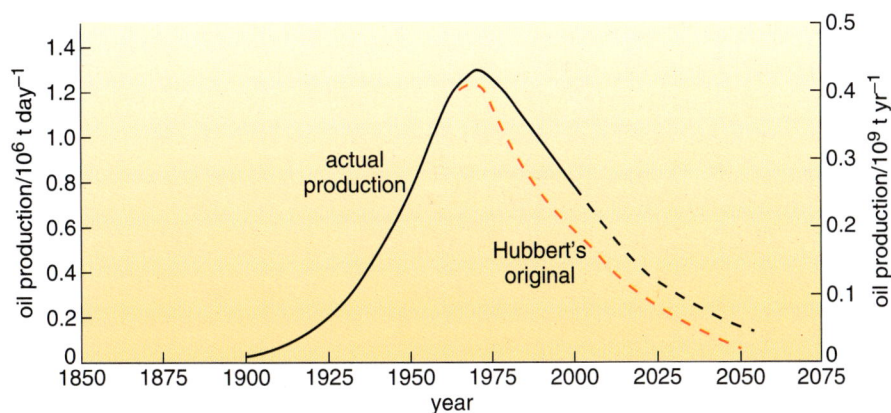

Figure 6.6 US oil production profile, 1850–2050. Actual production (solid curve) closely matches Hubbert's 1956 prediction (dashed red line). The continuation of the actual production beyond 2000 represents a realistic projection of US oil production.

The Hubbert example is relatively simple — considering a single energy source in a single, politically and economically stable country. Any attempt to forecast global energy *demand* must consider a variety of influences, including:

- availability and suitability of different energy sources;
- technological advances in energy production and use;
- economic factors;
- political trends, tensions and timescales;
- environmental pressures.

These points raise several important questions. For example, what proportions of fossil fuels will generate electricity in power stations that are only 30–50% efficient? What proportions will drive transport? If oil is to be used as a fuel, for how long can it also continue to supply the petrochemicals industry — another important role? Can we actually locate and extract the necessary reserves in the amounts required? To what extent can we reduce our dependence on environmentally damaging fossil fuels, either by conserving energy, or substituting alternative sources of power? These are just some of the questions to bear in mind as you read on.

One way of trying to visualize how energy resources might be used is by building 'scenarios', or imagined pictures of the future, which try to account for the factors listed above.

6.2.2 Global energy scenarios

Scenarios for forecasting future energy use tend to conform to three main types:

1 *Historical growth scenarios* assume that energy consumption will continue to rise along the same path as it has done historically.

2 *Technological 'fix' scenarios* reflect efforts to reduce energy demand by developing energy conservation, or more economic methods of producing and using energy, in order to sustain growth.

3 *Zero/negative growth scenarios* imply that society would make the voluntary or enforced decision that it had reached the end of growth.

- Could fossil fuels alone supply global consumption along a historical growth trend?

- Not indefinitely: Earth's fossil fuels are a finite resource, which would become depleted more and more rapidly in the face of rising consumption, no matter how vast they might seem today.

As early as 1972, a group of scientists and other professionals (the 'Club of Rome') warned that the exponential increase typical of historical growth scenarios for exploitation of most resources, including energy, was environmentally unsustainable. However, it was not their warning that forced a zero-growth scenario upon the world for a couple of years during the 1970s, but economic factors. Oil prices quadrupled during the oil crisis of 1973–74 following the Yom Kippur War (6–25 October, 1973), when the Organisation of Petroleum Exporting Countries (OPEC) cartel, dominated by Middle Eastern oil producing countries, cut oil supply dramatically. A further oil crisis occurred in 1979, when the Shah of Iran was overthrown and replaced by a Shia Muslim theocracy, soon to be followed by the Iran–Iraq war (at the time Iraq and Iran were the second and third largest oil-producing countries). Figure 6.7a shows details of the variation in oil production, from OPEC and other areas, in relation to oil-price changes for the period between 1973 and 2005. The effect of these major crises can be seen as sudden increases in oil price, and in the case of the Iranian Revolution as the beginning of a large decline in OPEC production, while non-OPEC production continued to rise. You might care to look for signs of other major developments in world history in Figure 6.7, as the 20th gave way to the 21st century — oil is *the* central feature of the world economy.

Although oil production in the US peaked at around 1975 (Figure 6.6), production in other parts of the world has continued to grow, as Figure 6.7b shows, albeit with considerable variations among OPEC countries and those outside OPEC. Several of the changes relate to important events, as do changes in oil price (Figure 6.7a).

Question 6.1

Which of the three main types of forecasting are depicted on Figures 6.4 and 6.7b for the following periods?
(a) 1940–1970
(b) 1973–1975 and 1979–1983
(c) 1985–2000

The latter half of the 20th century clearly illustrates how unforeseen political events can radically affect energy scenarios, and not only in the short term. The oil crises of the 1970s boosted oil exploration in regions outside the Middle East (e.g. Alaska, the North Sea, the Niger Delta and the Gulf of Mexico) that is

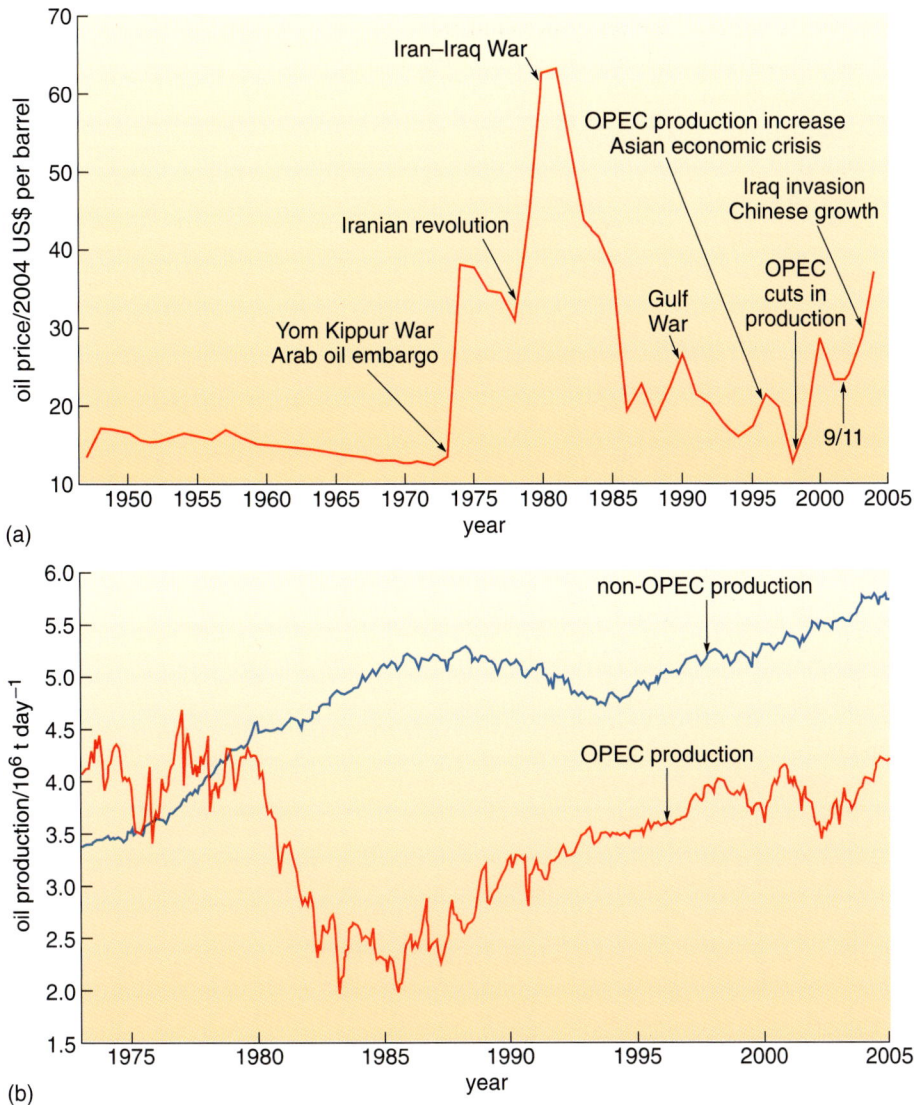

(a)

(b)

Figure 6.7 (a) Changes in the world price of oil from 1947 to 2004 — given here in terms of the purchasing value of the US dollar in 2004 — relative to major world events. (b) Oil production for OPEC and non-OPEC countries from January 1973 to February 2005.

partly reflected in the growth in non-OPEC production after 1973, as well as encouraging energy conservation and efficiency measures. Demand for oil was reduced so successfully that the oil price fell sharply in 1986 to about $17 a barrel (price in 2004 US$). While the world's reliance on Middle Eastern oil decreased, so did the incentive to develop alternative, renewable energy sources. However, the influence of the vast Middle Eastern oil reserves (Figure 6.3) seems set to increase once again, prompting predictions of an OPEC stranglehold on global supply as early as 2008. Equally, the huge gas reserves of the Russian Federation promise to give Russia economic and political leverage, especially in Europe where dependence on gas supplies from the east is growing.

Since 1973 the world price of oil has closely followed major events that either originated in or affected the Middle East (Figure 6.7a). At the time of writing (late 2005) the world price had risen to more than US$60 per barrel, and production from the main US oilfields in Louisiana had shut down following Hurricane Katrina. None of the events shown in Figure 6.7 would have been predictable only a few years before they occurred.

The World Energy Council scenarios

Among numerous attempts to forecast the future of global energy systems, one of the most recent and comprehensive was produced in 1998 by the World Energy Council (WEC) and the International Institute for Applied Systems Analysis (IIASA), a leading 'think tank' based in Austria. There are six WEC scenarios, grouped into three main cases, which compare predicted primary energy use in 2050 with a base year, 1990 (IIASA and WEC, 1998). These scenarios are summarized in Table 6.1, and their predictions are compared in Figure 6.8a.

Case A scenarios involve high economic growth and a medium improvement in the efficiency of primary energy use. Scenarios in Case C involve lower economic growth, a steady de-emphasis of fossil fuels and high improvement in efficiency. Case B employs an intermediate strategy as regards the blend of energy sources, the same economic growth as C and low improvements in efficiency. As well as the differences in economic growth, the different improvements in the efficiency of energy usage result in the three future trends for primary energy use shown on Figure 6.8a. As you might imagine, there is a huge range of possible scenarios that involve varying rates of economic growth, different blends of targeted energy sources, and various achievements in improving efficiency, of which the WEC settled on six that seemed realistic.

The inclusion of energy efficiency measures in the different WEC scenarios, unsurprisingly strongest for the Ecologically Driven case, is critical. However, the High Growth scenarios display a stronger commitment to energy efficiency than the Middle Course scenario, acknowledging that sustainability may not be incompatible with growth.

● Which of the scenarios from Cases A and B would you judge to be the most sustainable?

○ On the basis of the figures in Table 6.1, scenario A3 ('non-fossil future') should be the most sustainable, as it has the lowest proportion of fossil fuel use and the highest contribution by renewable alternatives. However, this would ultimately depend on the total energy consumption of each scenario.

Table 6.1 World Energy Council global energy scenarios (1998).

Case (economic growth per year; efficiency improvement)	Scenario description	Blend of primary energy sources in 2050/%				
		Coal	Oil	Gas	Nuclear	Renewables
A: High Growth (2.7%; medium)	A1: ample oil/gas	15	32	19	12	22
	A2: return to coal	32	19	22	4	23
	A3: non-fossil future	9	18	32	11	30
B: Middle Course (2.2%; low)	mixture of sources	21	20	23	14	22
C: Ecologically Driven (2.2%; high)	C1: new renewables	11	19	27	4	39
	C2: renewables + new nuclear	10	18	23	12	37
Base year: 1990	mainly fossil fuels	24	34	19	5	18

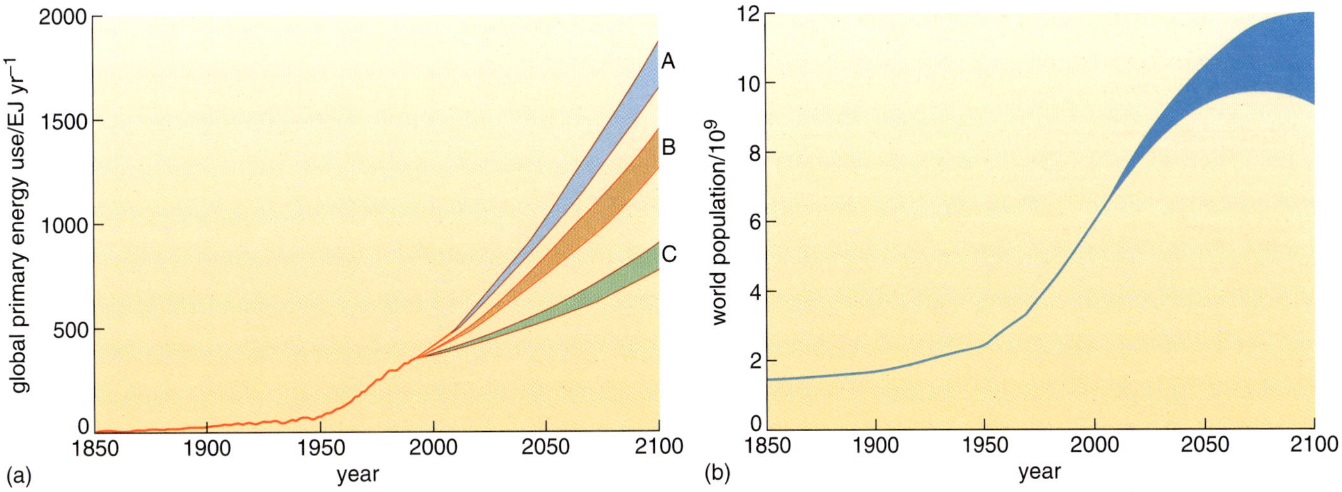

The WEC went further with the three 'probably sustainable' scenarios (A3, C1 and C2) in terms of their decreasing demands on non-renewable sources, by assigning specific contributions to the different kinds of alternative energy sources. The details of their modelling are shown in Figure 6.9. Note that the models rely most heavily on solar (25 to 38% by 2100) and biomass (18 to 25%) resources, with geothermal, wind, wave and tidal (i.e. other) sources having only the same weight as hydro (3 to 5%). Hydropower, as you will recall from Chapter 5, has the least potential for further development. If appropriate technologies can be adopted sufficiently, the likelihood of solar energy becoming the single most important contributor by 2100 poses no problems except for the amount of the Earth's surface that needs to be used. This surface could be that

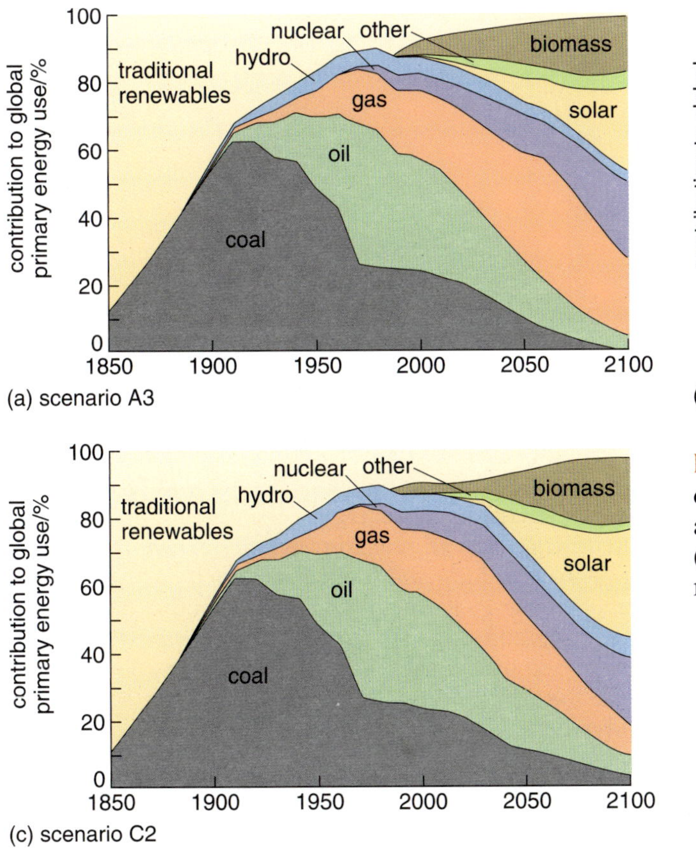

Figure 6.8 (a) Global primary energy use: historical development (1850–1990), and projections to the end of the 21st century, based on the three WEC scenarios (Table 6.1). (b) Global population growth 1850–1990 and projections to 2100.

Figure 6.9 Models of the relative contributions of different energy resources to global primary energy use according to WEC scenarios: (a) A3; (b) C1; (c) C2. (See Table 6.1.) *Note*: 'Other' refers to other renewable resources.

209

which would otherwise be non-productive desert or part of it could be incorporated into buildings and even roads. There is, however, a far less tractable issue concerning the massive adoption of another alternative energy source — biomass.

Question 6.2

Even for Case C, Figure 6.8a predicts annual primary energy use at the end of the 21st century to be about twice that at the start. If 25% is to be supplied by biomass, what proportion of the Earth's land surface would be needed to grow the vegetable matter needed? (You should refer back to Chapter 5.)

Figure 6.8b predicts a doubling of human population by 2100, which places at least twice the current demand on fertile land for food production. Any modelling that is constrained within arbitrary limits, inevitably conflicts with factors that lie outside its remit.

Other scenarios: Shell and Greenpeace

In 1995, Shell International Petroleum published two energy-supply scenarios ('Sustained Growth' and 'Dematerialization'), which contained similar elements to the WEC scenarios. In 2001, however, Shell published two further long-term scenarios: 'Dynamics as Usual' and 'Spirit of the Coming Age' (Shell International, 2001). The former envisages 'an evolutionary progression from coal to oil to gas to renewables (and possibly nuclear)…', whereas the latter is more radical, charting the rise of a new technological system based on hydrogen, aided by developments in fuel cells and sequestration of carbon dioxide. This system, widely known as the hydrogen economy, is described in Section 6.4.3. Shell's analysis of the global market shares of different fuels in these two scenarios is shown in Figure 6.10, tracking trends from 1850 through into the future.

Although superficially rather different, the scenarios shown in Figure 6.10 display several similarities, including a gradual decline in fossil fuels in the 21st century, and a global market share for all renewables of about one-third by 2050. Interestingly, the two earlier Shell scenarios (1995) both predicted a much higher market share (around 50%) for all renewables by 2050, demonstrating how marked changes in forecasts can occur on a relatively short timescale. Another indication of the frailty of forecasts was ironically also provided by Shell, which reduced its own estimates of the company's oil reserves on four separate occasions in 2004, by over 20% overall (see

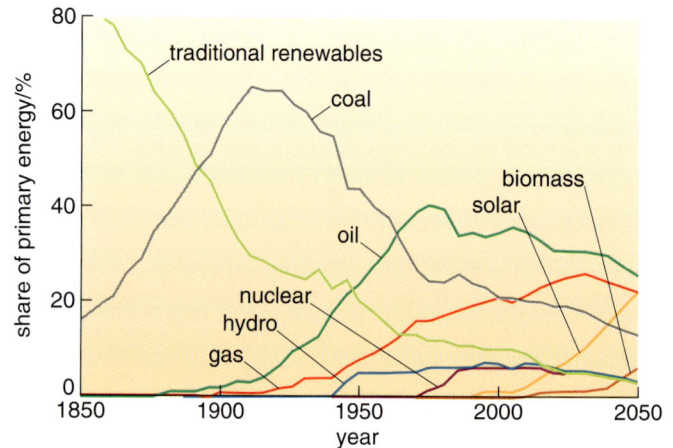

(a) 'Dynamics as Usual' scenario

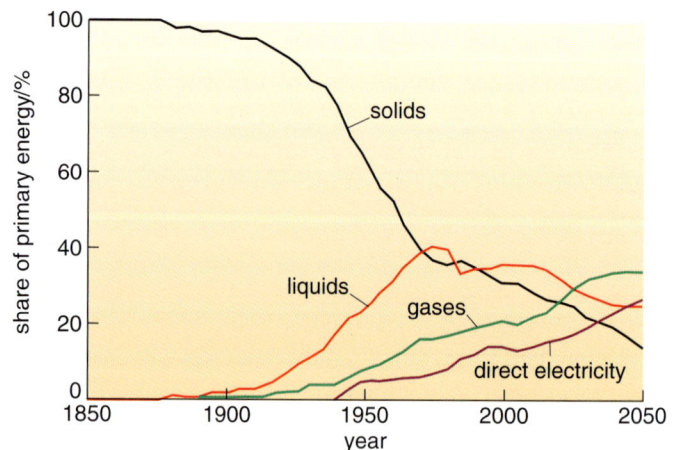

(b) 'Spirit of the Coming Age' scenario

Figure 6.10 Share of primary energy for different fuels in Shell International's energy scenarios (2001). (a) In the 'Dynamics as Usual' scenario, supplies evolve from high- to low-carbon fuels and towards electricity as the dominant energy carrier, driven by demands for security, cleanliness and sustainability. (*Note*: The curves for hydro, nuclear and traditional renewables converge after 2020.) (b) In the 'Spirit of the Coming Age' scenario, supplies evolve from solids through liquids to gas (methane and then hydrogen), supplemented by direct electricity from renewables and nuclear.

Box 3.4). Time will tell whether these re-evaluations heralded a hastening of fossil fuel depletion.

One organization, Greenpeace, spurred by warning signs that conventional energy resources may be unsustainable, commissioned a study by the Stockholm Environment Institute in the early 1990s that proved to be even more optimistic than the preceding examples (Lazarus et al., 1993). With similar assumptions on population increases and economic growth (to allow comparison with other scenarios), the Greenpeace scenario develops to a situation free from fossil fuels by 2100, dominated by solar and wind power (around 75%), with supplementary contributions from biomass, hydropower and geothermal sources. Nuclear power is phased out rapidly (by 2010), whereas the fossil fuels decline gradually through the century. This scenario requires improvements in energy efficiency that actually reduce global demand around 2030, before it rises again towards the end of the century, and hydrogen is used as a transport fuel and to store energy from intermittent sources such as wind.

In all these 'scenario-based' forecasts, as well as the tendency for the 'unthinkable' to happen (as it commonly does — Figure 6.7), it is important to temper such predictions by bearing in mind the difficulties with adopting alternative energy resources (Section 5.8). Achieving the modelled percentage shares of primary energy supply begins at a very low base level, and demands sustained high rates of growth in the alternatives sector; indeed, growth rates that far surpass any that have characterized global economic history during the last century.

6.2.3 Forecasting UK energy demand

In some ways, predicting demand for energy — or indeed any major resource — is more difficult for a single country than for the world (Section 6.2.1). Political decisions and regime changes have a dis-proportionally greater effect on energy usage within a single country, and the typically long lead times of major projects carry more significance. Lead times of 5–10 years are common for major developments such as power stations, oil fields, mines and quarries, all of which require substantial initial investment that depends on a guaranteed return at the culmination of the project. So, although lead times present a difficulty to forecasters, they are also one of the most important reasons for accurate forecasting in the first place.

In 2003, the UK Department of Trade and Industry published a White Paper entitled *Our Energy Future — Creating a Low Carbon Economy*. This Paper presented a general scenario for the UK energy system in 2020, which you can think about in the context of the UK potential for the alternative energy sources as discussed in Chapter 5:

'• Much of our energy will be imported, either from or through a single European market embracing more than 25 countries.

• The backbone of the electricity system will still be a market-based grid, balancing the supply of large power stations. But some of those large power stations will be offshore marine plants, including wave, tidal and windfarms. Generally smaller onshore windfarms will also be generating. The market will need to be able to handle intermittent generation by using backup capacity when weather conditions reduce or cut off these sources.

- There will be much more local generation, in part from medium to small local/community power plant, fuelled by locally grown biomass, from locally generated waste, from local wind sources, or possibly from local wave and tidal generators. These will feed into local distributed networks, which can sell excess capacity to the grid. Plant will also increasingly generate heat for local use.

- There will be much more micro-generation for example from CHP [combined heat and power] plant, fuel cells in buildings, or photovoltaics. This will also generate excess capacity from time to time, which will be sold back into the distributed network.

- New homes will be designed to need very little energy and will perhaps even achieve zero carbon emissions. The existing building stock will increasingly adopt energy efficiency measures. Many buildings will have the capacity at least to reduce their demand on the grid, for example by using solar heating systems to provide some of their water heating needs, if not to generate electricity to sell back to the local network.'

(Department of Trade and Industry, 2003, pp. 18 and 19)

The predicted shift away from fossil fuel dependence towards renewable energy sources is familiar by now, but the other striking theme in this scenario is the increasing emphasis on local power generation, as opposed to regional or national systems. This is intriguing, predicting a return to more 'primitive' times when the onus was on the individual to harness energy for their family or village, rather than on centralized provision. Alongside this localized generation comes an expectation that energy efficiency and conservation will rise markedly, through design of dwellings and appliances, resulting in a decrease in overall demand for energy.

⬤ What other prediction about demand is notably absent from the scenario as summarized above?

◯ There is no mention of consumers having to reduce their demand voluntarily, i.e. by reducing their implied standard of living. In fact, this point is seldom made in politically driven scenarios, for obvious reasons.

One group of scenarios, driven primarily by concerns over environmental damage, especially climate change, centres on large reductions in demand for energy. For example, the UK Royal Commission on Environmental Pollution published four energy scenarios in the year 2000; of these, three forecast reductions in total energy demand of between 36% and 47% by 2050. These reductions were aimed mainly at achieving 60% reductions in the 1997 level of carbon dioxide emissions by 2050, a target set around the same time by the Intergovernmental Panel on Climate Change. Such scenarios reflect the groundswell of scientific data, and increasingly public and political opinion, that environmental change is the most pressing problem facing human society in the 21st century.

6.3 Environmental consequences of fossil fuel combustion

Water vapour, carbon dioxide, sulphur dioxide, nitrogen oxides, carbon monoxide, ash, soot and fuel particles are all released into the atmosphere when fossil fuels burn. The types of emissions produced by burning any given fossil fuel are similar, whether the fuel is burned to generate electricity in power stations, to provide domestic heat, or for transportation. Chemical and physical conditions of combustion are different for each of these applications, so the *blend* of emissions varies depending on usage as well as the type of fuel. To give one example, atmospheric smoke concentrations in London have fallen more than tenfold since 1960 with the switch away from coal for home heating and electricity generation, but nowadays over 70% of the particulates in London air come from the exhausts of diesel-fuelled commercial vehicles.

Water vapour is one of the most significant emissions from fuel burning in terms of quantity, but since it is naturally part of the water cycle there is little concern over introducing more water vapour into the atmosphere. Three effects stemming from the changing composition of the atmosphere currently cause concern: acid rain; a decline in air quality; and, most significantly, global warming through enhancement of the greenhouse effect.

6.3.1 Fossil-fuel burning and global warming

The amount of carbon dioxide produced by combustion varies from fuel to fuel, depending on its ratio of carbon to hydrogen. Natural gas produces the least carbon dioxide. Burning 1 t of natural gas (e.g. methane, CH_4) in a power station releases about 2.75 t of CO_2; burning 1 t of petrol (that contains hexadecane, $C_{16}H_{34}$) in a car engine releases over 3 t of CO_2. Coal produces about 20% more, since coal contains between 60% (lignite) and 90% (bituminous) of carbon by mass.

● Why is the mass of CO_2 released greater than that of the fuel burned?

○ Carbon combines with oxygen in proportion to the relative atomic mass of the two elements (C = 12; O = 16) and the relative numbers of atoms in the molecule, so that every 12 t of carbon combines with 32 t of oxygen to produce 44 t of CO_2, i.e. burning 1 t of carbon produces 44/12 = 3.67 t of CO_2.

Burning fossil fuels emits 5×10^9 t of carbon into the atmosphere every year.

● How does the yearly amount of carbon released through burning fossil fuels compare, as a percentage, with that fixed by land plants?

○ Figure 1.9 showed that 60×10^9 t of carbon is fixed by plants each year. The carbon flux from burning fossil fuels is approximately 8% of this.

Until the 1960s, the atmosphere was considered to be large enough to absorb large quantities of carbon-based gases and dilute them to harmless levels. It has become clear that this is not the case. Rapid increase in carbon-based atmospheric gases interferes with the energy budget of the Earth itself by enhancing the 'greenhouse effect' (Box 6.2).

Box 6.2 The greenhouse effect

All objects emit electromagnetic radiation at wavelengths that depend on their temperature: the hotter an object is, the shorter the wavelength of radiation that it emits. The temperature of the Sun's surface is over 5500 °C, so it emits short-wavelength radiation (Figure 6.11) in the ultraviolet, visible and near-infrared parts of the spectrum. Much of the Sun's energy output is in the form of visible light.

Figure 1.7 summarized what happens to solar radiation received by the Earth. The crucial feature for climate is what happens to energy that is re-emitted. Because the Earth's surface temperature on average is about 16 °C, it emits radiation at much longer wavelengths than those of solar radiation.

Atmospheric gases absorb radiation at specific wavelengths that depend on the composition of each gas and on its concentration. Figure 6.11 shows schematically those parts of the electromagnetic spectrum absorbed by six important atmospheric gases. Nitrogen, which makes up 80% of the atmosphere, is an exception and is omitted; it does not absorb any infrared radiation. Dark-red bars signify absorption of incoming solar radiation, whereas the black bars show absorption of long-wave energy emitted from the Earth's surface.

Water vapour, carbon dioxide, methane, nitrous oxide (N_2O) and ozone (O_3) each absorb wavelengths in part of the long-wave infrared range emitted by the Earth. Taken together, atmospheric gases can absorb most wavelengths of terrestrial radiation, but water vapour and carbon dioxide make the most significant contribution, depending on their concentrations. The energy that they absorb raises atmospheric temperature. Almost all the mass of the atmosphere lies within 30 km of the Earth's surface, with half

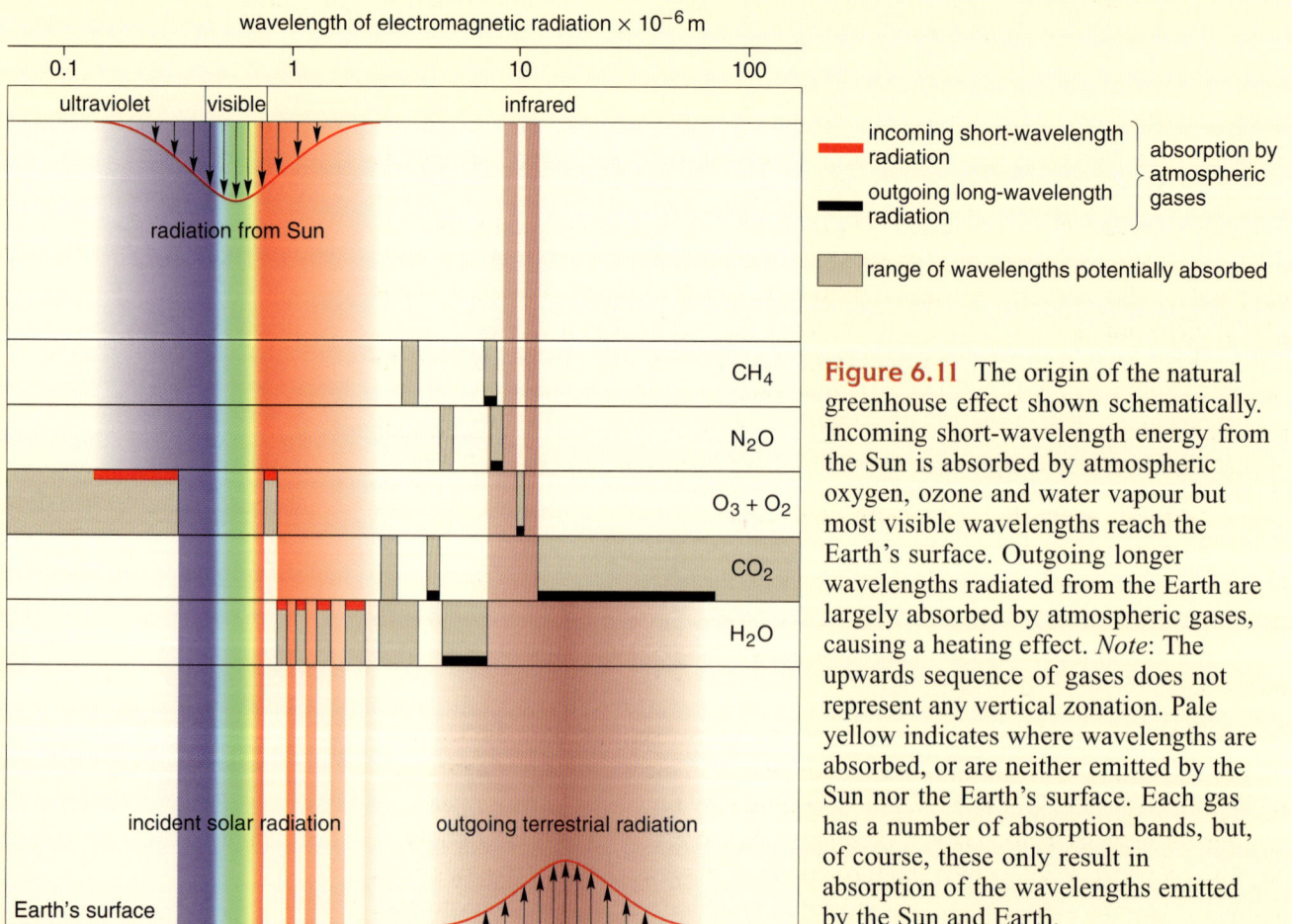

Figure 6.11 The origin of the natural greenhouse effect shown schematically. Incoming short-wavelength energy from the Sun is absorbed by atmospheric oxygen, ozone and water vapour but most visible wavelengths reach the Earth's surface. Outgoing longer wavelengths radiated from the Earth are largely absorbed by atmospheric gases, causing a heating effect. *Note*: The upwards sequence of gases does not represent any vertical zonation. Pale yellow indicates where wavelengths are absorbed, or are neither emitted by the Sun nor the Earth's surface. Each gas has a number of absorption bands, but, of course, these only result in absorption of the wavelengths emitted by the Sun and Earth.

comprising the lowermost 6 km. So the heating mostly affects the lowest atmospheric levels (with the exception of ozone, which heats up the stratosphere). As a result, the Earth's surface is some 33 °C warmer than it would be if it had no atmosphere. This phenomenon is the **greenhouse effect**, without which the Earth would be in a permanently ice-bound state.

There is a natural moderating effect to this 'greenhouse' heating. The warmer it is, the more the atmosphere eventually radiates long-wave radiation to space. Conversely, a cooler atmosphere radiates less, and absorbs more energy emitted by the surface and heats up. Before industrial greenhouse-gas emissions began, atmospheric heat gains and losses were roughly balanced at an average global temperature of 15 °C. The problem with these emissions is that this balanced temperature rises as concentrations of these gases increase; an enhanced greenhouse effect.

Some gases, such as methane, are far more potent absorbers of long-wavelength radiation than carbon dioxide and water. Each year, 108 t of methane is released into the atmosphere from natural gas venting at oil well heads, from leaking gas pipelines and during coal mining. One kilogram of methane released into the atmosphere can potentially cause eleven times more warming than 1 kg of CO2. But methane reacts with oxygen in a matter of years to form carbon dioxide, some of which dissolves in rainfall and in ocean water. Although the carbon cycle tends to achieve a balance (Section 1.4), that balance fluctuates as CO2 is added by volcanoes and the burning of biomass and fossil fuels, and reduced by various transfers of carbon to long-term storage (Section 1.4). Each gas therefore has its own warming potential, based on its potency and atmospheric lifetime. The greatest concern about global warming focuses on the large volumes of atmospheric CO_2 added through human activities, although there are fears that methane may be released from the vast natural stores of gas hydrate in ocean floor sediments (Section 3.7.2).

Global warming

Since the mid-19th century, human industrial and agricultural activity has caused a change in the concentrations of atmospheric gases. In particular, there have been dramatic increases in the production of carbon dioxide by combustion of fossil fuels. Figure 6.12 shows deviations in mean surface temperatures from recent average temperatures for two time periods. The first (Figure 6.12a) shows changes in the *global* mean annual surface temperature since 1860. The second (Figure 6.12b) covers the last 1000 years for the Northern Hemisphere, and has been dubbed the 'hockey-stick' trend from its shape; it represents the main evidence for anthropogenic global warming. Since about 1915, there has been an increase in the annual mean surface temperature in the Northern Hemisphere of about 0.7 °C.

Global warming seems to have a more significant effect on night-time rather than daytime temperatures. Since 1950 the minimum daily temperature over most of the landmass of the Northern Hemisphere has risen three times as fast as the maximum temperature. In the UK, nights are on average 0.84 °C warmer than they were in 1950, whereas days are only 0.28 °C warmer — a trend that applies to all northern continents and all seasons.

In North America and Europe, cloud cover has increased along with the warmer nights. Low-altitude clouds limit the loss of heat from the ground by radiation at night, but reflect sunlight and limit the warming effect by day. There are two reasons for increased cloud cover. One is particulate pollution (dust and smoke), since small particles encourage the formation of denser and more numerous clouds. The other is ocean evaporation: the enhanced greenhouse effect may be increasing evaporation from the oceans, leading to increased cloud cover over land.

(a)

(b)

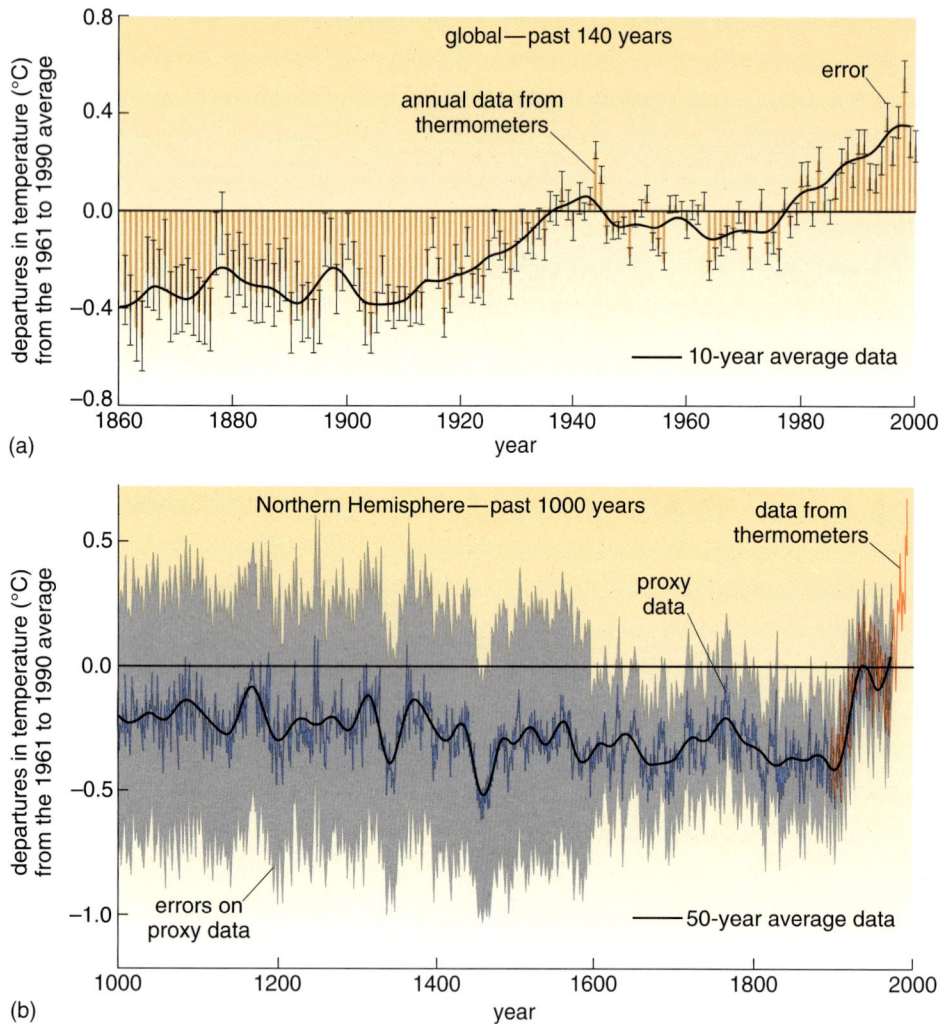

Figure 6.12 Deviations in mean surface temperature, relative to the average between 1961 to 1990: (a) globally for each year since 1860; (b) in the Northern Hemisphere since 1000. The curves are averages to show general trends. Red bars are temperatures based on thermometer readings; the blue bars show proxy estimates based on ice cores, tree rings and corals; greys show the errors on the proxy estimates.

The fossil fuels burned in the 200 years since the Industrial Revolution took many millions of years to accumulate. The sudden return of ancient carbon to the atmosphere upsets the outer Earth's energy balance, which now gives rise to much concern among scientists, environmentalists and politicians. With no change in the current 'blend' of energy sources, industrial greenhouse gas emissions may result in doubling of the pre-industrial CO_2 level by 2028. The global mean surface temperature could increase at about 0.3 °C per decade; faster than any rise seen over the last 10 000 years. By 2030, temperatures could be 0.7–2.0 °C higher than at present, and by the end of the 21st century the rise could be 6 °C.

Increased surface temperatures are likely to be accompanied by a rise in global sea-level, because of thermal expansion of the oceans and melting of land-based ice. A 'business-as-usual' scenario suggests a sea-level rise of 0.2 m by 2030 and 0.65 m by the end of the 21st century.

To stabilize CO_2 levels would require an immediate reduction of emissions from human activities. Yet even if fossil fuel burning stopped immediately, the temperature rise would continue because of the time lags in the gas–energy balance in the atmosphere. To eventually achieve a balance that does not destabilize global climate requires finding alternative sources of energy for transport and for electricity generation.

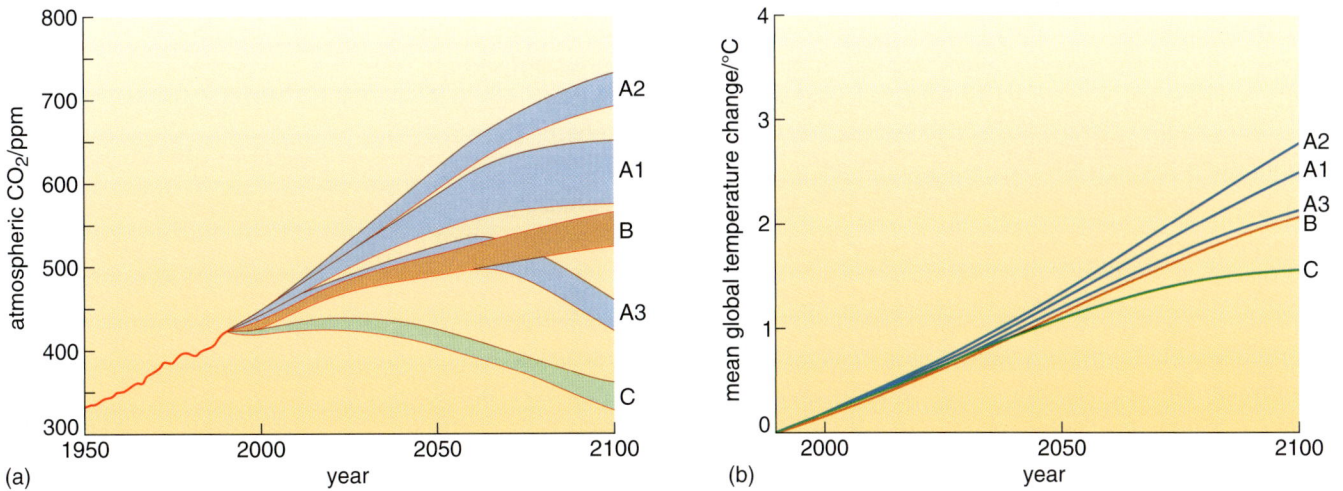

(a)

(b)

Figure 6.13 returns to the World Energy Council's scenarios of primary energy use during the 21st century (Figures 6.8 and 6.9), and shows the changes in atmospheric CO_2 concentration that would ensue from the three cases shown in Table 6.1. Whichever scenario is used, global mean surface temperature is predicted to rise over the next century by 1 to 2.7 °C (minimum for Case C to maximum for Case A2).

Figure 6.13 Changes in: (a) atmospheric CO_2 concentrations; (b) global mean surface temperature. Both are modelled to result from the WEC scenarios in Table 6.1.

6.3.2 Sulphur dioxide, acid rain and air quality

Emissions of sulphur dioxide (SO_2) and nitrogen oxides by burning fossil fuels cause globally significant environmental problems (Figure 6.14) due to **acid rain**. Sulphur dioxide reacts with water vapour to become first sulphurous acid (H_2SO_3) and then sulphuric acid (H_2SO_4) as a result of photochemical oxidation involving sunlight and other gases such as ozone (O_3), hydrogen peroxide vapour (H_2O_2) and ammonia (NH_3) which act as catalysts. Depending on the amount of moisture in the air, up to 80% of emitted sulphur dioxide may become acidic. Nitrogen oxide (NO_x) emissions also react with water vapour, to form dilute nitrous and nitric acid. If sulphur dioxide reaches the very dry stratosphere, rather than being 'rained-out', sulphuric acid forms minute droplets (aerosols) that have a very different effect (Section 6.3.3).

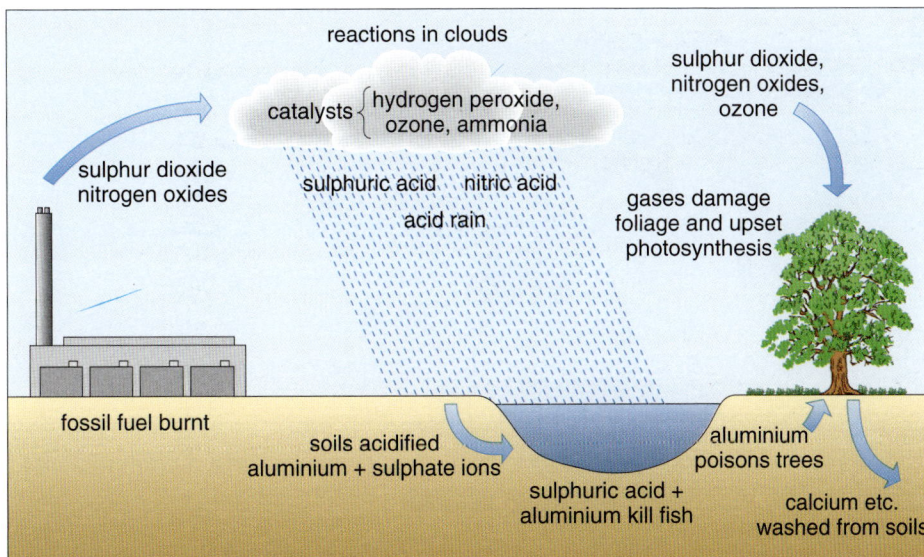

Figure 6.14 The chain of pollution caused by acid rain.

Acid rain increases the acidity of groundwater, which helps to dissolve metal ions from soil and rocks. Some of these, such as aluminium ions, are poisonous to plants and animals (Figure 6.14). Acid rain also attacks carbonate-rich soils and limestones exposed at the surface to release CO_2 into the atmosphere. Even if sulphur dioxide levels from power stations are strictly controlled, as they are in the USA and several European countries, NO_x emissions from increasing numbers of cars and power stations are capable of continuing the production of acid rain on about the same scale.

Current research suggests that even if SO_2 and NO_x pollution was halted today, the effects of acid rain outlined in Figure 6.14 would remain for decades. Acidity and aluminium would continue to poison lakes and streams. One UK investigation suggests that halving acid precipitation over the Scottish hills immediately would merely maintain the acidity of the local lakes at their present-day levels.

Activity 6.1

Figure 6.15 shows the pattern of global atmospheric fallout of sulphur (mainly as sulphates) during the 1990s and anticipated fallout during 2030 (based on atmospheric models). Analyse Figure 6.15, using the following tasks as a guide:

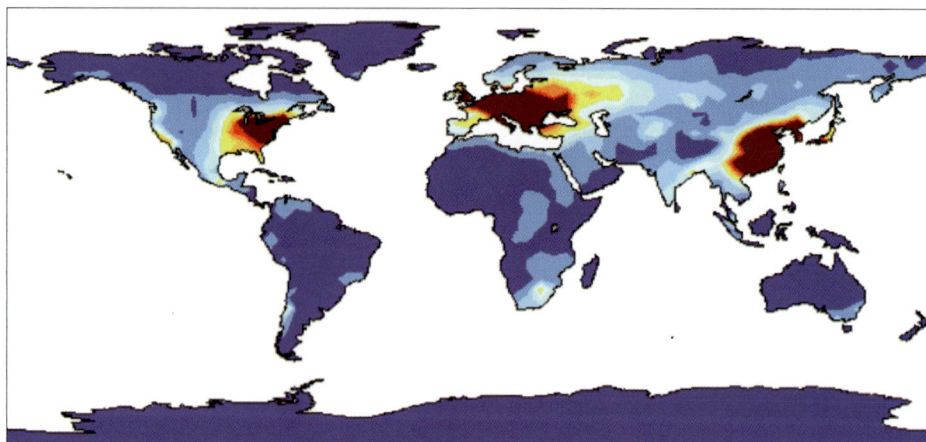

(a)

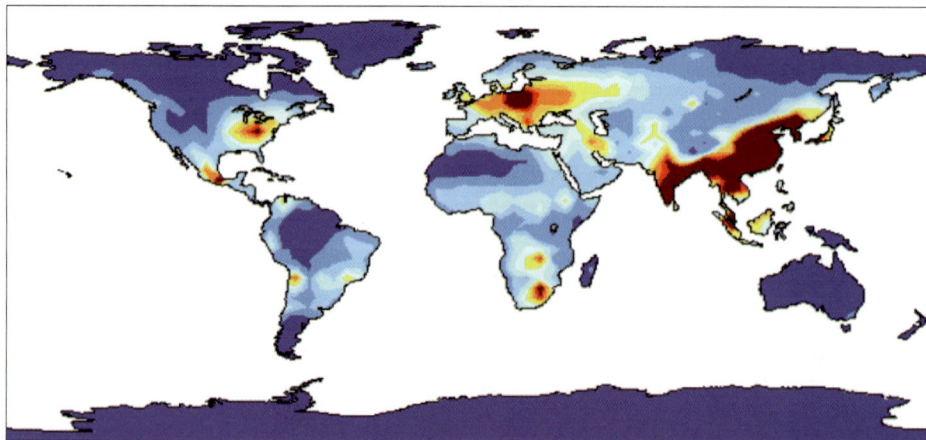

(b)

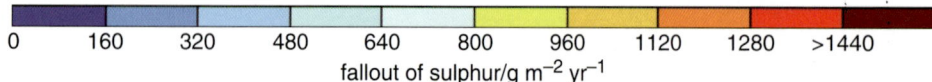

0   160   320   480   640   800   960   1120   1280   >1440

fallout of sulphur/g m^{-2} yr^{-1}

Figure 6.15 Global fallout of sulphur from acid rain (colours show ranges in g m^{-2} yr^{-1}), during (a) the 1990s and (b) estimated for 2030.

(a) Identify those regions that are likely to receive decreased sulphur pollution by 2030.

(b) Which regions are likely to expect an increased problem with acid rain?

(c) Suggest reasons for the main regional changes in acid rain between 1990 and 2030.

(d) Australia has one of the highest per capita energy consumption rates in the world. Why is acid rain so low there?

6.3.3 The effects of atmospheric aerosols

While many aspects of sulphur pollution are detrimental to the surface environment, another of its features may be counteracting global warming. There is growing evidence that global warming is at least partially offset by minute particles of dust (*aerosols*), including those of various sulphates, in the atmosphere. The effect is caused by emissions from power stations in Europe, North America and Asia, dust storms in the Sahara, burning tropical forests, iron and steel manufacture, and, importantly, by the tremendous growth in air transport.

Aerosols cool the atmosphere in two ways. They reflect and scatter solar radiation, slightly reducing the amount that reaches the surface. Sulphate aerosols also act as nuclei for condensation of water vapour, thereby encouraging the formation of clouds. Large numbers of aerosols produce small water droplets, of which reflective clouds are made. Clouds shade the ground during hot summer days. At night and in winter they warm the surface layers of the atmosphere by absorbing long-wavelength radiation emitted from the Earth's surface. Sulphur dioxide that enters the stratosphere also forms aerosol droplets of sulphuric acid that further reduce incoming solar radiation — but its source is explosive volcanism that is powerful enough to inject material to altitudes greater than 10 km.

In the USA, average daytime cloud cover has risen from a little less than 50% between 1900 to 1940 to above 58% since 1960. Whereas the average daytime temperature of much of the global landmass has risen since 1950, the areas with high SO_2 emissions have cooled. Average maximum daytime temperatures between June and November over land in the Northern Hemisphere *fell* by about 0.41 °C between the 1950s and 1990s. This is opposite to the trend shown by Figure 6.12b for average annual surface temperature over the whole of the Northern Hemisphere, and has been ascribed to the effects of atmospheric aerosols released by pollution.

This increased direct return of solar energy to space has been dubbed '**global dimming**' due to a decrease in solar radiation reaching parts of the Earth's surface, a trend observed until the early 1990s. Thereafter, growing controls on emissions from European and North American power stations have led to a 'brightening-up' of the skies — an increase in insolation. Ironically, reduced particulate pollution is likely to exacerbate the current global warming trend, since it had in part been offset by global dimming.

An important additional consequence of the combustion of fossil fuels is the effect the products of such burning have on air quality. In December 1952, a cold fog (smog) that contained high levels of smoke and sulphur dioxide hung over London for almost a week. It was one of the worst in a series of 'pea-souper' fogs that descended on London at that time, and was directly responsible for about 4000 deaths through bronchial infections and heart attacks. A public outcry led to smoke controls and wider use of smokeless solid fuel. At the time, doctors blamed high smoke levels in the fog, but more recent research suggests that the formation of highly acidic particles may have been important. The 1952 London smog had a pH of 1.6 — more acidic than lemon juice.

The advent of North Sea gas and the siting of power stations outside centres of population dramatically reduced the incidence of smoke-laden fogs. However, they have been replaced by a different form of air pollution — NO_x and smoke particles from vehicle emissions. A thick haze that built up from London traffic fumes during four windless days in December 1991 directly caused the deaths of 160 people. Two pollutants were exceptionally concentrated in the London air: NO_2 levels reached 423 parts per billion, the highest level ever recorded in the UK, and black smoke particles reached 228 $\mu g\ m^{-3}$.

Particularly on sunny summer days, the main respiratory irritant in 'modern' polluted air is ozone (O_3), only naturally abundant in the stratosphere. This ground-level ozone is formed from photochemical reactions involving both NO_x and traces of hydrocarbons in the air. Whereas the London smogs of the 1950s were the result mainly of domestic coal combustion, the NO_x that forms one of the main ozone precursors comes mainly from car exhausts and power station flues. While the consequences of ground-level ozone can have severe consequences for human health, the implications of this pollution can be even further reaching — natural ecosystems suffer and agricultural production falls. Already, rice yields in polluted parts of Asia are thought to be 10% lower than they would have been under 'ozone-clean' conditions, while soy bean crops are even more susceptible, with a projected 30% drop in yield by 2020.

6.3.4 Pollution solutions?

At the 1992 Earth Summit in Rio de Janeiro, it was agreed that the major industrialized nations would stabilize their CO_2 emissions from fossil fuels at 1990 levels by the year 2000. This decision was taken primarily to relieve possible global warming caused by the greenhouse effect, but such a policy would also stabilize emissions of other major contributors to atmospheric pollution and acid rain, such as oxides of sulphur and nitrogen. Since then, the Kyoto Protocol of 1997 committed signatory countries (but not the USA, Australia, India and China, which did not sign) to reducing annual greenhouse emissions to 5% below 1990 levels by 2012. In terms of acidification and air quality, the 1999 Gothenburg Protocol for the abatement of acidification, eutrophication (water pollution caused by excessive plant nutrients, such as NO_x) and ground-level ozone requires of its signatories the following reductions (relative to 1990 levels) by 2025: sulphur emissions by at least 63%; NO_x emissions by 41%; volatile organic carbon emissions by 40%; and ammonia emissions by 17%.

In Section 6.2 and in Figure 6.8 (C scenarios) you saw the dramatic potential effect of replacing fossil fuels by alternative renewable and nuclear sources, with a drive towards more efficient usage of energy. You should also have noted the caveats regarding the high pace of developing and deploying suitable technologies to achieve that (Section 5.8). An alternative view is that humanity should reduce its overall energy consumption rapidly and significantly. That means we all would have to adopt alternative ways of doing things: greater energy conservation at home and at work; less travel overall; better use of mass transport; different lifestyles and expectations, and so on.

Renewable energy sources must be developed if we are to wean ourselves off fossil fuels. Wind power is growing in the UK and in Europe — but not without some controversy over visual impact on the landscape — and solar technologies are improving in terms of efficiency and cost. Nuclear power could meet the demand for energy while these new technologies mature, without CO_2 being produced.

Another approach, however, may provide part of the answer to reducing CO_2 emissions from power stations — **CO_2 sequestration**. This experimental technology seeks to remove CO_2 at source, i.e. from within a power station, before it has a chance to escape into the atmosphere (Figure 6.16). In other words, the CO_2 is directly transferred to long-term storage, rather than being added to the carbon cycle (Section 1.4). Theoretically, such sequestration could exceed natural rates of carbon burial. Probably the simplest solution is to pump CO_2 into porous but sealed-off rocks that once held oil or gas. It is possible that some CO_2 could be induced to form carbonate minerals in the host rock, i.e. a solid and therefore more stable form of storage. Another possibility is exploiting the liquefaction of CO_2 at high pressures (Section 3.5.2, Figure 3.14) and injecting liquid CO_2 into deep ocean basins. Like methane, CO_2 can also form a solid gas hydrate in sea-floor sediments (Section 3.7.2); a further possibility for sequestration. There are of course immense technical challenges that need to be resolved. Such technological 'fixes' would also place a heavy economic burden on conventional CO_2 emitting power stations, making implementation unlikely in rapidly developing regions such as China and India. Moreover, this approach does nothing to alleviate CO_2 emitted outside power stations, especially from increasing road and air transport (now considerably greater than from electricity generation).

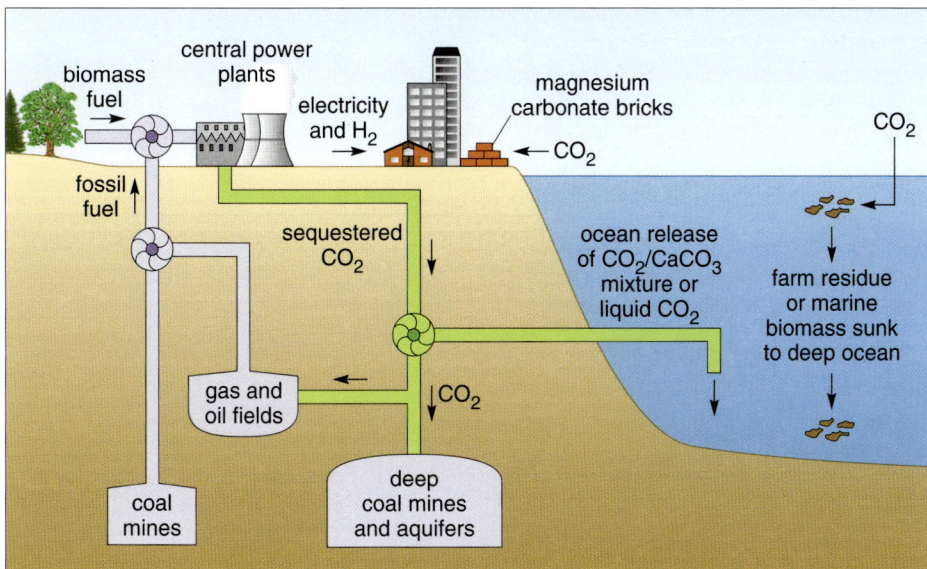

Figure 6.16 Diagram showing various options for the sequestration of carbon dioxide.

No technologies are currently envisaged that could selectively extract CO_2 from air in the volumes necessary to make any appreciable difference to its overall composition. There is one simpler means, however, photosynthesis is free and efficient, but needs encouragement. Planting enough trees to cover an area the size of Australia is believed to have the potential to reduce atmospheric CO_2 to levels compatible with a stable climate. Other biological methods of sequestration include 'fertilizing' open oceans to encourage phytoplankton growth, whose death, sinking and burial on the ocean floor would fix carbon in long-term storage, albeit at the possible expense of upsetting ecosystems.

6.4 Prospects and possibilities for the world's energy future

Before speculating on what might lie ahead for the global energy scene, it may help to review the principal characteristics of the range of energy resources available.

6.4.1 Currently available sources

Fossil fuels (Chapters 2 and 3) are the principal global energy source at the start of the 21st century, but on combustion they all emit CO_2, which contributes to global warming, and varying amounts of SO_2, along with nitrogen oxides, which all cause acid rain. Furthermore, they are a finite resource.

Coal is the most abundant of the fossil fuels but has the lowest 'energy density' and the greatest pollution potential. Commercial coal contains 5–10% or more of non-combustible mineral impurity (ash), around 1% of sulphur, and other elements in trace amounts, such as chlorine, uranium and arsenic. So-called clean coal technologies improve combustion efficiency and reduce pollution (see Section 6.4.2). As we saw in Box 6.1, however, the UK Government moved away from coal during the late 1980s and early 1990s, despite the fact that coal remains the UK's most abundant energy raw material. On a global scale, coal represented around 22% of primary energy consumption in 2002 (Figure 6.2), and there is enough coal to sustain demand at present levels for a couple of centuries, even without any new reserves being found.

Oil is a more versatile fuel than coal, with a higher 'energy density' and low to zero ash content. It still produces sulphur and nitrogen oxide emissions as well as CO_2 (though less than coal per unit of energy obtained). Motor vehicle exhaust gases can cause high levels of pollution in urban areas, even when reduced by catalytic exhaust systems. Global oil reserves (Chapter 3) are sufficient for several decades at current (2005) consumption levels, though finding and exploiting major new reserves will be increasingly difficult and costly.

Gas has the highest 'energy density' of all three fossil fuels and is widely held to be a 'clean' fuel, because it leaves no ash and contains virtually no sulphur; but burning gas still produces nitrogen oxides as well as CO_2 (though less than coal or oil per unit of energy obtained). Global reserves of natural gas (Chapter 3) are also sufficient for several decades at current (2005) levels of demand.

Nuclear fission (Chapter 4) is by far the most concentrated energy source currently available, and the actual generation of nuclear power releases no CO_2,

SO_2, nitrogen oxides or other chemicals. However, unless very strictly controlled, radioactive by-products can contaminate air, water and soil over wide areas for centuries to millennia. Known global reserves of uranium are sufficient to last the lifetime of reactors so far constructed, and well beyond (Section 4.2.2).

Biomass, mainly in the form of fuel woods and charcoal, represented around 10% of global primary energy consumption in 2002, largely in the developing world (Figure 6.2). Biomass is renewable if it is consumed at the same rate as new plants are grown, in which case the CO_2 released during combustion would be offset through uptake by growing plants. Biofuels may also be derived from wastes, many of which are biological in origin (e.g. dried yak dung in Tibet, sugar cane waste and rice hulls in India).

Alternative energy sources (Chapter 5) are all 'clean' as well as renewable, the 'fuel' costs nothing and contributes neither to global warming nor to acid rain. Their drawbacks include low energy density so that very large installations are needed for substantial power generation, erratic geographic distribution, intermittency of supply, and in the main they are not transportable, so power stations must be built near the energy source. There are also environmental costs of constructing the plant (e.g. the materials needed for wind turbines). Nevertheless, their combined potential is considerable, and many are adaptable to local power generation.

6.4.2 Developing current energy systems

Cleaner fossil fuels

A variety of techniques have been developed for manufacturing liquid or gaseous hydrocarbon fuels from coal or crude oil (in some cases to redress a deficit of a particular fuel in a single country). An historic example is 'town gas', mainly hydrogen and carbon monoxide with minor amounts of methane, which is produced by destructive distillation of coal in the absence of air (pyrolysis, Section 5.3) and with steam injection. The process was associated with the production of coke, coal tar and other industrially useful materials. In the UK, town gas was a major domestic and industrial fuel from 1804 to the 1970s, when it was rapidly replaced by natural gas from the North Sea. It was also produced from oil in the UK between 1960 and 1975.

Plentiful supplies of oil and gas have curbed research into developing these technologies; after all, why bother with complex conversions when the stuff simply comes out of the ground virtually ready to use? However, the large reserves of coal may offer a source of oil by liquefaction, as well as more sophisticated gasification. Production of oil from coal was first developed on a commercial scale in Germany during the 1930s and 40s, and has operated in South Africa since the mid-1950s (Figure 6.17), prompted initially by the country's lack of oil, abundant coal deposits, and increasingly

Figure 6.17 The Sasol plant in South Africa that converts coal into oil products.

isolated political stance. Although three plants remain operational, the real costs of production are not competitive with crude oil, unless oil prices continue to rise. Although the South African plants employ a two-stage process (gasification followed by catalytic synthesis), research currently favours the development of a different, relatively low-temperature, catalytic process where the coal is first dissolved in a suitable solvent before treatment with hydrogen.

Other non-conventional sources of liquid petroleum are estimated to represent a vast, untapped resource (they include tar sands, oil shales and heavy oil — see Section 3.7). However, these are not clean fuels, and in the main they require extensive (and expensive) processing to produce petroleum comparable to crude oil. The difficulty in their exploitation, and their potential environmental effects, have retarded their development thus far, and may continue to do so unless the world remains addicted to petroleum as an energy source beyond the lifetime of conventional reserves. Methane hydrates, currently residing in sea-floor sediments and some terrestrial settings, represent huge potential resources (Section 3.7.2). They also present a major potential hazard to climate, should warming of deep water destabilize them. Yet their formation involves another gas hydrate that incorporates carbon dioxide, and methane-hydrate exploitation could be combined with sea-floor CO_2 sequestration.

Renewable sources of energy

The potential and problems of individual renewable sources have been described in Chapter 5, so this section focuses on a few general issues concerning their future exploitation. It seems inevitable that these relatively clean energy sources will see extensive development during the 21st century. Only hydropower has been exploited close to capacity so far. Investment and environmental disruption remain barriers for further hydro schemes, both on a large scale in developing countries and on a small scale anywhere. This is true for many of the renewable energy sources, though some countries have fostered major renewables industries (e.g. wind power, offshore in Denmark and onshore in Germany).

Ironically, the *geographic distribution* of renewable energy potential is just as uneven as that of fossil and nuclear fuels, so that some countries have a far greater endowment than others. The UK, for example, is well-placed for wind, wave and tidal power, but much less so for hydro, solar and geothermal resources. In future such global inequalities in energy resources will have to be smoothed out (as today) by international trade, probably of generated electricity.

Another widespread difficulty with many renewable sources is intermittency of supply combined with *unpredictability*; the fact that the amount of electricity generated often varies on short timescales. Some sources are unpredictable (wind and wave energy), others are cyclical (tidal, solar), and others are more constant (geothermal). In designing an electricity supply system that can guarantee to meet demand, a delicate balancing act would be required. It is unlikely that renewable sources alone could provide a stable electricity supply, especially as most countries will not have the full range of renewable options. Thus, one of the most important features of future energy supply systems will be **integration**, that is, the ability to manage a complex mix of different energy sources as predicted for 2025 in Figure 6.18 to produce a stable power supply that can track the demand curve (see Figure 5.21). Sources that are designed to supply electricity continuously (mainly nuclear, tidal, waste, biomass and geothermal) will be harnessed to provide the **base load**, i.e. the minimum continuous demand for electricity between about 3.00 and 6.00 am during summer nights (Figure 5.21). Other sources can be brought on-stream to cope with periods of increased demand, which varies throughout the year and also during each day. The maximum demand, or *peak load*, occurs only briefly around 18.00 pm on winter evenings (Figure 5.21), for about 300 hours in a year. The brief, increased demand then can be satisfied by reliable, rapid-response power stations using oil or hydropower from pumped storage schemes. Demand during daytime hours, winter and summer, is mainly satisfied by power stations that burn coal and natural gas, supplemented by renewable sources that are less predictable (wind and wave) or which vary during the year (solar and hydropower). Although coal and natural gas still dominate the schematic electricity management scheme shown in Figure 6.18, their overall use is reduced from that in the early 21st century by the deployment of alternative energy sources.

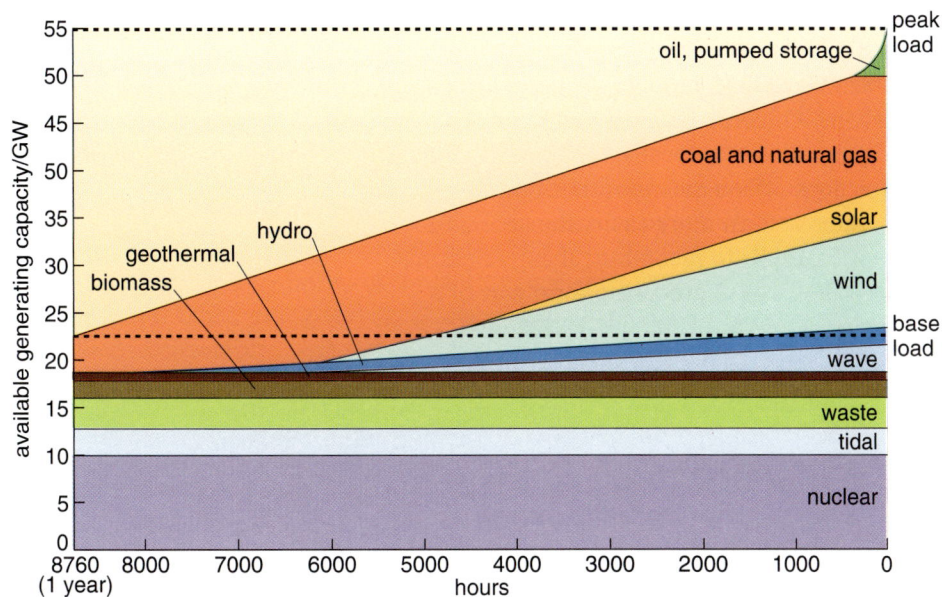

Figure 6.18 A prediction of the amount and duration of use of different types of generating plant in the UK over a year (8760 hours) around 2025, and the order in which they might be used as the power demand rose from the continuous base load — at the left. Reading the graph from left to right, as more power is required this is supplied from a blend of coal and gas power stations, and more unpredictable and intermittent renewable sources. This figure uses optimistic projections for the development of renewable technologies, and assumes steady investment in the nuclear industry. 'Waste' represents electricity generated from various waste products (e.g. sewage sludge, landfill gas, incineration of municipal rubbish).

Research at Oxford's Environmental Change Institute suggests that if intermittent renewable sources contributed a significant proportion of electricity (>20%) — and they will have to for a sustainable future with minimal global warming (Figures 6.9 and 6.12) — substantial backup generation capacity would be needed. Historically, this has been in the form of fossil-fuel power stations, including the most expensive 'spinning reserve', where turbines are actually rotated without generating but ready to supply electricity instantly. Various measures, however, can reduce the need for this fossil-fuel backup:

- distributing installations (e.g. wind turbines) widely, on the premise that the wind is always blowing somewhere;
- using a range of different intermittent energy sources, especially those that are partially complementary (e.g. sunny weather often means light winds, and vice versa);
- matching energy sources to periods of high demand (e.g. winter demand is matched by increased wind and wave potential);
- replacing fossil-fuel backup power stations with other storable energy sources.

This latter point highlights the last major issue with renewables — *storage*. Whereas fossil fuels are easily transportable storehouses of energy, renewable sources have to be harnessed and the energy used immediately. If some of the surplus energy from renewables produced at times of low demand (e.g. solar power in the summer, or night tides) could be stored ready for release when demand rose, many of the problems of renewable supply could be solved. The Dinorwic pumped storage scheme (Section 5.5) is the only example of such a storage scheme in the UK at the time of writing (2005), but there is potential for similar schemes in the form of tidal reservoirs. Of course, such geographically fixed means of storing surplus energy are most suited to local and at most regional energy planning. Another possible solution does offer a means of transporting stored energy. It involves electrolysis of water to form more easily transportable hydrogen and oxygen, and either the use of fuel cells to recombine them, releasing electrical energy, or burning hydrogen as a fuel. These processes form part of a futuristic economic system based on hydrogen.

6.4.3 Future energy systems

The hydrogen economy

The simplest use of hydrogen as a fuel depends on the reaction:

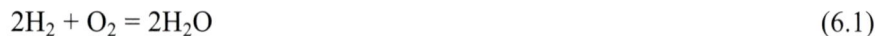

$$2H_2 + O_2 = 2H_2O \tag{6.1}$$

which takes place when hydrogen is burnt in air. The reaction is highly exothermic, yielding 1.21×10^8 J kg^{-1}, roughly twice the calorific value of petroleum products (0.5×10^8 J kg^{-1} for petrol). Hydrogen is thus a non-polluting fuel with a high 'energy density'. There are just two problems, the density of liquid hydrogen is ten times less than that of petrol, and hydrogen boils at -253 °C (20 K), so there are major, but not insoluble, difficulties of transport and storage.

Hydrogen could therefore be used as an energy storage medium to smooth out the vagaries of unpredictable fluctuations in energy supply from renewable sources. Figure 6.19 illustrates the storage concept; in this case wind energy is used to electrolyse water into its component gases — the reverse of Equation 6.1 and so an endothermic process. The energy used is thus 'stored' in the gases and can be released when they are recombined. Burning the hydrogen is one way of doing this; another way is to feed hydrogen to the anode and oxygen to the cathode of a **fuel cell**, where they are recombined into water. As a result of ionization processes ions migrate between the electrodes of the cell and electrons flow in the external circuit, producing a usable direct electric current. The inverter shown

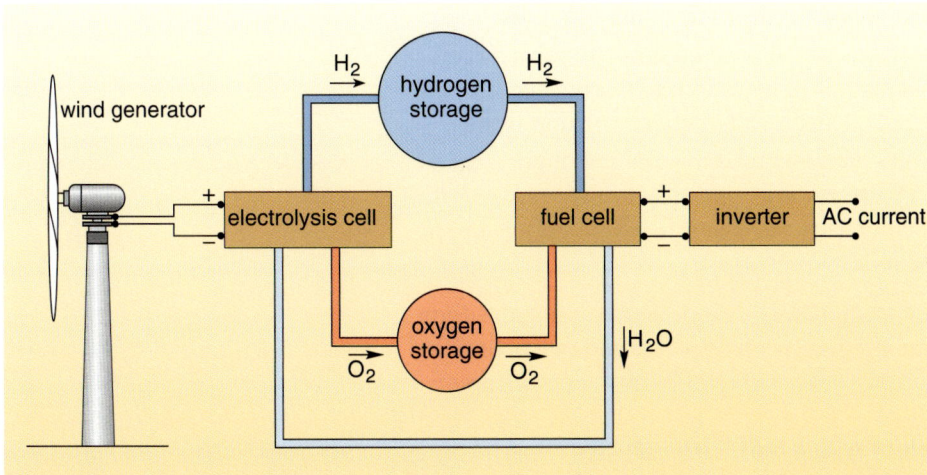

Figure 6.19 Storage of unpredictably fluctuating energy supplies from wind generators (or from any other source) may be achieved by electrolysis of water into its component gases followed by their recombination in a fuel cell at times when electrical energy is required.

in Figure 6.19 is required to convert this into alternating current for use on an electricity supply grid.

Electrolysis is just one of several methods by which hydrogen for fuel cells could be produced; other processes include the thermochemical splitting of water by high-temperature chemical reactions and the biochemical liberation of hydrogen by some plants and algae (during photosynthesis) or bacteria (by fermentation, for instance of putrescible waste). All three processes could form part of a futuristic integrated system based on hydrogen as the energy carrier (Figure 6.20) — the **hydrogen economy**. Rather than being liquefied, gaseous hydrogen could be stored in high-pressure vessels or as metal hydrides, or in the longer term, perhaps, in depleted aquifers or oil wells. The hydrogen economy offers a promising alternative to the present fossil fuel economy, but considerable technical development of fuel cells and both storage and transport of hydrogen is required before such a system is viable. And, of course, there must be a decisive and dramatic shift to increasing the use of renewables as primary energy sources.

Figure 6.20 Elements of the hydrogen economy. Energy sources are shown at the top, conversion methods in the middle and the storage, distribution and use of hydrogen at the bottom.

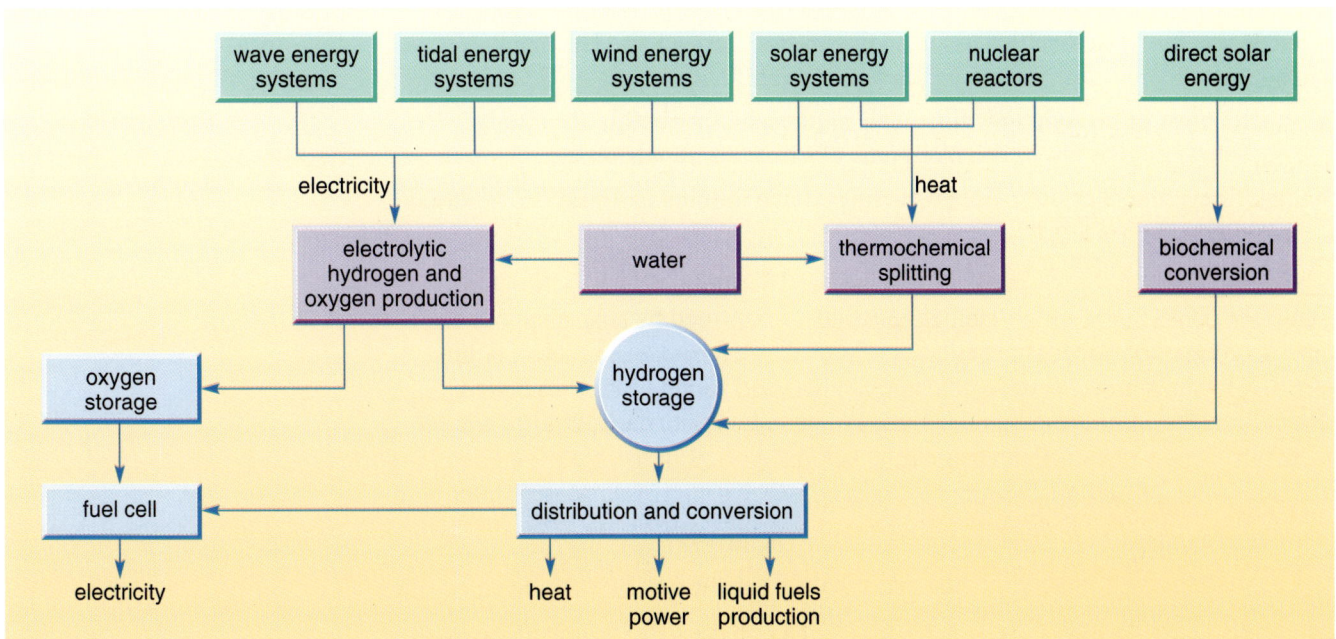

Nuclear fusion

When very high speed nuclei of light atoms collide and coalesce to form larger nuclei — **nuclear fusion**, any surplus mass is converted into energy according to the Einstein relation, $E = mc^2$ (Chapter 4). Uncontrolled nuclear fusion is the basis of the hydrogen bomb and powers the Sun and stars, but can it be controlled and harnessed? From the 1950s onwards, lavishly funded research has pursued this goal, achieving the first controlled release of fusion energy from the Joint European Torus (JET) experiment in 1991.

The fusion reactions which have been studied are those of deuterium and tritium (the heavy isotopes of hydrogen, containing, respectively, one and two neutrons plus a proton) which fuse to produce either helium-3 or helium-4. For example:

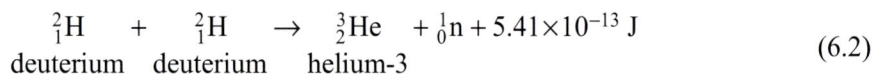

$$\underset{\text{deuterium}}{{}^{2}_{1}\text{H}} + \underset{\text{deuterium}}{{}^{2}_{1}\text{H}} \rightarrow \underset{\text{helium-3}}{{}^{3}_{2}\text{He}} + {}^{1}_{0}\text{n} + 5.41 \times 10^{-13}\ \text{J} \tag{6.2}$$

or,

$$\underset{\text{deuterium}}{{}^{2}_{1}\text{H}} + \underset{\text{tritium}}{{}^{3}_{1}\text{H}} \rightarrow \underset{\text{helium-4}}{{}^{4}_{2}\text{He}} + {}^{1}_{0}\text{n} + 3.0 \times 10^{-12}\ \text{J} \tag{6.3}$$

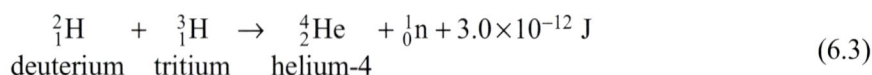

Deuterium (hydrogen-2, ${}^{2}_{1}\text{H}$) comprises 0.015% of the hydrogen atoms in natural waters; the oceans contain 4.2×10^{13} tonnes of deuterium. In theory, this could produce 3.4×10^{12} EJ of energy if extracted and fused, a factor of 10^7 greater than the entire fossil fuel bank.

Tritium (hydrogen-3, ${}^{3}_{1}\text{H}$) does not occur naturally, but is produced in nuclear fission reactors that use water as a moderator or coolant (Section 4.1.2). It is also a by-product of the nuclear fusion process itself, when lithium atoms that form part of the containment system of the reactor capture neutrons liberated by fusion; tritium use in nuclear fusion could conceivably extend this source of energy almost indefinitely. However, tritium is radioactive with a half-life of 12 years, and bombardment of fusion reactor vessels by neutrons and other subatomic particles would give rise to a radioactive waste problem — though on a much smaller scale than that associated with fission reactors. The main technical problems in sustaining fusion reactions are:

- nuclei must be brought to within 10^{-15} m of each other before the strong nuclear attractive forces can overcome electrostatic repulsions;

- to achieve the necessary kinetic energies, temperatures of about 10^8 °C are required.

The current design for a proposed International Thermonuclear Experimental Reactor (ITER) is for a high-temperature plasma confined by suspension in an intense doughnut-shaped magnetic field (known as a torus). Six partners (China, the EU, Japan, Russia, South Korea and the US) will fund ITER, with India expressing strong interest, but in early 2005 the reactor site had still to be agreed upon. Fusion research has three major objectives:

1   *break-even*: when total output power = total input power. This was demonstrated at the JET experiment in the UK in 1997;

2   *burning plasma*: where the plasma is mainly self-heated by particle collisions, with little external power needed;

3   *ignition*: when the plasma generates so much energy that no input power is
    required.

A reactor would probably need only to achieve the first two aims to be
commercially viable, and ITER is intended to demonstrate burning plasma for the
first time. However, a commercial prototype reactor is unlikely to be operational
until at least 2050, assuming that the research funding is maintained at current
levels.

6.5 Managing energy use in the future

> Anyone who believes exponential growth can go on forever in a finite world is
> either a madman or an economist.
>
> (Kenneth Boulding, *c.* 1980)

The final years of the 20th century brought increasing concerns over the use of all
resources, including energy, and the rise of international initiatives to address the
problems. The 1992 Earth Summit at Rio de Janeiro drew up a 'sustainable
development plan' showing how resources, transport, trade, biological diversity,
agriculture and fisheries could all be managed to maintain the quality of life for
future generations. Among other recommendations, the industrialized nations agreed
(in principle) to stabilize emission of carbon dioxide (from fossil fuels) at 1990 levels
by the year 2000. (This was not achieved.) Discussions at Rio were followed by the
1997 Kyoto Protocol, which aimed for 5% below 1990 CO_2 emission levels by 2012.

- Would achievement of the Kyoto Protocol aims stop the increase in the
  concentration of atmospheric CO_2?

- No. Stabilization of CO_2 emissions at even 5% below 1990 levels merely
  means that the atmospheric concentration of CO_2 would increase by almost
  the same amount each year as in 1990 — instead of a net annual addition of
  some 10^{10} t of CO_2 it would be 9.5×10^9 t.

Although some nations have been reluctant to commit to environmental initiatives,
growing numbers of people in the affluent societies of Western Europe, North
America and Australasia have begun to 'think globally, act locally', initiating and
supporting programmes of materials recycling, energy conservation and efficiency,
waste reduction, and so on. The ultimate aim is for 'sustainable development'
(Sheldon, 2005), that is, development within our ecological means, which modern
humans abandoned when they began consciously modifying their environment to
build our modern civilization.

To put it more explicitly, sustainable development must eventually involve:

- phasing out extraction of non-renewable resources;
- increased use of renewable resources;
- recycling all manufactured materials;
- releasing all anthropogenic wastes at rates commensurate with natural cycles.

In the early 21st century world, the main priority is to decrease fossil fuel
consumption. Using alternative, renewable energy sources will help, and in some
cases, using recycled and biodegradable materials — though a full energy audit

may reveal that more energy is required for recycling some products than for manufacturing them anew from raw materials. (More commonly, it is the high relative financial cost of recycling that deters such schemes.) Less equivocal is the benefit of energy conservation. This can take place either on the *supply side* or the *demand side*. Demand side measures are very diverse, and may involve approaches that are either technological or social; we do not consider them here. Supply side measures involve increasing the efficiency of power generation and distribution; as an illustration, less than half the energy in the gas fuel for the most efficient UK power stations in the early 2000s is actually available to the electricity customer. Much of this unused energy takes the form of waste heat, which could be used to heat buildings, as in Denmark.

Efficiency has been a theme throughout this book, but mainly applied to efficiencies of conversion, as in solar PV electricity generation. The theoretical maximum efficiency for this promising technology is limited to around 30% by physics, and is currently about 15%. Yet efficiency applies to all aspects of human energy use, a revealing example being the use of electricity to pump water; the most fundamental need of a modern society. Say the electricity was generated at a coal-fired power station using 100 arbitrary units of primary energy. Energy losses there are around 70%, so only 30 units enter the transmission grid. Transmission is very efficient (91%), pump motors operate at around 88%, and pumps themselves at around 75%. Once water is flowing through all the pipelines and valves to the user, distribution is about 47% efficient in energy terms, partly due to constrictions to the flow of a viscous fluid, and partly due to leaks. The net result of this chain of inefficiency is that the pumped water contains only 9.5 of the original 100 primary energy units. Transportation is very much worse. After more than a century of development, car engines deliver no more than 13% of fuel energy to the wheels, of which more than half heats the tyres, road and air. But the efficiency in terms of useful work, taking people back and forth, is a pathetic 1%, since 95% of the mass transported is the vehicle itself! More or less the same happens with every means of using energy to do useful work (see Section 1.2).

Technological measures involve improving the efficiency of energy use and effectiveness of conservation in a variety of ways:

- reducing heat loss from buildings, by improving insulation, window glazing, etc.;

- making more efficient appliances such as boilers, fridges, lightbulbs, computers, photocopiers, pumps, and other industrial, commercial or domestic machines;

- improving the efficiency of transport vehicles, and developing vehicles that run on alternative fuels, for example hydrogen in fuel cells (Figure 6.21) or biofuels;

- improving control systems so power is consumed only when needed, and at the lowest efficient output levels;

- recycling waste heat produced by some industrial processes (e.g. kilns) for lower temperature applications (e.g. drying raw materials or products);

- using less materials (e.g. thinner metals in car shells), or materials that are less energy-intensive (e.g. plastic, rather than steel, car bumpers).

The last point is part of a wider range of both technological and social measures grouped under the term **de-materialization**, which means using less material (and thus less energy) — either in production or consumption. One example would be product packaging; this can in many cases be reduced drastically at source without affecting the quality of the product, but in addition, consumers buying the product could re-use elements of the packaging rather than simply disposing of it. Such small social adjustments may seem trivial, but they can effect powerful changes in society. One illustration is the comparison between the consumerist, 'throw-away' attitudes of the late 20th century, and the ingenious practicality of people during World War II, when nothing was wasted unnecessarily. It could be argued that the only difference between that time of crisis and now is that the threat of war was perceived as being closer to home, because modern society cushions individuals in the developed world from the impacts of their consumerism.

Figure 6.21 This bus is powered by a fuel cell running on hydrogen gas.

In addition to de-materialization, social approaches to energy conservation would also involve rearranging our lifestyles, both individually and collectively, to reduce the energy required for a particular service. The author is fortunate enough to live within walking distance of a town centre, schools, and other amenities, but many newer towns (such as Milton Keynes) are designed with lower population densities, so that most journeys are impractical without using a car or bus. As Figure 6.22 shows, car and van use in the UK soared from 1952 to 1990, and is still increasing, partly reflecting a trend of fewer people per vehicle.

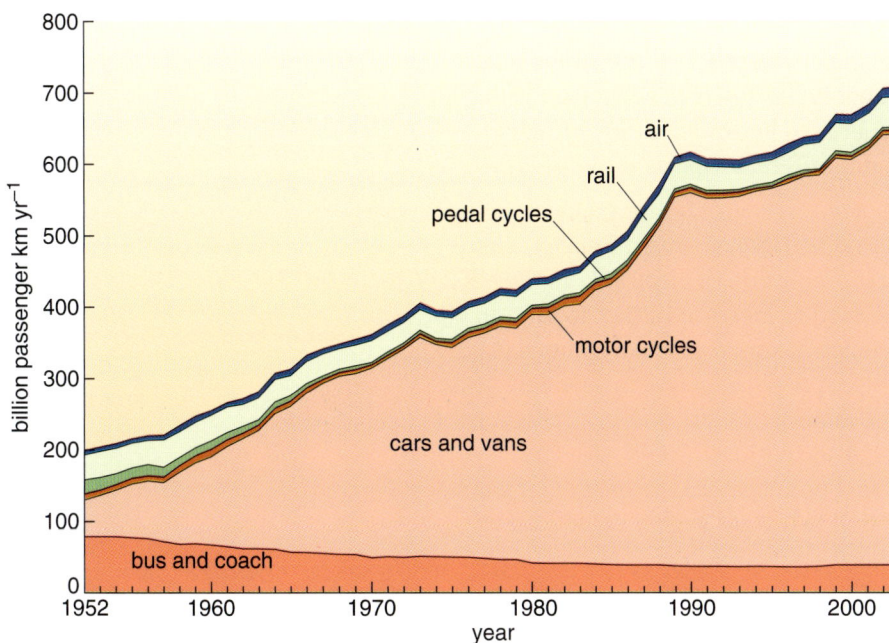

Figure 6.22 Annual passenger-kilometres travelled in the UK, 1952–2000, by transport mode. *Note*: Air travel data refers to internal flights only.

In 2005, there were few signs that this trend was reversing, even with the introduction of such measures as bus lanes, congestion charging (e.g. Inner London) and multiple-occupancy vehicle lanes (e.g. Leeds). The implication is that people in developed countries such as the UK are finding it difficult to shake off our addiction, as a society, to fossil fuels, and the lifestyle that cheap, plentiful energy has so far brought us. More worryingly, rapidly developing countries such as China and India are experiencing the same, dramatic rise in vehicle use and accompanying urban pollution. International air travel, not included in Figure 6.22, is a burgeoning problem, with air traffic, airports and even the size of airliners all increasing, while prices on many routes actually fall, encouraging demand. In effect, any savings in energy due to conservation measures are presently more than offset by increases in vehicle numbers and usage, and politicians are loth to upset potential voters by curbing consumer demand (for instance, by more draconian fuel taxes).

A further complication is known as the **rebound effect**. This is the tendency for individuals or organizations, once they have saved money by implementing energy saving measures, to spend that 'extra' money on additional energy-consuming activities, such as providing higher quality services. For instance, a householder who installs better loft insulation could save on heating bills. However, they may simply heat the house to a warmer temperature, or for longer periods, and use the same amount of energy as before. Alternatively, they may splash out on an overseas holiday with the money saved from the lower bills, using energy-intensive air travel that offsets any energy savings in the home. One way governments can counter the rebound effect is to give incentives for citizens to spend such financial savings in ways that are energy-frugal rather than energy-intensive. However, ultimately the responsibility lies with the individual concerned, and how much they share prevailing environmental concerns — in fact, how much they really desire '… to maintain the quality of life for future generations'.

Activity 6.2

The UK is fortunate in having access to a wide range of energy sources, and the numerous options for future energy supply has fostered considerable debate on what the best 'energy mix' should be. Printed below are brief, edited extracts from various sources that highlight different aspects of the energy debate.

> Since 1990, the demand for electricity in the UK has grown by 25%. Gas power stations now generate the largest proportion of electricity in the UK, and the share of gas generation is expected to increase further to 50% in 2012, following the closure of older nuclear stations and coal stations. At the current time, NGT [National Grid Transco] is not anticipating any new nuclear plants. The UK will soon become a net importer of gas, and in 2013/14, 66% of gas requirements will be imported. Of the current 22 GW gas-fired generation, only 6 GW has a back-up fuel option should the supply of gas be halted. It seems to me that an interruption to the gas supply could lead to significant shortfall of generation.
>
> (Simon Griew, National Grid Transco, 2004)

A serious situation faces this country if we do not address the looming energy crisis. By 2020, this country will be 90% dependent on gas for its energy needs and 70% of that will have to be imported from politically unstable regions of the world such as Russia, Ukraine, Iran and Algeria, through pipelines wide open to terrorist attack. We will be a net importer of energy and all the time sat on millions of tonnes of the coal reserves with which this nation is blessed. Our balance of payments will suffer directly as a consequence. Clean coal technologies are now available and are being further developed and used not only in the US, but in Australia, India, China and many other coal producing countries. In the UK we pioneered the research into clean-coal technology, which was abandoned by the last Tory government in its haste to butcher the mining industry that served, and still could serve, this nation well.

(Steve Kemp, National Secretary, National Union of Mineworkers, 2005)

AMEC is working closely with long-term customers like BP to develop new sources of supply or extend the productive lives of existing fields around the world — often in hostile environments. These oil and gas recovery projects are increasingly vital as the world's stocks begin to dwindle or become unreliable — just look at the recent global petrol price panic sparked by terrorism in Saudi Arabia.

No matter what your view about global warming and the burning of fossil fuels, recent events have shown how vital it is that we extract as much oil and gas as is technologically possible and so keep our carbon-based economy viable until new sources of energy are developed.

(Chris Bond, Oil and Gas Technology Director, AMEC, 2004)

The only renewable source currently capable of supplying a significant amount of electricity is hydropower, and there are few remaining opportunities for large hydropower schemes. The UK is aiming for 10% renewable generation by 2010. If this were to be fulfilled by wind power, then turbines would have to be installed at a rate of 40 per week between now and then; the current rate is 2 per week.

If carbon emissions are to be controlled, nuclear power will have to play an ever increasing role in electricity generation. New reactors produce just 10% of the nuclear waste the old ones account for. If we unlocked our coal it would transform the prospects for using fossil fuel, so carbon sequestration is the key to the future, together with those new nuclear plants. I can't believe the UK will ever get far beyond generating 10% or so of its energy renewably — and that would be a heroic effort.

(Professor Ian Fells, New and Renewable Energy Centre (NAREC), 2004)

Let's all go nuclear, it's the only way. Already nuclear is becoming the grown-up solution. And climate change is the nuclear lobby's best weapon: only global warming is more dangerous than massive proliferation of nuclear power across the world.

Malcolm Wicks, the new energy minister, rebuts the myths and factoids now so successfully spread by the anti-wind-power lobby. No, turbines are not taking over the country: only some 800 hectares are needed to reach the 10% target. No, they are not unpopular: 80% support them and 66% would like some in their area. No, the intermittent wind dropping is no problem, since the farms are spread far across the county and existing back-up is quite sufficient. (Eyesores? The UK had 90,000 windmills in the 17th century.) But these myths are gaining ground, alongside the bigger myth that nothing but nuclear will do. However, new nuclear stations would take a decade to build at £2bn each. So it's hard to see this parliament commissioning more nuclear power.

Everywhere there are green shoots of what might be done, if serious money and political attention were devoted to it now. Take micro-generation. You can buy a small windmill to stick in the garden or on the side of your house for just £900: it plugs into an ordinary 13 amp domestic plug, cuts electricity bills by a third and can feed into the grid. Imagine if each adult were given a carbon quota. Those who want to fly a lot or overheat a big house would have to buy extra quotas from low energy users. It would have the interesting side effect of redistributing funds towards those too poor to use their energy ration.

(Polly Toynbee, 2005)

By mixing between sites and mixing technologies, you can markedly reduce the variability of electricity supplied by renewables. And if you plan the right mix, renewable and intermittent technologies can even be made to match real-time electricity demand patterns. This reduces the need for backup, and makes renewables a serious alternative to conventional power sources.

Wind (onshore and offshore) could realistically provide some 35% of the UK's electricity, marine and dCHP [domestic combined heat and power] each 10–15%, and solar cells 5–10%. In other words, more than half the UK's electricity could ultimately derive from intermittent renewables. The high proportion of wind is because the wind blows hardest in the winter, and in the evening — when demand is highest. The dCHP also produces more at peak times, when demand for hot water and heating is also strongest. Solar makes a smaller contribution, and produces nothing at night. But it is still important to have it in the mix as it kicks in when wind and dCHP production is lowest. A marine-based renewable system works best when it includes both tide and wave. The combination has lower variability, is better at meeting demand patterns, and makes better use of expensive transmission infrastructure.

(Oliver Tickell, quoting Graham Sinden, 2005)

Imagine you control the UK's energy policy for the next 50 years. Considering the extracts above, which of the following measures would you consider to be essential. Bear in mind that different groups may have different outlooks. One may focus on 'green' idealism, another on economic incentives. Equally, the issues might be addressed pragmatically or in the context of which is technologically possible.

- Government's carbon dioxide reduction targets;
- keeping the nuclear option open while developing technologies for clean coal and maximizing oil/gas recovery, and placing more emphasis on carbon sequestration;
- substantial investment in research and development of new and renewable fuel sources on a large scale, such as wind, wave, tidal, geothermal, and nuclear fusion;
- focusing investment on integrating alternative energy sources and improving energy storage and supply, including the use of hydrogen;

- stimulating more local community and micro-generation schemes, while modifying electricity transmission systems to incorporate these diverse, widely distributed generators on a commercial to domestic scale;
- deploying legislation, supported by financial incentives and levies, to encourage energy efficiency and conservation at all levels and reduce consumption of primary energy.

Having studied this book, you will now realize that although a great deal is known about how the variety of available energy resources form, how they can be found and exploited, a great deal less is known about the implications of using them. It would be no exaggeration to state that, at the outset of the 21st century and little more than two centuries since the start of the Industrial Revolution, humanity is at a decisive point in its history. Decisions and actions, centred especially on changing how we get the energy that we need, will need to be put into practice — sooner rather than later.

6.6 Summary of Chapter 6

1   Human society's consumption of primary energy has increased rapidly over the last few hundred years, although global per capita consumption of commercial fuels was roughly constant for the last few decades of the 20th century.

2   The pattern of discovery, and rapid depletion, of fossil fuel reserves during the 20th century, as predicted for the US by M. King Hubbert, has prompted many different organizations to attempt forecasts of future energy sources and demand. These forecasts generally take the form of scenarios that illustrate how energy usage develops according to different factors, such as policy decisions, technological advances, and predictions of reserves.

3   The coal pit closures announced by the UK government in late 1992 were a classic example of how a high-value resource can be rendered worthless overnight by economic and political factors unrelated to geological availability. Such abrupt changes make forecasting difficult, especially in the energy industries.

4   Future energy scenarios tend to conform to three main types: *historical growth*, which extrapolate past trends; *technological fix*, which predict advances in energy efficiency and conservation, and *zero growth*, which suggest that society will be persuaded or forced to accept the end of economic growth. Many scenarios incorporate elements of all these approaches.

5   Most scenarios constructed around the start of the 21st century are based on some degree of reduction in fossil fuel use, aiming to reduce society's reliance on these finite, environmentally damaging resources by harnessing other energy sources, such as nuclear and renewables. In addition, these alternatives may alleviate the situation where a small number of countries hold the bulk of the world's remaining oil and gas reserves.

6　The global dominance of fossil fuels for energy supply carries with it the release of the 'greenhouse' gas CO_2 at such a rate that it accumulates in the atmosphere. This is thought to be raising global mean surface temperature; global warming has characterized the period since the Industrial Revolution. Even with the most ecologically sound transformation of the global energy economy, mean surface temperature is predicted to rise by 1 °C by the end of the 21st century, and possibly by even more, given 'business as usual' and a growing global population.

7　Other emissions, particularly sulphur dioxide, have increased the acidity of rain in some areas, thereby having significant ecological outcomes.

8　Effects of 'greenhouse' warming may have been partly hidden by particulate emissions from industry, thus reducing solar energy received by part of the Earth's surface. Reduction in such emissions will accelerate global warming.

9　A UK government White Paper in 2003 outlined a vision of energy supply in 2020, with a central electricity grid supplied by wave, tidal and wind power backed up by fossil fuel and nuclear power stations. Local and micro-generation (e.g. by community biomass plants, photovoltaics, fuel cells, local wind installations) would be more prominent, along with energy conservation measures, especially in the domestic sector. There was no suggestion of voluntarily reducing consumption.

10　Future development of energy sources may be focused on several goals: exploiting largely untapped fossil fuel resources (e.g. tar sands); developing technologies for cleaner fossil fuel use; developing the exploitation of renewable energy sources; and overcoming problems such as intermittency and energy storage.

11　One proposed solution to energy storage is to evolve a system known as the hydrogen economy, where hydrogen is used as a medium to store and transport the energy generated by intermittent sources. Fuel cell technology would have to be advanced, but one benefit would be that hydrogen is a relatively 'clean' fuel.

12　Another potential alternative energy source is nuclear fusion, which derives energy from combining atoms of hydrogen isotopes (deuterium and tritium) to form helium. Because this process requires a very high temperature plasma, which is contained in a strong magnetic field, it has not yet been developed on a commercial scale.

13　Technological measures to reduce fossil fuel dependence and develop alternative energy sources can be supplemented by a range of technological and social strategies to conserve energy, including de-materialization. However, energy conservation can be offset by the rebound effect, if savings are then spent consuming more energy.

LEARNING OUTCOMES

When you have completed this book, you should be able to explain in your own words, and use correctly, all the **bold** terms printed in the text and defined in the Glossary. You should also be able, among other things, to do the following:

General

1   Perform simple calculations, use formulae and equations, plot and interpret graphs, and evaluate tables of data relating to different aspects of energy resources.

Chapter 1

1.1   Understand the difference between energy and power, and their units and prefixes.

1.2   State the relative contributions of different natural energy sources to the global energy budget.

1.3   Describe the contribution of photosynthesis to the carbon cycle, and distinguish the terrestrial and marine parts of the cycle.

1.4   Discuss the issues involved in concentrating, storing and transporting energy.

1.5   Recognize which energy resources have a low energy density, and which have a high energy density.

1.6   Distinguish between renewable and non-renewable energy resources.

Chapter 2

2.1   Identify the types of sediments found in coal-bearing rocks, and the environments in which the coals of the future might currently be forming.

2.2   Explain how coal is formed, and give the reasons for variations in both coal rank and type.

2.3   Explain the order in which the methods used in coal exploration are employed, from initial surveys and ground exploration, to borehole drilling and geophysical logging.

2.4   Calculate a stripping ratio, given volumes (or depths) of overburden and workable coal, and use this ratio to argue for either surface or underground mine extraction.

2.5   Explain, using appropriate examples, the different ways in which coal is extracted from exposed and concealed coalfields.

2.6   Explain how geological problems and environmental issues surrounding extraction affect surface and underground mines. Describe ways in which these problems can be overcome.

2.7   Outline the main geological periods in which coal formed and explain why not all countries possess coal reserves.

2.8   Understand the difference between coal resources and reserves.

2.9   Use the R/P ratio to show how long a country's reserves may last and explain why this ratio changes with time.

2.10   Explain the reasons for the decline in the UK's coal industry between 1980 and the present day.

Chapter 3

3.1   Given basic geological information for a petroleum play, recognize the main 'ingredients' (petroleum charge, reservoir, seal and trap) that contribute to its potential.

3.2   Understand the roles played by different means of exploration in contributing to defining a petroleum play, and its evaluation.

3.3   Describe the various options for petroleum production in different settings.

3.4   Discuss the various hazards to operators and the environment that are presented by exploiting petroleum reserves.

3.5   Understand the criteria used in assessing petroleum reserves globally and in the UK.

3.6   Discuss the conditions under which unconventional petroleum resources form, and the requirements for their future exploitation.

Chapter 4

4.1   Distinguish between energy produced by nuclear fission and radioactive decay.

4.2   Describe the principles behind nuclear 'burner' and nuclear 'breeder' reactors.

4.3   Understand the geoscientific principles underlying the enrichment of uranium in ore deposits.

4.4   Summarize and explain the hazards associated with nuclear wastes and their safe disposal.

4.5   Summarize the fluctuating fortunes of the nuclear power industry.

Chapter 5

5.1   Explain the principles that underlie the ability of various natural phenomena to deliver useable energy.

5.2   Outline the technologies that are used to harness the power of alternative energy sources.

5.3   Discuss the positive and negative aspects of alternative energy sources in relation to natural and human aspects of the environment.

5.4   Assess the potential of various energy releasing or redistributing phenomena in supplying humanity's requirement's for energy, in both geographic and numeric senses.

Chapter 6

6.1   Summarize the difficulties of forecasting energy demand.

6.2   Assess the importance of political and economic issues, as well as geological and environmental factors, in determining trends in energy use.

6.3   Outline some of the contrasting scenarios for energy supply in the 21st century, and discuss evolving technologies that could play a part in future energy systems.

6.4   Appreciate the environmental consequences of society's current energy use, and the challenges of developing sustainable energy supply.

REFERENCES AND FURTHER SOURCES OF INFORMATION

The information in this book has been obtained from a wide range of sources, too numerous to mention. However, specific reference is made in the text to the following:

Argles, T. (2005) *Minerals: Bulk Materials for Building and Industry* (Book 2 of S278 *Earth's Physical Resources: Origin, Use and Environmental Impact*), The Open University, Milton Keynes.

Bond, C. (2004) in *Keeping the wheels turning*, online article available from http://www.amec.com/about/about.asp?PAGEID=912 [last accessed January 2006].

Boulding, K. (*c*. 1980); quoted in Douthwaite, R. (1992) *The Growth Illusion: How Economic Growth Has Enriched the Few, Impoverished the Many, and Endangered the Planet*, Lilliput Press, Dublin.

Boyle, G., Everett, B. and Ramage, J. (eds) (2003) *Energy Systems and Sustainability*, Oxford University Press and The Open University.

BP (2004) *Statistical Review of World Energy 2004*, available online at http://www.bp.com

BP (2005) *Statistical Review of World Energy 2005*, available online at http://www.bp.com

Department for Transport (DfT) Transport Statistics page. Statistics on UK transport. Available online from http://www.dft.gov.uk/stellent/groups/dft_transstats/documents/sectionhomepage/dft_transstats_page.hcsp [last accessed January 2006].

Department of Trade and Industry (2003) Energy White Paper *Our Energy Future — Creating a Low Carbon Economy*, The Stationery Office, London. Available online from http://www.dti.gov.uk/energy/whitepaper/ourenergyfuture.pdf [last accessed January 2006].

Department of Trade and Industry (dti) Information and statistics on trends in UK energy industry. Available online from http://www.dti.gov.uk/energy/inform/energy_trends/index.shtml [last accessed June 2005].

Evans, D., Graham, C., Armour, A. and Bathurst, P. (editors and co-ordinators) (2003) *The Millennium Atlas: petroleum geology of the central and northern North Sea*, The Geological Society of London, London.

Fells, I. (2004) Energy Management Group of the Institute of Physics, November 2004 London, and BBC Online, Tuesday 27 July, 2004.

Gluyas, J. and Swarbrick, R. (2004) *Petroleum Geoscience*, Blackwell Science, Oxford.

Griew, S. (2004) 'Gas and electricity demand; drivers and prospects over the next ten years' in conference proceedings 'Energy Demand — Spiralling Out of Control' 17 November 2004, organized by Energy Management Group of the Institute of Physics, London.

Harris, S., Bridgeman, F. and O'Reilly, T. (eds) (2002) *Britain's Offshore Oil and Gas*, UK Offshore Operators Association. Available online at http://www.ukooa.co.uk/issues/storyofoil

International Institute for Applied Systems Analysis (IIASA) and World Energy Council (WEC) (1998) *Global Energy Perspectives*. Summary available online from http://www.worldenergy.org/wec-geis/publications/reports/etwan/supporting_publications/annex_2_chap1.asp [last accessed January 2006)].

Kemp, S. (2005) letter to *The Guardian*, Friday 20 May, 2005.

Lazarus, M., Bartels, C., Bernow, S., Greber, L., Hall, J., Hansen, E. and Raskin, P. (1993) 'Towards global energy security: The next energy transition', Tellus Institute, Boston. Published by Greenpeace International, *Towards a fossil free energy future*, Greenpeace, Amsterdam.

Sheldon, P. (2005) *Earth's Physical Resources: An Introduction* (Book 1 of S278 *Earth's Physical Resources: Origin, Use and Environmental Impact*), The Open University, Milton Keynes.

Shell International (2001) *Energy Needs, Choices and Possibilities — Scenarios to 2050*. Available online from http://www.shell.com/static/media-en/downloads/scenarios.pdf [last accessed January 2006].

Smith, S. (2005) *Water: The Vital Resource* (Book 3 of S278 *Earth's Physical Resources: Origin, Use and Environmental Impact*), The Open University, Milton Keynes.

The Royal Commission on Environmental Pollution (2000) 22nd Report: *Energy: the Changing Climate*, Chapter 9; The Stationery Office, London. Available online from http://www.rcep.org.uk/newenergy.htm [last accessed January 2006].

Tickell, O. (2005) quoting G. Sinden, Environmental Change Institute, Oxford, in *The Guardian*, Thursday 12 May, 2005.

Toynbee, P. (2005) *The Guardian*, Wednesday 25 May, 2005.

Webb, P. (2006) *Metals: Ore Deposits and their Exploitation* (Book 5 of S278 *Earth's Physical Resources: Origin, Use and Environmental Impact*), The Open University, Milton Keynes.

ANSWERS TO QUESTIONS

Question 1.1

The Sun supplies 1.7×10^5 TW to the Earth, of which 1.1×10^5 TW enters the Earth's system (0.3×10^5 TW to atmospheric heating + 0.8×10^5 TW absorbed at the surface, see Figure 1.7). 1.1×10^5 TW is equivalent to 1.1×10^{17} J s^{-1}, and since there are 3.15×10^7 s in a year, the Sun's annual energy supply is

$$3.15 \times 10^7 \text{ s} \times 1.1 \times 10^{17} \text{ J s}^{-1} = 3.5 \times 10^{24} \text{ J}$$
$$\text{or } 3.5 \times 10^6 \text{ EJ}$$

Question 1.2

The fossil fuel energy 'bank' would last $(10^{23}$ J$)/$ $(4.51 \times 10^{20}$ J yr$^{-1})$, or about 220 years, at this rate of consumption.

Question 2.1

As 10 m of peat are required to produce 1 m of coal (Section 2.1.1), a 30 m thickness of peat would eventually produce a 3 m coal seam.

A subsidence rate of 1 mm yr^{-1} is equivalent to 10^{-3} m yr^{-1}.

So, 30 m of peat would be deposited in

$$\frac{30 \text{ m}}{1 \times 10^{-3} \text{ m yr}^{-1}} = 3 \times 10^4 \text{ yr.}$$

This calculation assumes that neither sea-level nor the delta subsidence rate change during this time. It also assumes that all 30 m of peat are deposited before any further compaction occurs. In reality, the earliest formed peat will compact, making the calculated time span an overestimate.

Question 2.2

The most important chemical change shown by increasing rank is the increase in the amount of carbon at the expense of oxygen. The proportion of hydrogen present remains relatively constant at 6–9% by weight over much of the rank series until about 90% carbon, where a significant reduction in hydrogen occurs.

Question 2.3

A (e); B (d); C (b); D (c); E (a).

Question 2.4

If drilling a borehole costs £190 per metre, coring must cost £190 per metre × 2.5 = £475 per metre. The cost of drilling the borehole is therefore:

$$(800 \text{ m} - 130 \text{ m}) \times £190 \text{ m}^{-1} = £127\,300$$

The cost of coring the remainder is therefore:

$$130 \text{ m} \times £475 \text{ m}^{-1} = £61\,750$$

The total cost is therefore £61 750 + £127 300 = £189 050

Question 2.5

The coal seam covers 76 mm^2 squares, whereas the overburden covers 21 cm^2 squares, i.e. (21×100) mm^2 = 2100 mm^2 squares.

So, the ratio of overburden : coal is 2100:76, or 27.6:1. This stripping ratio is higher than the 20:1 limit, so this coal seam would have to be extracted by underground mining.

Question 2.6

Down to depth A

The coal seam covers 125 mm^2 squares, whereas the overburden covers 25 cm^2 squares, i.e. 25×100 = 2500 mm^2 squares.

So, the ratio of overburden : coal is 2500:125, or 20:1. This stripping ratio is equal to the 20:1 limit, so this coal seam could be strip mined to this depth.

Down to depth B

The coal seam covers 150 mm^2 squares, whereas the overburden covers 33 cm^2 squares, i.e. 33×100 = 3300 mm^2 squares.

So, the ratio of overburden : coal is 3300:150, or 22:1. This stripping ratio is higher than the 20:1 limit, so surface mining of the coal seam to this depth would not be economic.

Clearly, at greater depths, proportionately more overburden has to be removed per volume of coal, and this steadily increases the stripping ratio.

Question 2.7

Gradual changes include seam thinning and splitting, whereas sudden changes include faulting and washouts. Gradual changes result in the deterioration in the quality of coal produced because more impurities are extracted as the seam thins. Sudden changes result in the face being halted because the seam is suddenly absent.

Question 3.1

The threshold depth is about 2.5 km. At this depth the temperature is said to be 120 °C, so the geothermal gradient is 120 °C/2.5 km = 48 °C km^{-1}.

Question 3.2

The average porosity of the Permian sandstone reservoirs is (19 + 18 + 13 + 13 + 12)/5 = 15%, whereas the Palaeocene–Eocene reservoirs average (20 + 27 + 29)/3 = 25%. The simplest explanation for this difference is that younger reservoirs tend to have higher porosities because they usually occur at shallower depths and are less compacted than their older counterparts.

Question 3.3

Source rocks only develop in sedimentary basins within continental crust, so areas formed by oceanic crust or crystalline rocks should be avoided. Remote regions (e.g. Antarctica) and inaccessible sedimentary basins (e.g. those beneath thick piles of lava or mountainous terrain) will inevitably be very expensive to explore. Some countries may be rejected for political or humanitarian reasons. It is salutary to remember that more than half of the world's countries produce no oil.

Question 3.4

(a) The Lower and Middle Jurassic play is the most productive, having yielded 2016 million toe. (b) About 50 wells were drilled before the Upper Jurassic play was discovered and a further 100 before the largest discovery was made. (c) The Palaeocene play has been targeted about 150 times and still no major discovery has been made. It seems unlikely that this play will have much impact.

Question 4.1

In one year, 1 kg of uranium-238 undergoing spontaneous radioactive decay will produce 3×10^3 J, whereas 1 kg of coal is equivalent to 2.8×10^7 J. So it would take $(2.8 \times 10^7 \text{ J yr}^{-1})/(3 \times 10^3 \text{ J}) \approx 9300$ years, or about 10^4 (10 000) years, for ordinary radioactive decay of uranium to produce the energy equivalent of the same mass of coal.

Question 4.2

(a) In a burner reactor the chain reaction depends entirely on fission of uranium-235 nuclei, which capture only slow neutrons (Section 4.1.1). But fission of a uranium-235 nucleus produces both fast and slow neutrons. The fast neutrons must therefore be slowed down so that they can be captured by other uranium-235 nuclei. (ii) Although a large number of neutrons is required to keep the chain reaction going (uranium-235 constitutes only one atom in 40 even in enriched uranium fuel), their action must be controlled. An unlimited number of neutrons would cause the fission of all the uranium-235 at once, creating a nuclear explosion. Excess neutrons above a certain number must therefore be absorbed.

(b) Fast breeder reactors depend on the ability of fast neutrons to transform uranium-238 to plutonium-239 (Equation 4.3), and to sustain the fission of plutonium. Because the moderator in burner reactors is designed to slow down fast neutrons produced in the chain reaction, it is not used in a breeder reactor.

(c) In burner reactors, most of the fast neutrons are moderated or absorbed — but it is not possible to remove them all. Those that remain can be captured by uranium-238, which forms the bulk of the fuel elements in all reactors, and plutonium is formed. The breeder reactor is expressly designed to produce plutonium, because it depends on the action of fast neutrons only, which turn uranium-238 into plutonium-239.

(d) Even in rich ore deposits, the uranium is 'diluted' by other elements in the minerals of the ore and surrounding rock. Nearly all neutrons are absorbed by surrounding rocks, which effectively act as 'control rods'.

Question 4.3

(a) Almost all supplies from the late 1940s to the late 1960s were for military use. Indeed much of production until the mid- to late-1980s went into military stockpiles.

(b) Prices rose sharply from $50 kg^{-1} to almost $300 kg^{-1} (in terms of the US$ in 2004) in 1974–75 (peaking in 1979), while production took 5–6 years to reach its peak in 1980–81. The lag represents the *lead time* required to expand production at existing mines and to bring new mines on stream.

(c) Uranium prices fell in the 1980s (from over $250 kg^{-1} to about $30 kg^{-1}), partly because the anticipated rate of growth in nuclear power did not happen and plans for new power stations were cancelled; partly because artificial influences on the market by the US government had ceased and the end to the Cold War cut demand for military uranium and indeed, by the 1990s, stockpiles were starting to be run down.

(d) Uranium prices rose from 2003 to 2004, partly because stockpiles and ex-military supplies were running down, and partly because the future for nuclear energy was looking more promising as a means to satisfy increasing energy requirements without increasing CO_2 emissions.

Question 4.4

(a) True. See discussion of incompatible elements.

(b) True. See comments on hydrothermal vein and unconformity deposits.

(c) False. Underground mining of veins generally will be more expensive in energy and staff costs than surface mining of sandstone-hosted uranium.

(d) True. See comments on roll-front type uranium ores.

(e) False. Yellowcake is impure uranium trioxide.

(f) True. In its reduced form, uranium is insoluble.

Question 4.5

Clay-rich sedimentary rocks occur in the Mesozoic–Tertiary rocks of Dorset, Hampshire, London and the Yorkshire–Lincolnshire–Home Counties areas, and as mudstones and shales among the Palaeozoic rocks of central Wales, the Lake District, the southern uplands of Scotland, and Northern Ireland.

Appropriate salt deposits of Permo-Triassic age underlie the Midlands, particularly the Cheshire area, Dorset and the North Sea (Argles, 2005).

Palaeozoic and Precambrian crystalline igneous and metamorphic rocks are generally found in the upland regions of Wales, Northern Ireland, northern England and Scotland north of the Midland Valley.

Question 4.6

(a) New fuel rods consist of pure uranium oxide or metal which are radioactive but with a long half-life; spent fuel rods contain a few per cent of highly radioactive fission products, some of which have short half-lives, and so are intensely radioactive (see discussion of Figure 4.18).

(b) Uranium stockpiles that existed in the early 1990s were large because of: release of weapons-grade uranium at the end of the Cold War; cutbacks in nuclear power programmes; uranium overproduction from mines. So there was less need to recycle uranium-235 that remained 'unburned' in spent nuclear fuel. The UK's fast breeder research programme at Dounreay was terminated, so there was no longer a UK market for reclaimed fissionable material from spent fuel rods. Several customers began to reconsider using THORP.

(c) Vitrification has two aims: first to concentrate HLW into a small volume for disposal, and second to lock the wastes into a relatively stable glass, which forms part of the multiple containment system resistant to groundwater penetration (Figure 4.19).

Question 5.1

(a) Shales and most metamorphic rocks have low thermal conductivity. (b) Many granites have high concentrations of uranium (Section 4.2.1), and also thorium and potassium, distributed through large volumes. Large granitic intrusions therefore contain higher total masses of heat-producing isotopes than most other crustal rocks. The fact that other rocks, such as those containing uranium ores, coal and some shales, have higher concentrations of these isotopes has little effect on overall heat production, because they generally occur in small volumes relative to those of granite intrusions.

Question 5.2

The geothermal gradient must produce a temperature difference of $150 - 10 = 140$°C. So the minimum depth of drilling needs to be 140 °C/37 °C km^{-1} = 3.78 km.

Question 5.3

The wind power available, from Equation 5.5, is

0.5×1.2 kg m$^{-3} \times 300$ m$^2 \times 10^3$ m^3 s$^{-3} = 1.8 \times 10^5$ W

For an efficiency of 40% (0.4), the power output of the turbine at 10 m s^{-1} is therefore 7.2×10^4 W = 72 kW.

Question 5.4

Substituting the given values into Equation 5.6 gives:

$N = 0.8 \times (9.8$ m s$^{-2}) \times (10^3$ kg m$^{-3}) \times (60$ m^3 s$^{-1}) \times (5$ m$)$

$= 2.35 \times 10^6$ W = 2.35 MW

Question 6.1

(a) The period 1940–1970 was dominated by gas and oil production curves that steepen with time, representing a scenario of historical growth, which was checked abruptly in the early 1970s.

(b) For the two short periods, 1973–1975 and 1979–1983, oil and gas production followed a zero or even negative growth scenario, due to the two separate political crises.

(c) The gradients of the oil and gas production curves between 1985 and 2000 are still positive (rising production), but they do not tend to steepen with time as they did earlier in the century; if anything, there is a trend towards curves that are becoming less steep with time. This is characteristic of technological 'fix' scenarios. Minor fluctuations in the curves' gradients reflect the interplay between technological and political factors.

Question 6.2

You saw in Section 5.3 that about 11% of all land would be needed to grow enough biomass totally to supplant global primary energy use at the start of the 21st century, so this would rise to 22% by the end of the century. If biomass were to contribute 25% of global primary energy by 2100, $0.25 \times 22\% = 5.5\%$ of all land would be needed to grow it, and most of that would have to be in biologically productive areas (Figure 5.15).

COMMENTS ON ACTIVITIES

Activity 2.1

(a) Using Equation 2.1, you should have been able to calculate the values in Table 2.4. Your graph of s against d should be similar to Figure 2.39.

Table 2.4

d/m	$\sqrt{d}$	$\sqrt{d}+4$	$1/(\sqrt{d}+4)$	t/m	$4t/\text{m}$	$s=[4t/(\sqrt{d}+4)]/\text{m}$
50	7.07	11.1	0.09	3	12	1.08
150	12.3	16.3	0.06	3	12	0.74
250	15.8	19.8	0.05	3	12	0.60
350	18.7	22.7	0.04	3	12	0.53
450	21.2	25.2	0.04	3	12	0.48
550	23.5	27.5	0.04	3	12	0.44
650	25.5	29.5	0.03	3	12	0.41
750	27.4	31.4	0.03	3	12	0.38

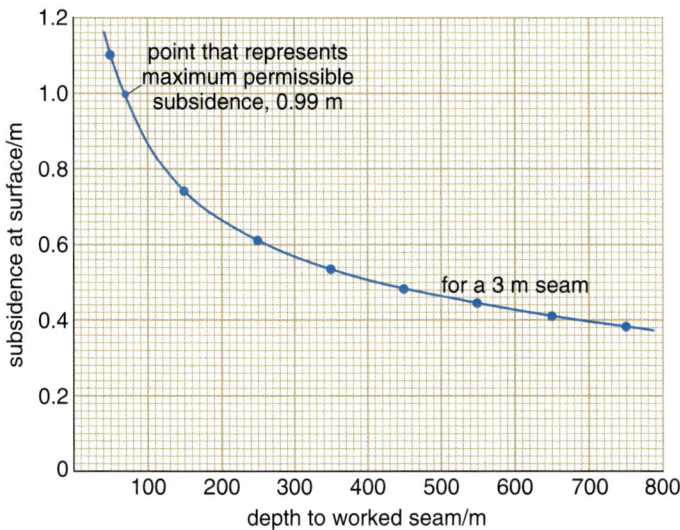

Figure 2.39 Graph of estimated subsidence against depth to worked seam for a seam of 3 m thickness.

(b) The graph shows that a 3 m seam can be worked at depths of about 60 m and deeper if it is to produce less than the 0.99 m of surface subsidence demanded at Selby.

Activity 6.1

(a) Europe and North America are likely to receive less sulphur-rich acid rain pollution by 2030.

(b) Asia is the major region likely to experience an increase in acid rain pollution by 2030, although other regions such as Africa, the Middle East, South and Central America are also anticipated to experience increases in acid deposition.

(c) Europe and North America will probably experience less pollution, due mainly to the introduction of pollution abatement legislation. Between 1990 and 2030 technological developments that allow reduced emissions of sulphurous compounds will lead to less pollution, even though the economies of these regions are predicted to continue to grow. In Asia and, to a lesser extent, in Africa, South and Central America and the Middle East, economies are predicted to grow by 2030. The accompanying rapid demand for energy that will mostly be met by the consumption of fossil fuels would result in an overall increase in sulphurous acid rain pollution, even if emission abatement legislation were introduced there in the intervening time.

(d) Although the per capita energy consumption is high, the population density of Australia is low, so energy use per unit land area is likely to be lower than in more densely populated developing regions such as Asia.

Activity 6.2

There is no 'right' answer to this, and as the extracts demonstrate, opinions vary widely on such a broad range of issues. The following remarks are included to show the sort of considerations that might arise in a debate of this kind.

An idealistic approach might be to invest heavily in development of alternative energy sources, including methods of storing and supplying such energy (perhaps even devising a practical 'hydrogen economy'). This investment could operate in tandem with draconian taxes and other disincentives (personal carbon quotas?) on fossil fuel consumption (especially by the transport sector), to drive the switch from fossil fuels to renewables. But just how much energy could renewables reasonably provide? How 'green' are massed ranks of wind turbines anyway? And what might this approach do for your political or economic future?

A more subtle way to induce a shift from fossil fuels to renewables might be to employ a range of incentives, investment and education measures to encourage development not only of renewable technologies, but also energy efficiency — at all levels — including conservation, home improvements, and perhaps local community schemes for energy generation and supply. Gradual integration of renewable energy into the existing distribution system would be implemented. You might have to accept a gradual transition over several decades, or even longer, for society to adapt to using different modes of energy. In the meantime, global warming becomes a stark, undeniable reality — but will the comfortable society of an industrialized nation feel threatened enough to change its lifestyle?

A more pragmatic view would be to accept that fossil fuel consumption will only start to abate when prices are forced higher by depletion of reserves, or political pressures, or both in tandem. The emphasis could be more on developing the cleanest possible technologies for burning fossil fuels, and exploiting marginal resources such as tar sands in a more environmentally sound way than at present. After all, necessity is the mother of invention: when the time comes, technologies for renewable energy or nuclear fusion will arise to meet any energy crisis. But can you really rely on such ingenuity? Who, initially, will pay the price if this strategy fails?

A different technological strategy might be to follow a nuclear route: revitalize the nuclear fission generation industry to ameliorate the damaging effects of CO_2 emissions, while at the same time pouring money into research and development of nuclear fusion. Fossil fuels will tide us over the long lead times on nuclear power station construction, and by then surely the problems of nuclear waste disposal will have been solved. This approach would give society breathing space to encourage the development of renewable technologies on a more realistic timescale. But what would be the consequences of a major nuclear accident?

ACKNOWLEDGEMENTS

Among the many people who helped in various ways during the preparation of this book, the editor would particularly like to thank colleagues on the S278 Course Team — Peter Webb especially — for their constructive suggestions and attention to detail in the proofs.

Grateful acknowledgement is made to the following sources for permission to reproduce material within this book.

Figure 1.2 National Railway Museum/Science and Society Picture Library; *Figures 1.5a, 3.17 and 3.18* BP (2004) *Statistical Review of World Energy 2004*, http://www.bp.com; *Figure 1.5b* Adapted from NASA material; *Figures 1.8, 3.8a and 6.21* Martin Bond/Science Photo Library; *Figure 1.12* NASA's Earth Observatory; *Figure 1.13b* © David Sanger Photography/Alamy; *Figure 1.13c* Steve Allen/Science Photo Library; *Figure 2.1a* Department of Earth Sciences Image Library, The Open University; *Figure 2.1b* Professor R. A. Spicer, The Open University; *Figure 2.2b* Norman Tomalin/Alamy; *Figure 2.2c* John Watson, The Open University; *Figures 2.3, 5.15, 5.18b and 5.19a* NASA; *Figures 2.4 and 2.35* Thomas, L. (2002) *Coal Geology*, John Wiley and Sons Ltd; *Figure 2.6* Kevin Church, The Open University; *Figure 2.7* Skinner, B. J. (1976) *Earth Resources*, Figure 13, Prentice-Hall, Inc.; Englewood Cliffs, NJ; *Figure 2.8* Cossons, N. (1987) 'Coal', *The BP Book of Industrial Archaeology*, 2nd edn, David and Charles Publishing Inc; *Figure 2.10* WesternGeco; *Figures 2.11, 2.17, 2.22, 2.23 and 2.31* MRP Photography; *Figure 2.12 and 2.15* James A. Luppens, USGS; *Figure 2.14* Scottish Coal Company Ltd; *Figure 2.21a* Arthur Meyerson Photography; *Figure 2.21b* Tejada Photography Inc; *Figures 2.24 and 2.25* Guion, P. D. and Fulton, I. M. (1993) 'The importance of sedimentology in deep-mined coal extraction', Figures 30 and 31, *Geoscientist*, **3**(2), The Geological Society; *Figure 2.27* The Selby Coalfield, British Coal; *Figure 2.30* Dunrud, C. R. and Osterwald, F. W. (1980) 'Effects of coal mine subsidence in the Sheridan, Wyoming area: a summary of geology, subsidence, and other effects of past and present mining as related to the environment, coal resource management, and land use', US Geological Survey Professional Paper 1164, 49 p; *Figure 2.34* Based on a map by Professor A. J. Smith; *Figures 2.37 and 2.38* UK Energy Brief July 2003, Department of Trade and Industry. 'Crown copyright material is reproduced under Class Licence Number C01W0000065 with the permission of the Controller of HMSO and the Queen's Printer for Scotland'; *Figure 3.1* Tarling, D. H. (1981) *Economic Geology and Tectonics*, Blackwell Science Ltd.; *Figure 3.2* Tissot, B. P and Welte, D. H. 'From kerogen to petroleum' in *Petroleum Formation and Occurrence*, Springer-Verlag Berlin; *Figures 3.3a, 3.3b, 5.13a and 5.14b* © Copyright Andy Sutton Photography; *Figure 3.4* Chapman Bounford; *Figure 3.5* Pegrum, R. M. and Spencer, A. M. (eds) (1990) 'Hydrocarbon plays in the North Sea', in Brooks, J. (ed.) *Classic Petroleum Provinces*, Special Publication No. 50, Geological Society, London; *Figure 3.6* © 2002 United Kingdom Offshore Operators Association http://www.oilandgas.org.uk; *Figures 3.7 and 3.16* Evans, D., et al. (2003) *The Millennium Atlas: Petroleum Geology of Central and Northern North Sea*, Geological Society, London; *Figure 3.8b* John Mead/Science Photo Library;

Laherrère, Jean, *Forecasting Future Production for Past Discovery*, OPEC Seminar, 28 September 2001. Courtesy of the author; *Figure 6.5* 'Mining', in *The Guardian*, 14 August 1993, with additional data adapted from Department of Trade and Industry; *Figures 6.7a and 6.7b* Copyright © 1996–2005 by James L. Williams; *Figures 6.8, 6.9 and 6.13* The World Energy Council, London, UK; *Figure 6.10 Exploring the Future Energy Needs, Choices and Possibilities: Scenarios to 2050*, Global Business Environment, © Shell International 2001; *Figure 6.12* The World Meteorological Organisation; *Figure 6.14* Pearce, F. (1987) 'Acid rain' in *New Scientist*, **116**(1585), 5 November 1987, © IPC Magazines, 1987; *Figure 6.17* Courtesy of Sasol Ltd; *Figure 6.22* Department of Transport, Local Government and Regions (DTLR) 2001, Transport statistics, Great Britain (27th edn), Crown copyright material is reproduced under Class Licence Number C01W00000 with the permission of the Controller of HMSO and the Queen's Printer for Scotland.

Every effort has been made to contact copyright holders. If any have been inadvertently overlooked the publishers will be pleased to make the necessary arrangements at the first opportunity.

GLOSSARY

Items in this Glossary are printed in **bold** in the main text, usually where they are first mentioned. Terms printed in italics below are defined elsewhere in the Glossary.

acid mine drainage (AMD)   Surface and groundwater containing sulphuric acid produced by oxidation of iron sulphide, which escapes to the environment from mine workings and waste heaps.

acid rain   Rain in which significant amounts of sulphur and/or nitrogen oxides are dissolved, thereby releasing hydrogen ions and lowering the pH of rainwater.

active solar heating   The use of mirrors to focus solar radiation, thereby concentrating heat to create high temperatures, suitable for cooking and thermal electricity generation.

adit   A shallow angled tunnel into hillsides to mine coal or minerals.

advance face   A variant of *longwall* underground mining, whereby the coal face is progressively cut into a block of coal away from a permanent access road.

alpha (α-)particles   Subatomic particles consisting of two neutrons and two protons, emitted in the radioactive decay of some unstable nuclei with high *mass number* (>160) and large *atomic number* (>60), including uranium.

anoxic (anoxia)   Conditions that are oxygen poor (lacking oxygen). Not all the fixed carbon in dead plant tissue returns to the atmosphere as CO_2: some may be retained as carbon-enriched residue and yet more converted into *hydrocarbons*.

anthracite   High-rank coal that contains more than 90% carbon and less than 10% volatiles.

atomic number   The number of protons within the nucleus of an atom of a particular element. By definition, atoms of the same element all have the same atomic number.

base load   In electricity generation, the minimum continuous demand for electricity on a supply network (e.g. the UK National Grid).

base-line study (environmental)   An assessment of the original, natural state of the environment before development takes place.

bell pit   A shallow vertical shaft down to a coal seam or ore deposit, with workings outwards in all directions from the bottom of the shaft (no longer used).

bench mining   A way of working coal seams in which overburden is stripped off each seam, leaving shallow dipping 'benches' of profitable coal.

biogas   Hydrocarbon gas, mainly methane, formed by anaerobic digestion of biomass, usually human or animal waste.

biomass   Wood, and other plant materials that can be used as fuels.

bitumen   Solid form of naturally occurring petroleum that consists of high molecular weight hydrocarbons. Sometimes known as tar, although that can also be produced from coal during the coking process.

bituminous coal   Coal containing over 10 % volatiles and variable carbon content. The most common rank of British coal.

burner reactor   A nuclear reactor that relies on the fission of uranium-235.

calorific value   The amount of heat liberated per unit mass by burning a fuel under controlled conditions.

cap rock   An impermeable rock layer that *seals* petroleum in rocks beneath it, to form a *trap*.

carbohydrates   Organic compounds of carbon, hydrogen and oxygen, with the general formula $C_nH_{2n}O_n$. Glucose, one of the simplest carbohydrates, has the formula $C_6H_{12}O_6$.

carbon cycle   The movement of carbon between the major natural stores of carbon (called 'reservoirs') on the Earth. The two major divisions of the cycle are the *terrestrial* and *marine carbon cycles*.

chain reaction   In nuclear physics, a reaction in which neutrons released by fission of atomic nuclei — especially those of uranium — initiate fission in other nuclei, so that successive fissions are linked in a 'chain'. Many millions of chain reactions occur simultaneously in a reactor core.

channel-fill deposits   Sedimentary structures that form when erosive drainage channels cut down into the underlying sediments and become filled with coarse sand.

CO₂ sequestration See *gas sequestration*.

coal Carbon-rich material formed from decayed, compacted and chemically changed vegetation that has been buried. *Note*: Not all coal is black, that which has been least compacted and chemically changed is brown in colour.

coalification The geological and chemical processes whereby dead vegetable matter is compressed and loses water and hydrogen to become carbon-rich coal of different grades.

coke A solid material containing over 95% carbon, which is produced when coal is heated (in the absence of air) to distil off the volatile constituents. It is used in smelting as a reducing agent and as a fuel for melting the ore and producing a temperature sufficient for reduction to occur.

combination traps (petroleum) Traps with both structural and stratigraphic components.

compaction The forcing closer together of grains in a sediment as it becomes buried, especially in fine-grained sediments.

concealed coalfields Coalfields where the coal-bearing strata are hidden beneath younger strata.

conduction Heat transfer through a material along a thermal gradient.

control rod Rod made of a neutron absorber (e.g. boron, cadmium), which can be moved in and out of a nuclear reactor *core* to vary the amount of neutron absorption and thus control the *chain reaction*.

convection Heat transfer by the buoyant rise of hotter, less dense fluids — in the Earth of hot, ductile rock.

core (drilling) A cylinder of rock obtained by drilling; also called 'drill core'.

core (reactor) The 'heart' of a nuclear reactor, which contains the fuel rods and *moderator*, through which the coolant fluid circulates to extract heat, and into which the *control rods* can be inserted to control the *chain reaction*.

creaming curve A plot showing the varying rate of increase in petroleum reserves as exploration of an oilfield or *play* progresses.

critical mass The amount of fissionable material required for nuclear fission chain reactions to be self-sustaining.

critical point The highest temperature and pressure at which a substance can exist in both liquid and gaseous states. At higher values of P and T the substance takes the form of a *supercritical fluid*. In the case of pure water, the critical point is at 374 °C and a pressure of about 150 atmospheres.

crude oil Liquid *petroleum* that occurs naturally in *reservoirs* within sedimentary rocks.

decay curve (radioactivity) The exponential relationship between the fraction of the original radioactive atoms of an unstable isotope remaining and the number of *half-lives* that have elapsed.

decommissioning Disposal of an installation after its useful lifetime has expired.

de-materialization A range of technological and social measures aimed at using less material (and hence less energy), in either production of goods or consumption of resources.

digestion (biomass) Production of hydrocarbon gases by anaerobic bacterial metabolism of semi-liquid human and animal wastes.

direct solar energy use Exploitation of *insolation* either by using it as a source of heat or by *photovoltaic* conversion to electricity.

disseminated magmatic uranium deposits Low-grade uranium deposits consisting of uranium-rich accessory minerals dispersed in granitic rocks and pegmatites.

drift mines Early mines that were simply trenches following the coal seam.

drill cuttings Chips of rock produced by drilling that return to the surface in the drilling lubricant fluid, which are recovered to yield information about rock variations down the hole.

energy The capacity to do work, measured in joules (J).

energy density The amount of energy stored by a resource relative to the volume that it occupies.

energy efficiency The ratio of the useful output of work to the input of energy suppled — usually expressed as a percentage.

enriched uranium Uranium in which the content of uranium-235 has been artificially increased to more than the 0.7% found in natural uranium (usually to ~2.3%), for use in nuclear reactors (e.g. AGR, PWR).

enthalpy   The total energy content of a physical system.

environmental impact assessment (EIA)
Ecological study aimed at assessing the likely effects of developing physical resources on the pre-existing environment, as defined in a *base-line study*, and means of mitigating the effects.

exposed coalfields   Coalfields in which the coal-bearing rocks occur at the surface.

fast breeder reactor   Nuclear reactor that allows *fast neutrons* to convert uranium-238 into plutonium-239, thereby generating more fissionable fuel than they use.

fast neutrons   The higher-energy neutrons emitted in nuclear fission; they can be slowed down by the *moderator* in a nuclear reactor.

fast-response system (electricity generation)
Power plant that can be started and stopped quickly, such as gas-fired and hydroelectric plants or a *solar PV cell*.

fault traps (petroleum)   Formed when an inclined bed of reservoir rock is brought into contact with impermeable strata by movement along a fault.

fermentation (biomass)   Production of combustible alcohols from the action of yeasts on the sugars in plant materials.

fission (nuclear)   See *nuclear fission*.

fission energy   The energy released by the conversion of mass into energy on fission of heavy unstable nuclei.

flash (steam)   The spontaneous transformation of superheated hot water to steam as geothermal fluid travels up the borehole from high-pressure aquifers to low pressure at the surface.

force   The means by which the direction or speed of movement of an object is changed; equal to mass × acceleration.

fossil fuels   Combustible resources derived from ancient organisms, especially coal, oil and natural gas.

fuel   Materials capable of liberating energy by nuclear and chemical reactions, including combustion. They commonly have a high *energy density*.

fuel cell   Electrolytic cell in which hydrogen and oxygen combine to form water and release electrical energy (heat is also released).

fusion (nuclear)   See *nuclear fusion*.

gas hydrates   Crystalline hydrocarbons that form by the combination of methane and/or carbon dioxide with water under the low-temperature and high-pressure conditions of the deep ocean floor.

gas sequestration   Disposal into long-term storage of the carbon dioxide produced by the use of fossil fuels, either by using petroleum *reservoirs* or other secure sites from which the gas cannot escape.

gas venting and flaring   The deliberate release and burning of natural gas that comes up wells along with oil, because collecting the gas for sale is not economic.

gasification (biomass)   See *pyrolysis*.

geothermal gradient   The rate at which temperature increases with depth in the Earth.

global dimming   Reduction in global atmospheric temperature because of the reflection and scattering of incoming short-wave solar radiation by fine-grained particles and droplets (aerosols) in the atmosphere, so that it returns back to space.

gravity surveys   Measurements of variations in the Earth's gravitational field, which can show areas underlain by rocks of different densities.

'greenhouse' effect (enhanced)   Absorption of long-wave radiation emitted from the Earth's surface by atmospheric gases, such as carbon dioxide and methane, thereby retaining heat in the atmosphere. Global warming that is caused by increased emissions of some gases to the atmosphere by industrial burning of fossil fuels is an enhanced form of the greenhouse effect.

ground-source heat pump   A means of exchanging heat from low-enthalpy geothermal sources — generally close to the surface — to fluids that can be used for space heating.

half-life   The time taken for half of the atoms in a sample of a radioactive isotope to decay.

heat flow   The passage through the Earth of heat that has been generated by decay of radioactive isotopes — mainly those of uranium, thorium and potassium — which occur naturally in minerals of the crust and mantle.

heavy oil   Oil with high viscosity and density that occurs in *oil sand* deposits.

hot dry rock (HDR) system   A means of exploiting the geothermal energy from hot crystalline rocks that are impermeable and contain very little groundwater. An HDR system involves an artificial fracture system that acts as a heat exchanger when water is pumped through.

humification   The process of conversion of dead vegetation to *peat* through the action of anaerobic bacteria.

hydrocarbons   Organic compounds, such as *methane*, that contain only hydrogen and carbon.

hydrogen economy   Energy system based on the electrolysis or biochemical splitting of water (using alternative energy sources) to produce hydrogen, which can then be stored and used as a transportable fuel.

hydropower   Conversion of the *potential energy* of water in rivers or reservoirs to electricity, by using the *kinetic energy* that is released when it flows to turn turbines. Because rainfall stems from water vapour evaporated from the ocean surface, hydropower is an indirect form of solar energy.

hydrothermal vein uranium deposits   Uranium mineralization deposited from hydrothermal solutions in fissures to form veins.

***in situ* leaching (ISL)**   A means of extracting soluble metals, such as uranium, from deeply buried ore deposits by pumping a solvent into the deposit through boreholes.

insolation   Solar radiation that reaches the Earth's surface.

integration (electricity supply)   The management of a complex mix of different energy sources to produce a stable electricity supply that can track the demand curve.

intermittency of supply (electricity)   Unpredictable duration and power of a means of electricity generation, such as wind and solar energy.

isotope   Versions of an element that each have the same number of protons but different numbers of neutrons. Each isotope of an element has identical chemical behaviour. Most elements have several naturally occurring isotopes. Some isotopes are radioactive.

joule   The SI unit of energy and work. A joule (J) of work is done when a force of one *newton* moves an object through a distance of one metre.

kerogen   A general term for organic material in petroleum *source rocks*; the name is derived from the Greek for 'wax producer'.

kinetic energy   Mechanical energy that exists by virtue of movement. For example, a car engine turns chemical energy (petrol) through heat into movement (engine parts and road wheels).

longwall mining   An underground mining technique that involves a wall being cut into the coal seam, extraction advancing away from or towards the shaft.

magnetic surveys   Measurements of the Earth's magnetic field, which can show variations due to the occurrence of rocks with different levels of magnetization, mainly due to oxides containing iron and titanium.

marine carbon cycle   That part of the carbon cycle that involves the circulation of carbon in seawater through biological activity and inorganic processes.

mass number   The total number of protons and neutrons in an atomic nucleus.

maturation   The process — involving chemical reactions, time and burial temperature — that transforms natural *kerogen* into crude oil and natural gas.

methane CH_4; a gaseous *hydrocarbon* that is the main component of *natural gas*.

mire   Poorly drained land that is waterlogged, thereby encouraging the *anoxic* conditions that allow preservation of dead vegetation to form *peat*.

moderator   Material containing nuclei of low atomic mass (small atomic number), such as hydrogen (in water) or carbon (graphite), which absorb some of the energy of *fast neutrons* by scattering them, thus transforming them into the slow neutrons that are more likely to induce fission in uranium-235.

natural gas   Gaseous form of *petroleum* that consists mainly of methane, with other low molecular weight hydrocarbon gases.

net to gross   The ratio or percentage of porous and permeable rock thickness to that of the overall *reservoir rock*.

newton   The *force* that gives a mass of one kilogram (kg) an acceleration of one metre per second per second (m s^{-2}), and is therefore equivalent to 1 kg m s^{-2}.

non-renewable energy   Energy resources that are replenished naturally over extended timescales of thousands or millions of years. As they are exploited faster than they are replenished they are considered as non-renewable.

nuclear fission   The breakdown of a radioactive isotope in which the atomic nucleus splits into different, lower mass combinations of protons and neutrons that form other isotopes. Of the naturally occurring elements, only uranium and thorium have fissionable nuclei.

nuclear fusion   The combination of the nuclei of two atoms to form another nucleus of higher mass. In the context of nuclear energy, nuclear fusion refers to the fusion of hydrogen isotopes that contain one proton and one neutron (deuterium) or one proton and two neutrons (tritium) to form helium atoms.

oil sand   A sedimentary sandstone whose pores are saturated with *heavy oil* or *bitumen* that is too viscous to flow easily — sometimes known as a tar sand because of the tarry consistency of the oil.

overburden   Waste rock that overlays a valuable resource, e.g. coal.

passive solar heating   Exploiting *insolation* to heat materials and spaces, often incorporating some means of heat storage.

peat   Compressed layers of decaying plant material; part of the initial stages of coal formation.

permeability   A measure of the degree to which fluid passes through a porous rock.

petroleum   A complex mixture of hydrocarbons and lesser quantities of other organic molecules containing sulphur (S), oxygen (O), nitrogen (N) and some metals. Petroleum is derived from the chemical alteration of *kerogen* in source rocks.

petroleum charge   An concept concerning the formation, migration and accumulation of petroleum in a body of sedimentary rocks. It involves consideration of *source rocks*, *maturation*, geological history and migration paths in a sedimentary basin.

petroleum play   A model that can be used to direct petroleum exploration. Plays consolidate what is known about *petroleum charge*, *reservoirs*, *seals* and *traps*, together with geological structure, and enable the chances of petroleum accumulations to be assessed at specific stratigraphic levels within a sedimentary basin.

photosynthesis   A chemical reaction in green plants in which carbon dioxide from the atmosphere combines with water to form carbohydrates, using the energy of solar radiation.

photovoltaic (PV) effect   The result of electrons being freed from impurities in *semiconductors* by the quantum energy of light photons, so that the electrons move to generate both electrical potential and current. The principle behind solar cells.

pillar-and-stall   A once-used underground mining method that involved a rectangular system of tunnels to extract coal from blocks or stalls, the roof being supported by pillars of unmined coal — an important and continuing cause of surface subsidence.

play fairway map   Depiction of areas where geological conditions imply a high likelihood of petroleum occurrences, as deduced from a *petroleum play*.

porosity   The void space in a rock, commonly expressed as a percentage.

possible reserves (petroleum)   A category of reserves for which there is a significant (>0 but <50%) probability that they are technically and economically recoverable. A definition used widely in the petroleum industry.

potential energy   Mechanical energy that exists by virtue of position or potential in a force field — usually that of gravity.

power   The rate at which energy is delivered or work is done, measured in watts (W). One watt is equivalent to one *joule* per second.

primary energy   The energy released by directly burning *fossil fuels* or consumed at source in electricity generation.

primary migration   The initial expulsion of *petroleum* from its *source rock*, before its later movement over long distances through a *reservoir rock* eventually to meet a *trap* and accumulate.

primary recovery   A phase of hydrocarbon production that relies only on the internal pressure within the reservoir rock to expel fluids; contrast with *secondary recovery*.

probable reserves (petroleum)   Reserves of a resource for which there is a better than 50% probability that they are technically and economically recoverable. A definition used widely in the petroleum industry.

proved reserves (petroleum) Reserves of a resource for which there is a better than 90% probability that they will be produced over the lifetime of a deposit. A definition used widely in the petroleum industry.

pumped storage schemes Hydroelectric power stations that use off-peak electricity to pump water back up to a holding reservoir. They are useful at times of peak demand, but are net consumers of electricity.

pyrolysis (biomass) Strong heating (effectively distillation) of organic matter in the absence of air, to produce liquid and gaseous hydrocarbons, leaving a solid residue.

quartz-pebble conglomerate uranium deposits Uranium deposits consisting of uranium ore minerals in river gravels older than about 2000 million years.

R/P ratio The ratio of reserves to annual production for any resource, effectively the number of years that proven reserves would be expected to last at current rates of production. The same concept as *reserves lifetime*.

radioactive decay The spontaneous disintegration of unstable nuclei, involving the emission of subatomic particles (α-*particles*, β-particles (electrons)) and energy (γ-rays).

rank The position of *coal* in a series of increasing carbon content; an indicator of the maturity of the coal. Low-rank coals (lignites) contain a higher proportion of volatiles than high-rank coals (anthracites).

rebound effect (energy consumption) The tendency for individuals or organizations, having saved money by implementing energy saving measures, to spend that money on additional energy-consuming activities, such as providing higher quality services.

remote sensing Gathering data about the Earth's surface — mainly in the form of images and detectable geophysical properties — from a distance, either from satellites in the case of images or an aircraft for geophysical data.

renewable energy resources Supplies of energy on Earth from internal and external processes that are continually available, whether they are used or not. They include solar, tidal, wave, wind, hydro and geothermal energy.

reservoir rocks Sedimentary rocks — usually sandstones or limestones — that contain enough space within them to store hydrocarbons.

residence time (carbon) The average time that carbon stays in a reservoir before moving to another reservoir, estimated by the amount in the reservoir divided by the transfer rate.

respiration The process by which organisms combine oxygen from the air with *carbohydrates* and other substances to liberate energy, and release carbon dioxide and water back into the atmosphere.

retreat face A variant of *longwall* underground mining, whereby the coal face is worked backwards, once accessed by temporary tunnels, towards a shaft or permanent access tunnel.

sandstone-hosted uranium deposit Uranium mineralization formed by precipitation of reduced uranium from oxidizing percolating groundwaters at a boundary between oxidizing and reducing conditions in sandstones.

seals (petroleum) Impermeable rocks that prevent the flow of fluids out of a reservoir rock. They are also often called *cap rocks*.

seatearth A fossil soil underlying a coal in which coal-forming plants grew.

secondary migration The movement of petroleum through permeable pathways into a *reservoir rock* after it has escaped from the *source rock*, generally by moving upwards under gravity: it floats on any water in the reservoir rock.

secondary recovery A phase of hydrocarbon production, involving the injection of natural gas into the reservoir above the oil forcing the oil downwards, and/or injecting water below the oil to force it upwards.

seismic section An image of seismic reflectors that reveals geological structures in cross section at depth, using data obtained from *seismic surveying*.

seismic surveying A geophysical technique used extensively in petroleum and coal exploration and evaluation to identify rock sequences and structures. It is based on recording the time taken for sound waves to travel from a surface source down into the rocks, reflect off distinctive rock boundaries and return to surface detectors. The data can be processed to image the geological structures at depth, known as a *seismic section*.

seismic velocity The speed at which seismic waves (generally sound waves in exploration) pass through rock.

semiconductors   Poorly conductive materials doped with impurities, which act to increase electrical conductivity to a level intermediate between non-conductors and conductors. Bonded together in *photovoltaic cells*, they can generate an electrical current.

slow neutrons   Low-energy neutrons that are the most likely to cause fission in uranium-235; they are often referred to as 'thermal neutrons'.

slow-response system (electricity generation)   Power plant that requires lengthy periods to start or stop generation, such as nuclear, coal-fired and geothermal generators.

solar energy   Solar radiation that reaches the Earth's surface is potentially available as an energy resource as direct solar energy using *photovoltaic cells*.

solar PV cell   Wafer of semi-conducting materials that exhibit the *photovoltaic effect* so that it converts light into electrical energy.

solid fuel combustion   Applied to biomass fuels: burning of wood, peat and waste materials in solid form as a primary energy source.

source rocks   Sediments, usually shales, that contain enough organic carbon to generate petroleum under the right burial conditions. Source rocks usually contain at least 5% total organic carbon.

stratigraphic traps (petroleum)   Traps that occur where there is a barrier caused by lateral variation in *permeability*.

strip mining   A method of extracting nearly horizontal beds of a valuable resource by surface mining, in which the operations proceed parallel to the outcrop of the beds.

stripping ratio   The ratio of the amount of *overburden* to the amount of a valuable resource (e.g. coal) to be mined in surface operations.

structural traps (petroleum)   A trap where a barrier to petroleum migration formed as a result of tectonic deformation of strata.

supercritical fluid   The state in which a substance exists at temperatures and pressures above the *critical point*. It shows unique properties that are different from those of either gases or liquids below the critical point, such as being able to move more easily through rock,

like a gas, yet having a high capacity for dissolving other compounds, like a liquid.

surface mine   A mine operated directly from the surface by excavation of *overburden* and the resource.

terrestrial carbon cycle   That part of the carbon cycle involving biological and inorganic processes on land.

threshold (petroleum)   The subsurface temperature and depth at which petroleum starts to form from *kerogen*.

tonne oil equivalent (toe)   The chemical energy contained in one tonne of oil, it equates approximately to 4.19×10^{10} J. Often used as a convenient means of comparing the energy contents of different fossil fuels.

trap   A structure (tectonic or sedimentary) that involves a *seal* to constrain petroleum within a *reservoir* rock.

two-way travel time (TWT)   The time taken for a seismic wave to pass from a sound source to a reflector, and back again to a detector. The vertical or y-axis of a *seismic section*.

unconformity related uranium deposits   Uranium mineralization, often at high grade, deposited from hydrothermal solutions at erosional unconformities associated with faults, where oxidation conditions change.

underground mining   Operations in which a physical resource is mined entirely beneath the surface by systems of tunnels and working faces, which are accessed from shafts driven from the surface.

washout   A sedimentary feature formed where a channel has cut down into a coal seam and locally eroded it away.

watt   The SI unit of power. One watt (W) is equivalent to converting one *joule* of energy per second, or doing one joule of work every second.

wireline logging   Measurement of the physical properties of rock using instruments lowered down a borehole.

work   A force acting on an object that causes its displacement, equivalent to *energy*.

working head   The vertical distance through which the water falls in a hydroelectric power scheme.

INDEX

Note that **bold** page numbers refer to where terms defined in the Glossary are printed in **bold** in the text.

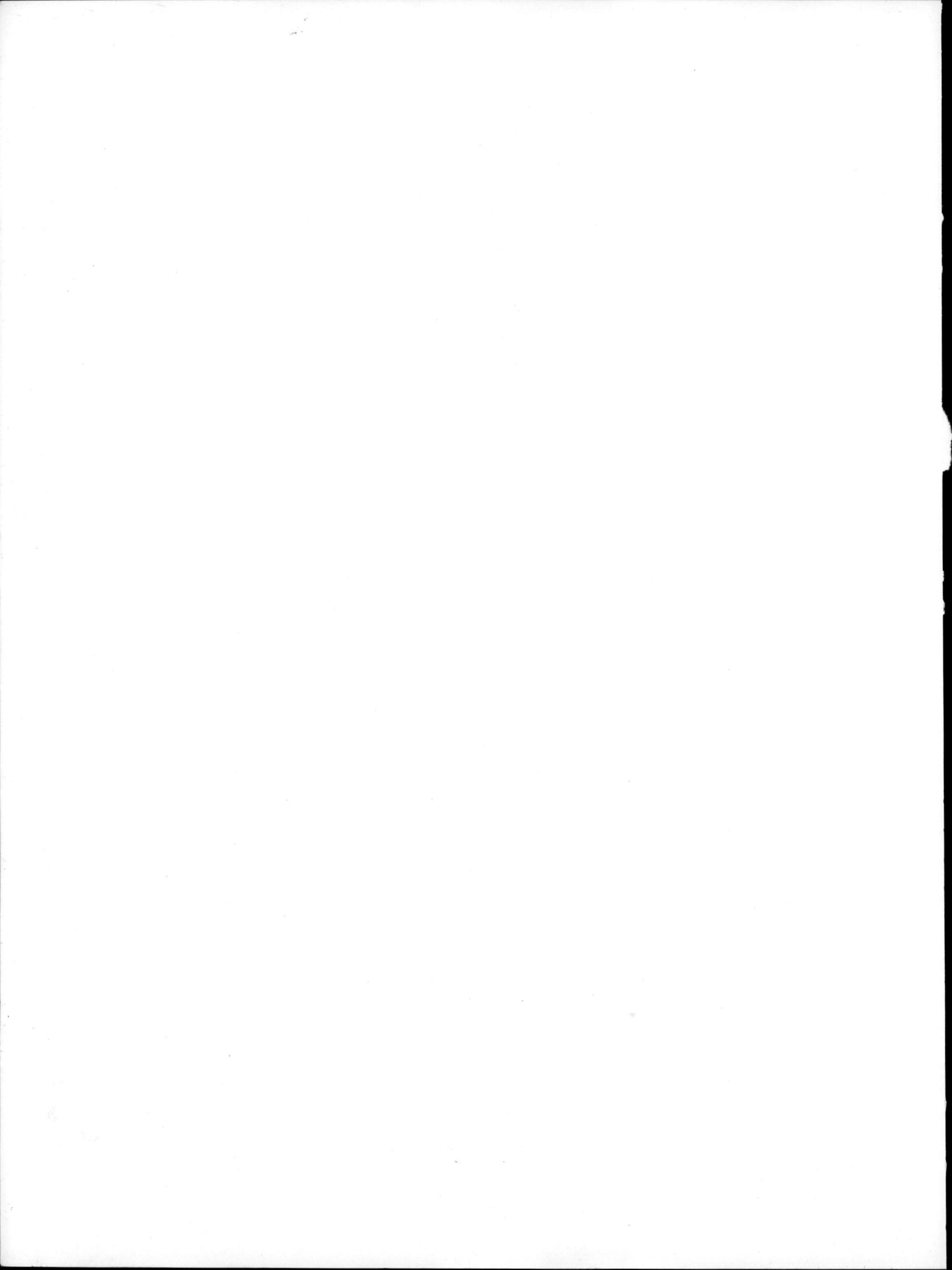